AF594954

Technologies of Coatings and Surface Hardening for Tool Industry

Technologies of Coatings and Surface Hardening for Tool Industry

Editor

Sergey N. Grigoriev

MDPI • Basel • Beijing • Wuhan • Barcelona • Belgrade • Manchester • Tokyo • Cluj • Tianjin

Editor
Sergey N. Grigoriev
University of Technology
Russia

Editorial Office
MDPI
St. Alban-Anlage 66
4052 Basel, Switzerland

This is a reprint of articles from the Special Issue published online in the open access journal *Coatings* (ISSN 2079-6412) (available at: https://www.mdpi.com/journal/coatings/special_issues/Coat_Surf_Tool).

For citation purposes, cite each article independently as indicated on the article page online and as indicated below:

LastName, A.A.; LastName, B.B.; LastName, C.C. Article Title. *Journal Name* **Year**, *Volume Number*, Page Range.

ISBN 978-3-0365-2452-8 (Hbk)
ISBN 978-3-0365-2453-5 (PDF)

Cover image courtesy of Moscow State University of Technology "STANKIN", Moscow, Russian Federation.

Contents

About the Editor

Sergey N. Grigoriev (Doctor of Engineering Science, Professor) is head of the department of High-Efficiency Processing Technologies of Moscow State University of Technology "STANKIN". The scientific team of the department was awarded at the most prestigious international scientific competitions and exhibitions under the supervision of Dr. of Eng. Sci., Prof. S. N. Grigoriev. The researchers and academic staff of the department are winners of grants of the President for the leading scientific schools in the field of Engineering and Technical Sciences, and are invited lecturers at conferences and editors of international scientific journals.

Preface to "Technologies of Coatings and Surface Hardening for Tool Industry"

The innovative **Coatings and Surface Hardening** technologies developed in recent years allow us to obtain practically any physical–mechanical or crystal–chemical complex property of the metalworking tool surface layer. Today, the scientific approach to improving the operational characteristics of the tool surface layers produced from traditional tools industrial materials is a highly costly and long-lasting process. Different technological techniques, such as coatings (physical and chemical methods), surface hardening and alloying (chemical-thermal treatment, implantation), a combination of the listed methods, and other solutions are used for this. The high efficiency of this approach may be explained by the fact that under the diversified operating conditions of the metalworking tool, in all cases, the most loaded is its surface layer. First, its properties precisely define the workability of the tool piece during the machining of a part. There is no all-purpose coating and surface hardening method; everything is very particular for each type of tool piece and its operating conditions. It should additionally be emphasized that the metalworking tool as an object of research was not chosen accidentally. Production experience shows that even with the most advanced machine tools, it is not possible to achieve high technical and economical rates of the machining process of the part with a low work resource of the tool. Diverse conditions of the tool operation bring on diverse injuries and failures of the technological system. At the same time, the wear rate of the tool is significantly higher than the wear rate of the machine tool parts and units. Therefore, the operational ability of the technological system as a whole depends on the metalworking cutting and die tool used. The roles of the tool multiply when it comes to machining composite materials, high-hardened steels, chrome-nickel, titan, or other hard-to-machine alloys. The scientific school of Moscow State **University of Technology STANKIN** is the leading scientific and educational center in the Russian Federation in the field of the science and technology of the machine-building industry, as well as other science-intensive manufacturing sectors, which introduces traditional and innovative approaches in the educational process for students and experienced specialists, and conducts research activities. The non-interrupted development of projects related to the critical stages of the modern technical progress in **Coatings and Surface Hardening for Tool Industry** by academic staff demonstrates the most current and innovative scientific results, which are published in high-ranking international scientific journals. The years of experience of the team of highly qualified scientists and specialists contribute to developing innovative products and technologies in the tool industry. The most remarkable results of scientific activities in the form of patented technologies, methods, techniques, and equipment are commissioned in the engineering enterprises for producing reliable and high-performance cutting tools. The most recent achievements of the department and invited colleagues in the field of **Tool Coatings and Surface Hardening** are presented in the Special Issue of *Coatings*.

Sergey N. Grigoriev
Editor

Article

Development of DLC-Coated Solid SiAlON/TiN Ceramic End Mills for Nickel Alloy Machining: Problems and Prospects

Sergey N. Grigoriev, Marina A. Volosova *, Sergey V. Fedorov, Anna A. Okunkova , Petr M. Pivkin, Pavel Y. Peretyagin and Artem Ershov

Department of High-Efficiency Processing Technologies, Moscow State University of Technology "STANKIN", Vadkovskiy per. 3A, 127055 Moscow, Russia; s.grigoriev@stankin.ru (S.N.G.); sv.fedorov@stankin.ru (S.V.F.); a.okunkova@stankin.ru (A.A.O.); p.pivkin@stankin.ru (P.M.P.); p.peretyagin@stankin.ru (P.Y.P.); a.ershov@stankin.ru (A.E.)

* Correspondence: m.volosova@stankin.ru; Tel.: +7-916-308-49-00

Abstract: The study is devoted to the development and testing of technological principles for the manufacture of solid end mills from ceramics based on a powder composition of α-SiAlON, β-SiAlON, and TiN additives, including spark plasma sintering powder composition, diamond sharpening of sintered ceramic blanks for shaping the cutting part of mills and deposition of anti-friction Si-containing diamond-like carbon (DLC) coatings in the final stage. A rational relationship between the components of the powder composition at spark plasma sintering was established. The influence of optimum temperature, which is the most critical sintering parameter, on ceramic samples' basic physical and mechanical properties was investigated. DLC coatings' role in changing the surface properties of ceramics based on SiAlON, such as microrelief, friction coefficient, et cetera, was studied. A comparative analysis of the efficiency of two tool options, such as developed samples of experimental mills made of SiAlON/TiN and commercial samples ceramic mills based on SiAlON, doped with stabilizing additives containing Yb when processing nickel alloys (NiCr20TiAl alloy was used as an example). DLC coatings' contribution to the quantitative indicators of the durability of ceramic mills and the surface quality of machined products made of nickel alloy is shown.

Keywords: diamond-like carbon coating; high-speed milling; nickel alloy; SiAlON; spark plasma sintering; roughness; wear resistance

Citation: Grigoriev, S.N.; Volosova, M.A.; Fedorov, S.V.; Okunkova, A.A.; Pivkin, P.M.; Peretyagin, P.Y.; Ershov, A. Development of DLC-Coated Solid SiAlON/TiN Ceramic End Mills for Nickel Alloy Machining: Problems and Prospects. *Coatings* **2021**, *11*, 532. https://doi.org/10.3390/coatings11050532

Received: 13 April 2021
Accepted: 28 April 2021
Published: 29 April 2021

Publisher's Note: MDPI stays neutral with regard to jurisdictional claims in published maps and institutional affiliations.

1. Introduction

At present, heat-resistant nickel alloys such as Inconel 718 type are widely used to manufacture critical parts operating under high thermal loads. For example, it can be nozzles and working turbine blades, fairings, et cetera. The high-performance characteristics of parts made of nickel alloys determine the difficulties in their machining, accompanied by increased heat and power loads on the cutting tool [1–3]. The machinability factor for high-temperature nickel alloys is in the order of 0.25 compared to the machining of C45 steel (according to EN 10083-2: 2006). Simultaneously, solid carbide end mills are the most popular and versatile tool for machining aircraft parts. A modern approach to solving the problem of increasing machining nickel alloys' productivity over the past years is the development and use for these purposes of end mills made of tool ceramics [4–6]. Today, large international companies, which are recognized leaders in cutting tools production (Kennametal, Mitsubishi Materials, Iscar, and others), produce solid ceramic end mills on an industrial scale [7–9]. Figure 1a shows an example of industrial use of solid ceramic end mills made of SiAlON material for high-speed milling a turbine blade made of nickel alloy on a multi-axis computer numerical control (CNC) machine.

The main advantage of using ceramic end mills when machining nickel alloys over carbide tools is higher heat resistance. Nickel alloys of the Inconel type begin to soften at cutting temperatures of 800 °C and above (Figure 1b), after which significantly lower power loads accompany their machining on the tool [1]. These temperatures correspond to cutting speeds above 350 m/min. Tool ceramics can be successfully operated due to their higher heat resistance in this high-speed milling mode. Sintered hard alloys at cutting temperatures over 800 °C lose their hardness rapidly (Figure 1b), which excludes the possibility of their use in high-speed milling nickel alloys using the surface layer plasticization effect [10,11].

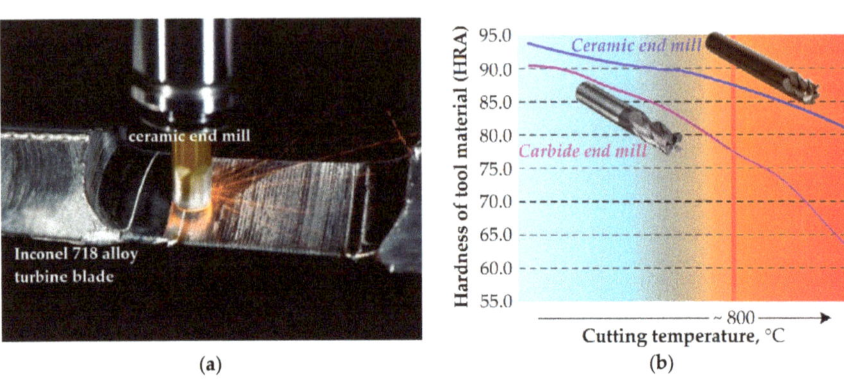

Figure 1. An example of industrial use of SiAlON solid ceramic mills in high-speed milling of a nickel alloy turbine blade (**a**) and the variation in surface hardness of end mills produced from ceramics and hard alloy (**b**) as a function of the cutting temperature.

Practice shows that among a wide variety of well-known technical ceramics brands for tool purposes for the solid ceramic end mills manufacturing, the most suitable ceramic material is SiAlON (silicon-aluminum oxynitride), which belongs to the class of ceramics based on silicon nitrides [12–15]. This ceramic consists of three or more phases: α-sialonic, β-sialonic, and amorphous or partially crystallized grain-boundary phases. Sintered ceramics based on α/β-sialons are characterized by a unique combination of even higher hardness than traditional silicon nitride while providing a high level of strength properties. The α-sialon phase has a high hardness that is retained at elevated temperatures, and the β-sialon phase has a high impact strength and fracture toughness. α- and β-sialons are ideally combined, and the ratio between these phases can be quite easily varied when preparing powder compositions (raw materials or precursors) for subsequent sintering [4–6]. It makes it possible to obtain a different set of physical and mechanical properties of the sintered ceramic necessary for the cutting tool's specific operating conditions [16–19].

Moreover, any ceramics, even those obtained using the most advanced technology, are structurally inhomogeneous materials that a priori contain certain defects (such as micropores). The sintering process of ceramic powders involves providing an increased temperature mode and is very technologically complex. In sintering, grain growth can occur; a significant amount of residual glassy phase can be formed, which worsens the hardness and strength of the material during high-temperature operation [20,21]. Excessive porosity can form at a low rate of grain-boundary diffusion and insufficient shrinkage of the powder composition, which sharply worsens the sintered ceramics' fracture toughness [22–24]. It should be borne in mind that a sintered ceramic workpiece, to obtain multi-edge tools, must be subjected to dimensional shaping processing by mechanical methods. One of the main machining methods is grinding with diamond abrasive tools by CNC grinding and sharpening machines. Grinding is also extremely difficult and energy intensive. Additional difficulties are associated with the reduced electrical and thermal conductivity of ceramics. The physical and technical nature of the diamond grinding process is such that it introduces additional defects (chips, microcracks) into the surface layer of a ceramic workpiece [22,25–27]. Therefore, if a workpiece containing cracks and pores was formed at the sintering stage, they will inevitably become local foci of microfracture and chips of the cutting part, which will reduce the tool performance. That is why leading researchers and technologists prioritize developments in improving technological processes for sintering ceramics and improving the ceramic powder composition to improve the basic physical and mechanical properties of ceramic blanks with a simultaneous improvement of their machinability.

A promising technological process for sintering SiAlON ceramics is the spark plasma sintering (SPS) method. The sintered powder composition and the used mold are heated by passing high-frequency low-voltage pulses of direct electric current through them. Compared to traditional hot pressing, SPS can significantly reduce the holding time at the maximum temperature (in the order of several minutes). Due to the direct transmission of electric current, it is possible to reduce grain growth rate since it allows high heating and cooling rates of the ceramics [28–30].

Large reserves for improving the physical and mechanical properties of ceramics based on SiAlON are introducing alloying components. For example, the introduction of up to 20 wt.% TiN nanoparticles into a powder composition based on α/β-sialons can provide an optimal ratio between hardness and impact toughness, improve electrical and thermal conductivity, crack resistance, and machinability [31–36]. Besides, stabilization during unstable high-temperature phases sintering, which is carried out by alloying with certain stabilizing elements based on rare-earth metal oxides, is of great importance. Today, Nd, Sm, Gd, Dy, Y, and Yb elements introduced in the form of nanoparticles into a powder composition with a volume of up to 7 wt.% are sufficiently studied and have proven a certain efficiency in SiAlON production [37–40]. World manufacturers, currently industrializing solid ceramic end mills, sinter powder compositions based on α/β-sialons with stabilizing additives Y_2O_3 and Yb_2O_3 (the latter option is recognized as more effective). The introduction of alloying rare-earth metals yttrium and ytterbium complicates powder composition preparation and sintering and increases the cost of an already expensive end product with apparent positive aspects. For example, the average selling price per unit for a one-piece ceramic four-flute endmill with a diameter of 10 mm is over 500 Euro. Such a high cost is partly offset by a manifold increase in productivity and durability during operation. Simultaneously, such a tool, compared to hard alloy, is not suitable for subsequent regrinding after reaching the limit wear by the cutting part. This constrains the broader distribution of solid ceramic end mills in mechanical engineering. Today, such a tool is not publicly available for manufacturing enterprises, and exploratory research aimed at finding approaches to simplifying solid ceramic end mills manufacturing technology and reducing their manufacturing costs is relevant.

One of the approaches to increasing the wear resistance of one-piece ceramic end mills can be various coatings deposition on their working surfaces based on complex nitrides and diamond-like carbon structures. This approach is based on the fact that under various loading conditions of the tool cutting part, its surface layer is the most loaded, and in many respects, this layer determines the wear resistance under the influence of external loads [41–44]. The coatings' effectiveness in increasing the wear resistance of solid ceramic mills in machining nickel alloys has not been experimentally studied until now. Simultaneously, in assessing the effectiveness of this approach to improve assembled turning tools and end mills equipped with ceramic plates, researchers do not have one single point of view [25,26]. Some experts are not inclined to consider coatings' deposition as a viable approach to improve ceramic tools' wear resistance. It can be assumed that such conclusions are the result of overestimated expectations from coatings applied to ceramics. One should not expect an effect comparable to that achieved for carbide tools (multiple increases in durability) from coatings. Initially, it is necessary to proceed from the fact that the role of coatings deposited onto ceramics is specific, and even a twofold increase in the ceramics' resistance should be considered a good result. Many experimental works demonstrate that with rationally chosen deposition technology, architecture, and coatings' composition, their particular possibilities for increasing resistance are quite real and significant [42–46]. For example, the authors of this work have achieved an increase in resistance during turning and milling hardened steels by 1.4–1.9 times in previous studies when vacuum-plasma coatings' deposition onto cutting inserts made of tool ceramics based on Al_2O_3 + TiC and Al_2O_3 + SiC [25,26,42], which was a consequence of the effect of coatings on the properties of the surface and the surface layer of ceramic plates. The changes were manifested in the form of a decrease in the friction coefficient between the tool and the processed material, some improvement in strength characteristics, smoothing and reduction of the height of surface microroughness, and a decrease ("healing") of ceramic plates' surface defects that were formed during diamond grinding [25,47]. Thus, there are prerequisites for the fact that the approach to increasing the wear resistance of solid ceramic mills, based on various coatings' deposition onto their working surfaces that can also have a positive effect.

Within the framework of this work, a set of problems was solved, which can be formulated as follows:

1. development and implementation of laboratory technology for the manufacture of solid ceramic end mills from SiAlON ceramics, including spark plasma sintering powder composition and subsequent diamond grinding of sintered ceramic blanks for shaping the cutting part of end mills of the required geometry;
2. the establishment of a rational ratio between the components of the powder composition α-SiAlON, β-SiAlON, and TiN during spark plasma sintering ceramic blanks and the choice of the sintering process optimal temperature, as the most critical parameter of the technological process (expensive stabilizing additives used in industry in the production of SiAlON ceramics were deliberately excluded);
3. using the example of promising Si-containing diamond-like carbon (DLC) coatings to show their effect on the properties of the surface and the surface layer of sintered ceramic samples; to carry out a comparative analysis of the effectiveness of two cutting tool options—developed samples of experimental end mills made of SiAlON/TiN and industrial samples of commercial ceramic mills based on SiAlON, alloyed with stabilizing additives containing Yb; and to establish the contribution of DLC coatings to the quantitative indicators of the resistance of ceramic end mills and the surface quality of workpieces made of nickel alloy (for example, NiCr20TiAl alloy, standard EN 10269).

2. Materials and Methods

2.1. Solid Ceramic End Mill Design and Geometry

Figure 2a shows a drawing, and Figure 2b is a general view of experimental solid ceramic end mills under development and field testing. Table 1 provides detailed information on the design and geometric parameters of the cutting tool. The experimental samples of end mills produced in this framework of the current study were compared with ceramic tools of a similar design and geometry, which are now commercially manufactured from SiAlON-based ceramics, including Yb-containing stabilizing phases (commercial ceramic mills), and adopted by the authors as a standard.

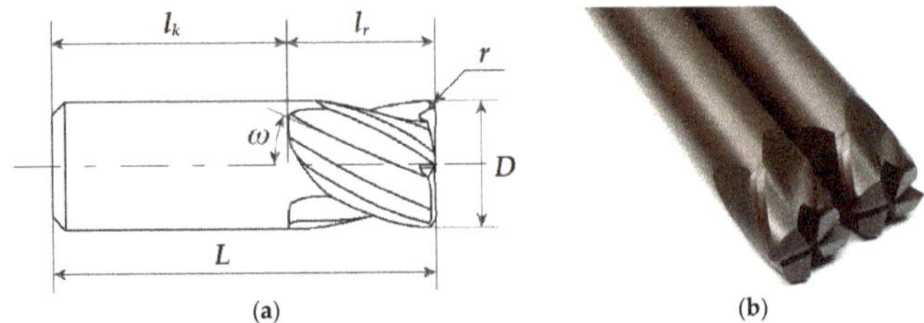

Figure 2. Drawing (**a**) and general view (**b**) of experimental samples of solid ceramic end mills made by the team of authors.

Table 1. The geometric design of solid ceramic end mills for development and testing purposes.

Tool Parameter	Value
End mill diameter D, mm	10
Number of teeth	4
Overall length L, mm	48
Fastener (shank) length l_k, mm	40
Cutting length l_r, mm	8
Vertex radius r, mm	1.2
Flute helix angle ω, degree	30

2.2. Technological Algorithm of Solid Ceramic End Mill Manufacturing

The technological algorithm for solid ceramic end mills manufacturing developed by the study's authors is visualized in Figure 3. It includes the following technological operation steps when the finishing operation of coating is not indicated on the diagram:

1. preparation of powder composition based on α-SiAlON, β-SiAlON, and TiN nanoparticles;
2. high-temperature spark plasma sintering of a powder composition in graphite dies, which defines the shape of a future workpiece (a ceramic disk), using the force action of the upper and lower punches and passing powerful DC pulses through the composition to be sintered;
3. cutting a sintered ceramic disk into quadrangular parallelepipeds on a disk abrasive cutting machine (dimensions are set based on the main overall dimensions of ceramic rods and allowances for subsequent grinding);
4. shaping a ceramic rod on a centerless grinding machine;
5. shaping a ceramic rod on a CNC tool sharpening and grinding machine: helical flute, undercut face, and toroidal cutting edge (margin);
6. deposition of vacuum-plasma coatings on solid ceramic end mills.

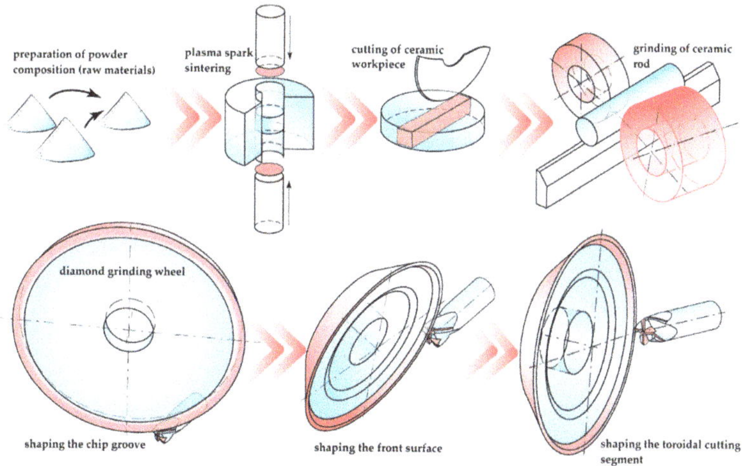

Figure 3. Technological algorithm for manufacturing solid ceramic end mills.

2.3. Powder Compositions' Preparation and Sintering Ceramic Blanks

Powders produced by Plasmotherm (Moscow, Russia) were used as initial ceramic precursors: α-SiAlON and β-SiAlON with a particle size of 1 ± 0.5 μm and TiN additives with a particle size of 15–175 nm. The colloidal method was used for mixing the ceramic powder composition since it allows mixing with minimal time and energy costs.

Three suspension options based on powder compositions were prepared in a ball mill for 24 h using isopropyl alcohol and ceramic grinding bodies in a polyethylene container. The choice of composition content was made taking into account the data of previous works conducted by authoritative researchers and based on the professional experience o the current study authors [22,31,32]:

1. Option 1: ceramic base α–β SiAlON 80 wt.% (α-SiAlON 90 wt.% + β-SiAlON 10 wt.% + TiN 20 wt.%; composition code 80% (90α10β) + 20% TiN;
2. Option 2: ceramic base α–β SiAlON 90 wt.% (α-SiAlON 90 wt.% + β-SiAlON 10 wt.% + TiN 10 wt.%; composition code 90% (90α10β) + 10% TiN;
3. Option 3: ceramic base α–β SiAlON 80 wt.% (α-SiAlON 70 wt.% + β-SiAlON 30 wt.% + TiN 20 wt.%; composition code 80% (70α30β) + 20% TiN.

The resulting suspensions were dried using a vacuum drying oven, VO400, at +50 °C for 24 h (Memmert GmbH + Co. KG, Schwabach, Germany). After drying, the powders were sieved using a vibration machine (NL 1015X/010, NL Scientific, Klang, Malaysia) sieve laboratory control of 2 ± 0.5 μm, and Si_3N_4 ceramic grinding bodies [48].

After completing all the preparatory operations, the powder compositions were subjected to spark plasma sintering on a technological unit manufactured by FCT Systeme GmbH (Effelder-Rauenstein, Germany) (Figure 4).

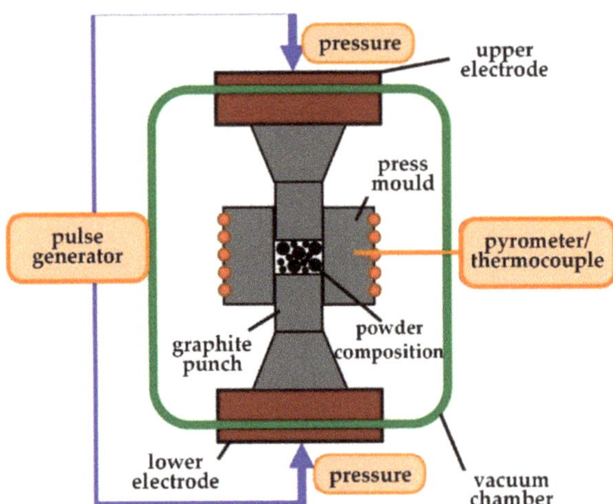

Figure 4. A schematic diagram of the ceramic composition's spark plasma sintering.

Ten disk-shaped samples of 3.0 mm thickness and 20.0 mm diameter were made from each version of the powder composition. Further, they were subsequently used to asses sintered samples' main physical and mechanical properties. The ceramic composition that exhibits the best results was subsequently used for coating and tribological tests. The sintering pressure was constant for all types of materials and amounted to 80 MPa, and the holding time at the maximum temperature was 30 min. The sintering temperature was varied from 1600 to 1750 °C with a pitch of 50 °C. These indicators were selected based on the analysis of technical literature [28,30,32] and preliminary experiments carried out by the authors of the current work with spark plasma sintering ceramics based on SiAlON. During sintering, the pressure value was considered rational because its effect on the final samples' mechanical properties was described in a previously published study by the authors of [49]. After determining the powder composition's optimal version and choosing a rational sintering temperature mode, disk-shaped ceramic blanks with 10.2 mm thickness and 80.0 mm diameter were made, from which ceramic rods were obtained.

2.4. Diamond Sharpening Solid Ceramic End Mills

Multi-stage diamond sharpening was performed on a Helitronic Micro multi-axis machine from Walter Maschinenbau GmbH (Tübingen, Germany) to manufacture solid end mills from ceramic rod-blanks with the required design and geometric parameters. The sequence of the main operations of forming ceramic cutters is illustrated in Figure 5.

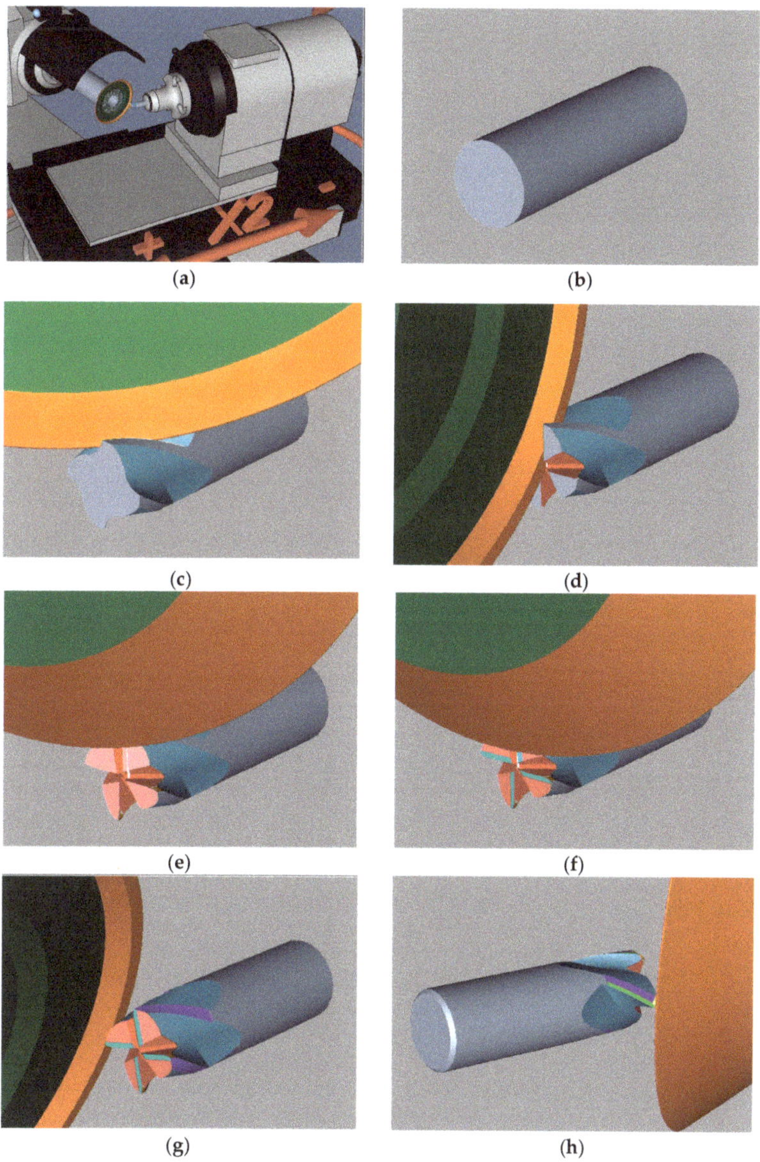

Figure 5. The main stages sequence for diamond sharpening solid ceramic end mill: (**a**) the relative position of the machine spindle and the workpiece when forming a ceramic end mill; (**b**) a ceramic rod; (**c**) grinding the helical chip grooves with a tapered disk; (**d**) grinding the rake face at the end with a cup-bevel wheel; (**e**) grinding the secondary relief surface at the end with a cup-bevel wheel; (**f**) grinding the primary relief surface at the end with a cup-bevel wheel; (**g**) grinding the secondary relief surface at the periphery with the cup-bevel wheel; and (**h**) grinding the secondary relief surface on the transitional radius section with the cup-bevel wheel.

The information on chosen cutting modes and the type of grinding wheels used are given in Table 2. The given technological operations and processing modes were selected based on ensuring high-precision execution of ceramic end mills, their profile, geometric dimensions, and high cleanliness of the tool surfaces (0.32–0.5 μm in R_a surface roughness parameter). Particular attention was paid to quality control of the cutting edges of ceramic end mills and minimization of micro-chips that significantly reduce the tool life. The manufactured end mills' control was carried out on a non-contact precision measuring device, Helicheck Plus, manufactured by Walter Maschinenbau GmbH (Tübingen, Germany).

Table 2. Modes of ceramic end mill diamond grinding and the type of used grinding wheels.

Diamond Sharpening Stage	Grinding Wheel Type	Cutting Speed, m·s^{-1}	Feed, mm·min^{-1}
Grinding the helical flute	Organic bonded diamond tapered wheel (1V1)	17	10
Grinding the rake face at the end–undercuts at the transitional radius section		20	7
Grinding the rake at the transitional radius section	Organic bonded diamond cup-tapered wheel (12V9)	22	50
Grinding the secondary relief surface at the end		25	30
Grinding the primary relief surface at the end		25	35
Grinding the secondary relief surface on the transitional radius section		25	30
Grinding the secondary relief surface at the periphery		25	30
Grinding the primary relief surface on the transitional radius section		28	40
Grinding the primary relief surface at the periphery		25	20

2.5. *Vacuum Plasma Coating Composition Selection and Deposition on Solid Ceramic End Mills*

The choice of coating deposition method for SiAlON ceramic samples was made in favor of the well-proven technology of physical coating deposition of an evaporated material from a vacuum-arc discharge plasma. This technology makes it possible to form multicomponent coatings of the required composition with high productivity and reproducibility [50–55]. The coating was carried out on a STANKIN-APP technological unit prototype (MSTU Stankin, Moscow, Russia). Figure 6 shows a schematic diagram of a technological unit. The unit provides deposition of various compositions' coatings by varying the composition of the gas mixture supplied to the vacuum chamber through a multi-channel gas injection system and replacing cathodes, the outer surface of which has the shape of a right cone frustum. The formation of plasma by a vacuum arc discharge at a reduced pressure of various gases is accompanied by the appearance of a micro-droplet fraction and a neutral atomic and molecular component of the cathode material's erosion products [56–61]. The unit is equipped with a plasma flow filtration system to eliminate this drawback. A powerful electromagnetic field deflects ions, and microdroplets and neutral particles are captured and removed from the chamber, making it possible to form higher quality coatings on the ceramic samples' surfaces.

Three variations of coatings for deposition on SiAlON ceramic samples, which should contact with nickel alloys, were chosen based on the results of the authors' last works [62–70]: (CrAlSi)N—single-layer coating; (TiAl)N—sandwich-type multilayer coating; and (CrAlSi)N/DLC—diamond-like carbon two-layer coating (this coating was developed by Platit AG, Selzach, Switzerland). The total thickness of the coatings was in the range of 3.5–3.9 μm. The ceramic samples were thoroughly cleaned in an ultrasonic tank using a soap solution at a temperature of 60 °C for 20 min and in alcohol for 5 min before coating deposition.

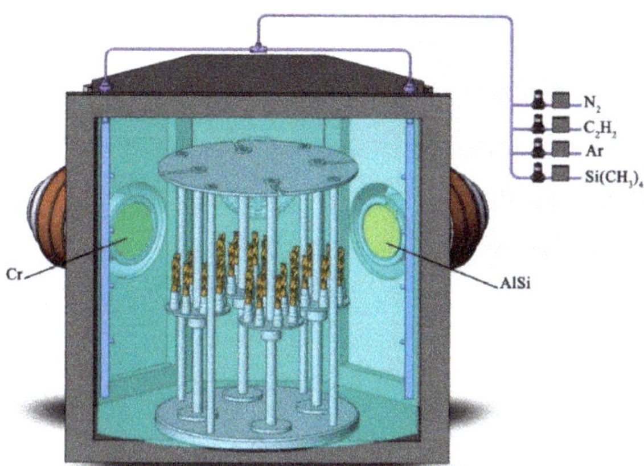

Figure 6. A technological unit's principal scheme for coatings' deposition on ceramic samples is using physical deposition of evaporated material from the vacuum arc discharge plasma.

Tribological tests were carried out on disk-shaped sintered specimens with three coating options to assess the prospects of the selected compositions' coatings to increase the wear resistance of SiAlON ceramics. Considering that the coated ceramic tool is expected to operate at elevated cutting temperatures, the tests were carried out both at room temperature and heated to 800 °C. The results of the high-temperature tests were of crucial importance for the selection of the coating, which was subsequently applied to the samples of solid ceramic end mills. The tests were carried out on a TNT tribometer by Anton Paar TriTec (Corcelles-Cormondrèche, Switzerland) using the "ball-on-disk" method with the rotation of the sample relative to a stationary counter body (nickel alloy ball), set at a distance relative to the rotation axis of the sample [71]. Table 3 shows the results of tribological tests with a friction length of 250 m, an applied load of 1 N, and a sliding speed of 10 cm/s.

Table 3. Results of preliminary tribological tests of 80% (90α10β) + 20% TiN ceramic specimens with different coatings.

Specimen	Total Coating Thickness, μm	Testing Temperature, °C	Average Coefficient of Friction (μ)	Wear Track Depth, μm	Ball Wear Diameter, μm
SiAlON without coating	–	20 800	0.49 0.81	1.4 2.2	520 1000
SiAlON with (CrAlSi)N coating	3.6 ± 0.1	20 800	0.45 0.96	0.33 2.4	195 1000
SiAlON with (CrAlSi)N/DLC coating	3.6 ± 0.1	20 800	0.12 0.4	0.29 1.7	165 860
SiAlON with (TiAl)N multilayer coating	3.8 ± 0.1	20 800	0.41 0.80	0.36 2.1	180 950

The coating choice was made in favor of a (CrAlSi)N/DLC two-layer coating based on the experimental data obtained (hereinafter referred to as the DLC coating) since it demonstrated the lowest coefficient of friction and wear-track depth in contact with the counter body during high-temperature heating. It makes it possible to expect that this coating will reduce the intensity of the frictional and adhesive interactions of the cutting tool during milling nickel alloys.

The DLC coating deposition process on solid ceramic end mills (experimental and commercial end mills), implemented in the unit shown in Figure 6, was based on several innovative solutions of the current work's authors and well-known solutions of Platit AG

(Selzach, Switzerland) and provided for the sequential implementation of the complex o the steps listed below:

1. heating of the tool: pressure in the chamber of 0.03 Pa, temperature in the chamber o 500 °C, table rotation speed of 5 rpm, and holding time of 60 min;
2. cleaning with argon ions: pressure in the chamber of 2.2 Pa, temperature in the chamber of 500 °C, rotation speed of the table of 5 rpm, negative bias voltage on the table of 800 V, arc current at the Cr cathode of 90 A, and holding time of 20 min;
3. sublayer (CrAlSi)N deposition: a gas mixture of 5% (Ar) and 95% (N_2), pressure ir the chamber of 0.9 Pa, temperature in the chamber of 500 °C, table rotation speed o 5 rpm, negative bias voltage on the table of 400 V, arc current at the AlSi cathode o 100 A, arc current at the Cr cathode of 100 A, and holding time of 70 min;
4. functional DLC coating deposition, including:
   (1) formation of a gradient layer: a gas mixture of 20% (Ar), 73% (N_2), and 7% ($Si(CH_3)_4$), pressure in the chamber of 1.5 Pa, temperature in the chamber o 180 °C, rotation speed table of 5 rpm, negative bias voltage on the table o 500 V, and holding time of 20 min;
   (2) formation of a diamond-like carbon layer: a gas mixture of 2% ($Si(CH_3)_4$), 55% (Ar), and 43% (C_2H_2), pressure in the chamber of 0.8 Pa, process temperature of 180 °C, table rotation speed of 5 rpm, negative bias voltage on the table o 500 V, and holding time of 100 min.

2.6. Evaluation of Physical, Mechanical and Surface Properties of Ceramic Materials and Coatings

The hardness of sintered ceramic compositions based on SiAlON was determined at a load of 2 kg by the Vickers pyramid indentation method on a QnessQ10A universal microhardness tester manufactured (Qness GmbH, Mammelzen, Germany).

The sintered samples' density was estimated using the hydrostatic weighing technique based on Archimedes' law comparing the mass of ceramic samples obtained in air and liquid under normal conditions. The measurements were carried out in distilled water on a GR-300 high-precision analytical balance (A&D Company, Tokyo, Japan).

The samples' strength characteristics were measured by 3-point bending tests at room temperature on an AutoGraph AG-X universal testing machine (Shimadzu, Kyoto, Japan The loading rate during the tests was 0.5 mm/min, and the maximum displacement wa 40 mm. Fracture toughness was also assessed using AutoGraph equipment using a simila loading pattern. A one-sided notch perpendicular to its longitudinal axis was made on ceramic samples' surface with a diamond disk. The sample was installed with a notch downwards and loaded.

The morphology and structure of sintered ceramic specimens and DLC coatings were studied using a VEGA 3 LMH scanning electron microscope (SEM) (Tescan, Brno, Czech Republic) equipped with an Oxford Instruments INCA Energy energy-dispersive X-ra spectroscopy (EDX) system. The sample atoms under study were excited and emitted X rays characteristic of each chemical element with an electron beam's assistance in the ED process. When studying the energy spectrum of the indicated radiation using a specialized program, data on the samples' qualitative and quantitative elemental composition were obtained. The program identifies an element and even reveals hidden elements in the sample using the point analysis mode. Transmission electron microscopy (TEM) and selected area diffraction (SAED) studies of the surface layer of ceramic samples with DLC coatings were carried out on JEM-2100F equipment manufactured by JEOL (Tokyo, Japan The samples were prepared according to the standard probe techniques using Opal 41 Jade 700, and Saphir 300 sample equipment (ATM, Haan, The Netherlands). An epox resin with quartz sand was used as a filler for SEM samples. The samples were coated with gold by Quorum 150T ES (Laughton, UK). An instrument controlled the thickness of th coating of less than ~30 μm. TEM samples were polished, cut off 100 nm by an ARTOS 3I Ultramicrotome (Leica, Wetzlar, Germany), and sputtered with gold.

An experimental technique and a special stand, described and tested by the authors in [25], were used to assess the effect of coatings on the resistance of sintered ceramic specimens to fracturing under the action of external loads. The stand was a device in the form of a massive parallelepiped with a groove and step. Sintered ceramic specimens in the form of 20.0 mm diameter and 3.0 mm thickness disks were installed on the step with a small gap. During the experiment, a force was applied to the punch, rigidly fixed on the INSTRON universal testing machine, until the plate broke. A mandatory requirement during the test was to ensure strict reproducibility of the test process, particularly the identity for the entire group of samples of the place of application of force from the punch to the ceramic sample. In the experiment's course, the diagram "force – the punch movement along the z-axis" was recorded, and the critical force at which the destruction occurred was recorded.

The surface microroughness of the sintered ceramic samples before and after DLC coatings' deposition was evaluated on a DektakXT stylus profilometer (Bruker, Billerica, MA, USA).

2.7. Performance Testing of Solid Ceramic End Mills

A cylindrical workpiece with a diameter of 40 mm from a heat-resistant nickel alloy NiCr20TiAl [72] was used as the processed material to carry out comparative performance tests of experimental ceramic end mills and commercial ceramic mills. The chemical composition of the workpiece is shown in Table 4. The hardness (HRC) of the processed material was 34 units, and the ultimate strength is 1150 MPa.

Table 4. Chemical composition (wt.%) of processed nickel alloy NiCr20TiAl.

Element	Ni	Cr	Ti	Al	Fe	Si	Co	Mn	Cu	C	Rest
Wt.%	73	20	2.5	1.0	1.0	0.6	0.5	0.4	0.2	0.07	0.73

The tests were carried out on a 5-axis turning-milling machining center CTX beta 1250TC (DMG, Hüfingen, Germany). The tool and part clamping system's high rigidity minimizes the risks of accidental brittle fracture of the cutting part of ceramic end mills during resistance tests. A SCHUNK GZB-S Ø20/Ø10 adapter sleeve and an SDF-EC bt40 Ø20 HSK-A63 hydraulic arbor to fix the end mills were used, which allow the tool to be securely clamped with high positioning accuracy.

The tool's tests were carried out according to the program written in the CAD/CAM system GeMMa-3D (LLC STC GeMMa, Zhukovsky, Moscow, Russia) when implementing the strategy of machining the plane of the workpiece end face with an end mill when the cutter moves along a spiral. Figure 7a shows a general view of the ceramic end mill's location relative to the nickel alloy NiCr20TiAl workpiece during testing, and Figure 7b visualizes the processing strategy. The tests were carried out under the following cutting modes: cutting speed V_c = 376.8 m/min (rotation frequency n = 12,000 rpm), feed S = 1500 mm/min, and feed per tooth S_t = 0.031 mm/tooth. The scheme of "dry" processing without cutting fluids was used.

The wear area's critical size h_f along the end mill tooth's flank face equal to 0.4 mm was taken as the criterion for the loss of performance (failure) of ceramic end mills (Figure 8). The tool's wear resistance was determined as the time of milling until the end mill reached critical wear (when this value was exceeded, the cases of spalling and chipping of the cutting part increased many times). Each end mill sample was researched optically every minute during milling. The performance tests continued until h_f reaches the range of 0.4–0.5 mm to demonstrate the dramatic tendency in tool wear. A metallographic optical microscope of the Stereo Discovery V12 model (Carl Zeiss Vision GmbH, Jena, Germany) was used to quantify wear with a measurement accuracy of ±0.025 mm, which is 5.00–6.25% of the measured value h_f and less than standard accuracy tolerance. The wear area was monitored on each of the four teeth of the ceramic end mills that were tested. The arithmetic mean values were calculated based on the data obtained, which were used to plot the wear curves of ceramic end mills.

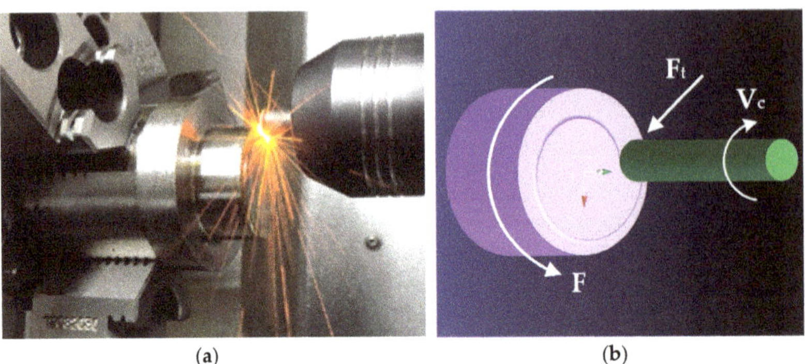

(a) (b)

Figure 7. Strategy for resistance testing of solid ceramic end mills: (**a**) general view of the ceramic end mill location relative to the NiCr20TiAl nickel alloy workpiece during testing; and (**b**) processing strategy visualization.

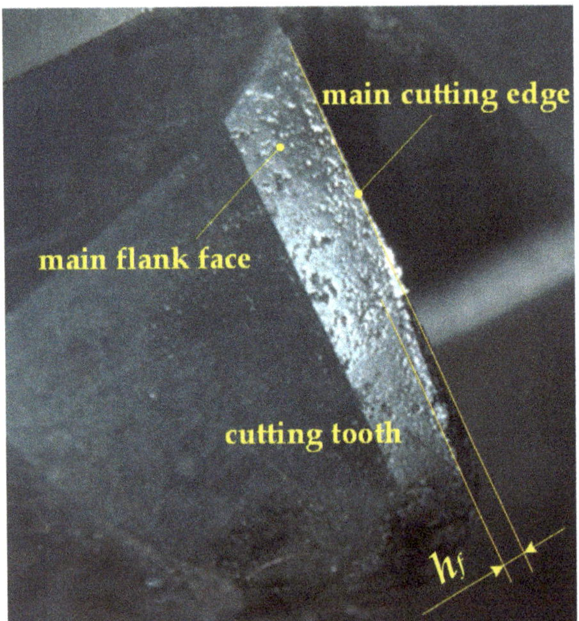

Figure 8. The wear zone location on the flank face of the ceramic end mill tooth and the h_f area (land) to measure the quantified wear.

A Surftest SJ-410 portable profilometer (Mitutoyo, Kawasaki, Japan) was used for a quantitative and qualitative assessment of a nickel alloy workpiece's surface roughness after machining with various ceramic end mills.

3. Results and Discussion

3.1. Structure and Properties of Sintered Ceramic Blanks

Figure 9 shows the experimentally obtained data on the effect of various powder compositions (1) 80% (90α10β) + 20% TiN, (2) 90% (90α10β) + 10% TiN and (3) 80%

(70α30β) + 20% TiN and spark plasma temperatures in the range of 1600–1750 °C on the basic physical and mechanical properties of sintered ceramic samples: percentage theoretical density (Figure 9a), hardness (Figure 9b), crack resistance (Figure 9c), and flexural strength (Figure 9d).

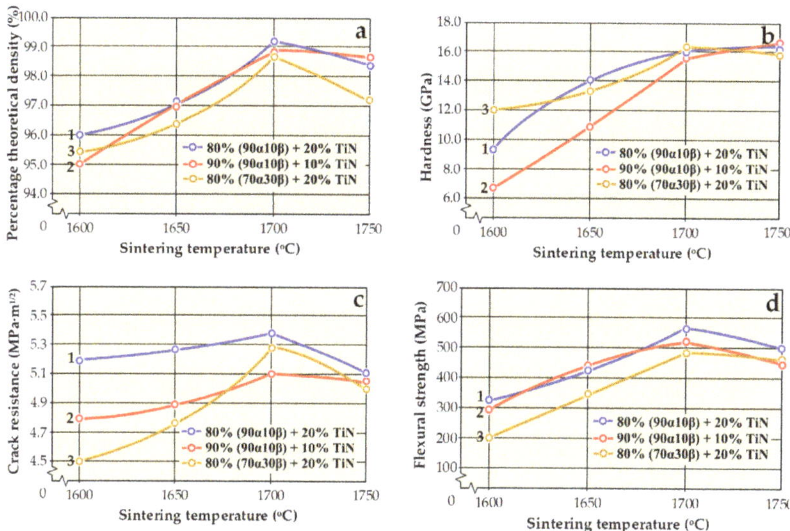

Figure 9. Dependences of the basic physical and mechanical properties of sintered ceramic samples made of various powder compositions such as 80% (90α10β) + 20% TiN (1), 90% (90α10β) + 10% TiN (2), and 80% (70α30β) + 20% TiN (3) from the spark plasma sintering temperature: (**a**) percentage theoretical density; (**b**) hardness; (**c**) crack resistance; and (**d**) flexural strength.

A general characteristic tendency of the temperature factor's influence at spark plasma sintering on the basic physical and mechanical properties is visible in Figure 9. The best values of the properties in all cases are achieved at a process temperature of 1700 °C. A further increase in temperature leads to a sharp decrease in the sintered material properties. The exception is hardness, which at 1750 °C remains at the same level for all specimens, Figure 9b. There is no doubt that a higher sintering temperature promotes recrystallization processes leading to rapid grain growth. The temperature of 1700 °C is a rational value for sintering all studied powder compositions based on SiAlON. Ceramics with a minimum number of pores are formed under these conditions judging by the fact that they have the maximum theoretical density (Figure 9a). It explains the maximum values at 1700 °C of crack resistance (Figure 9c) and flexural strength (Figure 9d).

Analysis of the basic physical and mechanical properties of the samples sintered from three versions of powder compositions shows that the best ratio of "theoretical density–hardness–crack resistance–flexural strength" in all cases had option (1): ceramic base α–β SiAlON 80 wt.% (α–SiAlON 90 wt.% + β–SiAlON 10 wt.%) + TiN 20 wt.% (composition code 80% (90α10β) + 20% TiN). The indicated samples at a temperature of 1700 °C have the following quantitative values of properties: theoretical density of 99.1%, hardness of 16.0 GPa, crack resistance of 5.3 MPa·$m^{\frac{1}{2}}$, and flexural strength of 550 MPa. It should be noted that the hardness of the experimental ceramics is in no way inferior to the industrially produced ceramics based on SiAlON, stabilized with Yb_2O_3 additives, but it is somewhat inferior to the best samples in terms of strength properties and crack resistance [30,37,39].

Summarizing the character of the experimentally obtained curves for various powder compositions (Figure 9), it can be concluded that the introduction of TiN particles in a volume of 10 wt.% during sintering (curve 2) does not provide an acceptable level of properties of ceramics based on α–β SiAlON (mainly the above refers to crack resistance). In contrast, the introduction of TiN into the powder composition in a volume of 20 wt.% (curve 1) provides the best combination of properties. In this case, the SiAlON ceramic base's rational composition is the following ratio between the phases: α–SiAlON 90 wt.% and β–SiAlON 10 wt.%. It was shown experimentally that an increase in the β-SiAlON content to 30 wt.% in a ceramic powder composition noticeably decreases the sintered material's strength characteristics (curve 3).

Table 5 shows the results of an elemental quantitative EDX analysis of an experimental sample of 80% (90α10β) + 20% TiN and the ceramic material, from which commercial ceramic end mills are made, intended for the processing of nickel alloys, taken in the current study as a standard. The data of an experimental sample of 80% (90α10β) + 20% TiN showed the best combination of physical and mechanical properties in studies. As follows from the presented data, the experimental ceramics contains the following content of elements: 37.4% Si, 19.1% Ti, and 12.6% Al. Commercial ceramic significantly differs in elemental composition and contains: 45.7% Si, 7.0% Yb, and 6.5% Al. Furthermore, we studied two fundamentally different types of ceramics based on SiAlON.

Table 5. Results of elemental quantitative EDX analysis of sintered ceramics based on SiAlON.

Specimen	Si	N	Ti	Al	O	Yb
Experimental 80% (90α10β) + 20% TiN	37.4	19.9	19.1	12.6	11.0	–
Commercial ceramics	45.7	35.0	–	6.5	5.8	7.0

Figure 10 shows the results of microstructural analysis of fractures of two types of ceramics based on SiAlON, which demonstrate significant differences in the experimental sample 80% (90α10β) + 20% TiN (Figure 10a,c,e) and specimens of commercial ceramics (Figure 10b,d,f). It is natural since sintered ceramics' structure depends mainly on the type and number of sintering components and additives and the technological process parameters. Comparing scanning electron microscopy (SEM) images of the microstructure of the fracture surface of ceramic samples shows that both samples contain a significant amount of the α-modification of SiAlON as the dominant phase. This phase has high homogeneity and fineness for a specimen of commercial ceramics. Experimental ceramics are characterized by equiaxed and non-equiaxed grains with an average grain size of 3–5 μm, while commercial ceramics have a grain size of 1.5–3.5 μm. Besides, both samples have a certain content of the β-modification of SiAlON. This phase is more characteristic and noticeable in commercial ceramics since it has elongated (oblong) particles.

(a)

(b)

Figure 10. *Cont.*

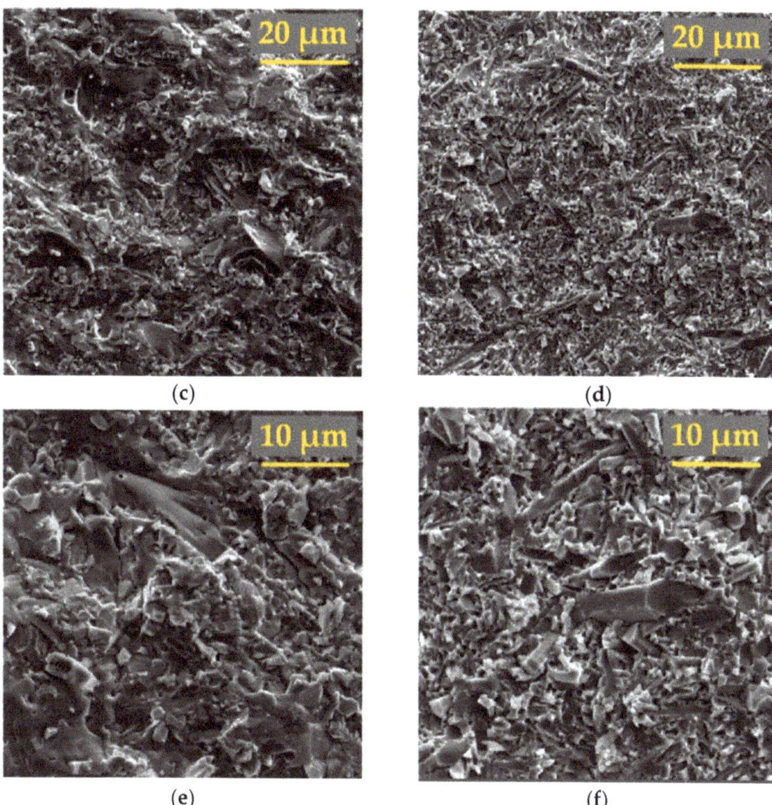

Figure 10. SEM images of fracture surface microstructures of sintered ceramic samples based on SiAlON: (**a**) experimental sample 80% (90α10β) + 20% TiN, ×1000; (**b**) specimen of commercial ceramics, ×1000; (**c**) experimental sample 80% (90α10β) + 20% TiN, ×5000; (**d**) specimen of commercial ceramics, ×5000; (**e**) experimental sample 80% (90α10β) + 20% TiN, ×10,000; and (**f**) specimen of commercial ceramics, ×10,000.

The elements distribution maps along the samples' fracture were obtained using EDX analysis for the possibility of a qualitative assessment of the uniformity of distribution in ceramic samples of the main elements and Ti- and Yb-containing phases (Figure 11). It can be seen that the Ti-containing phase is fairly uniformly distributed over the material volume after sintering in the experimental sample of 80% (90α10β) + 20% TiN (Figure 11a–c). Element distribution maps in a specimen of commercial ceramics (Figure 11d–f) demonstrate that the Yb-containing phase is uniformly distributed throughout the material structure. It can be concluded that the Yb-containing phase (stabilizing phase) is formed along the grain boundaries of ceramics based on SiAlON, and it is this that inhibits their growth during sintering and ensures the formation of a finer-grained and uniform structure in comparison with samples from experimental ceramics 80% (90α10β) + 20% TiN (Figure 10) based on the data of authoritative researchers [37–40].

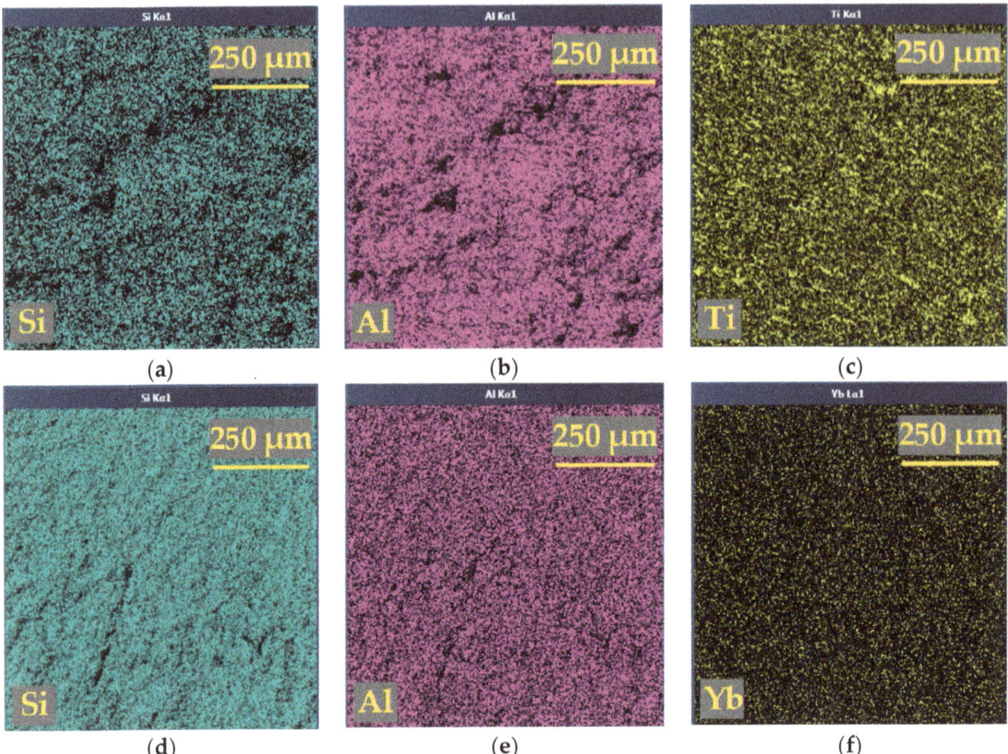

(a) (b) (c)

(d) (e) (f)

Figure 11. Distribution of chemical elements on the fracture surface of sintered ceramic material based on SiAlON samples, obtained by EDX analysis: (**a**) Si-content of experimental sample 80% (90α10β) + 20% TiN; (**b**) Al-content of experimental sample 80% (90α10β) + 20% TiN; (**c**) Ti-content of experimental sample 80% (90α10β) + 20% TiN; (**d**) Si-content of commercial ceramics specimen; (**e**) Al-content of commercial ceramics specimen; and (**f**) Yb-content of commercial ceramics specimen.

3.2. Influence of DLC Coatings on the Transformation of the Sintered Ceramic Blanks Characteristics

Figure 12 shows SEM images of the microstructure of thin sections of an experimental sample of 80% (90α10β) + 20% TiN ceramics (Figure 12a) and specimen of commercial ceramics (Figure 12b) after DLC coating deposition under identical technological conditions. Figure 12c presents an SEM image of a DLC coating's surface structure, and Figure 12d shows a TEM image of its volume structure. The coating thickness in total was 3.6 μm, including the thickness of 1.7 μm of the (CrAlSi)N sublayer and 1.9 μm of the functional DLC layer. It can be seen that the nitride sublayer has a columnar structure, while the outer DLC layer is characterized by an amorphous structure that microscopic methods cannot differentiate. The DLC layer does not have visible boundaries between grains in thickness, which are traditionally formed during the deposition of nitride coatings, while the surface of the DLC layer is a specific relief in the form of intergrown spherical segments of 0.2–1.5 μm in size. It should be noted that the formation of a characteristic relief is not affected by the ceramic base's surface layer's state before the coating deposition. The selected area diffraction (SAED) of the complex coating (Figure 12e,f) presents a complex phase of (CrAlSi)N sublayer and amorphous structure of DLC layers.

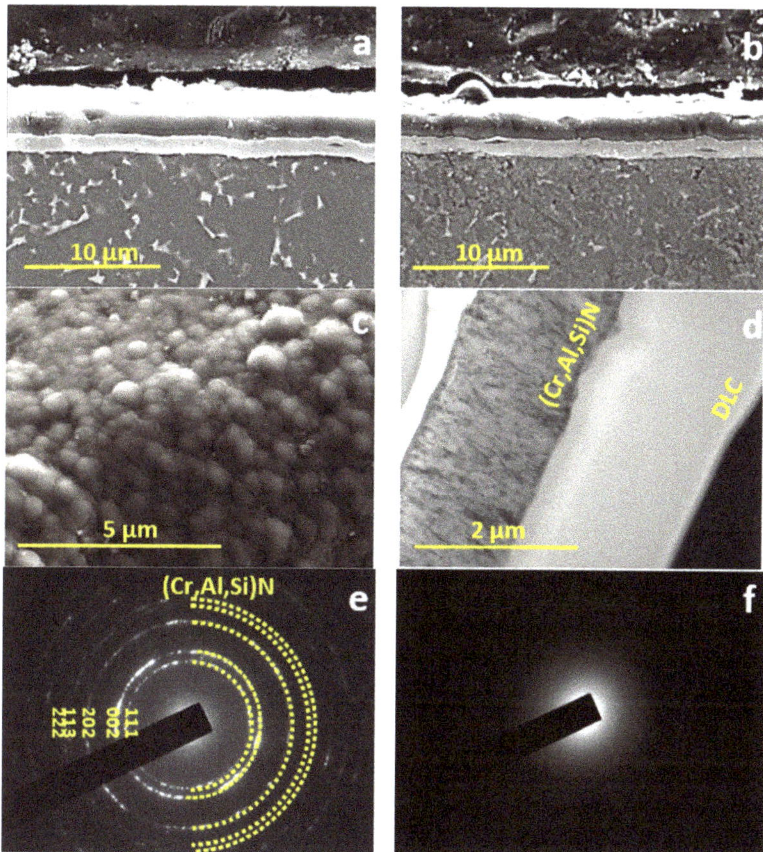

Figure 12. Microstructure of ceramic specimens with DLC coatings: (**a**) SEM image of a thin section of an experimental specimen 80% (90α10β) + 20% TiN ceramics with DLC coating; (**b**) SEM image of a thin section of commercial ceramics specimen with DLC coating; (**c**) SEM image of the DLC coating surface structure; (**d**) TEM image of the a two-layer DLC coating structure; (**e**) SAED of the (CrAlSi)N sublayer; and (**f**) SAED of the functional DLC layer.

Figure 13 shows the surface topographies of 80% (90α10β) + 20% TiN ceramic specimens before (Figure 13a) and after DLC coating deposition (Figure 13b). It is seen (Figure 13a) that the surface of the ceramic specimen includes a pronounced network of abrasive scratches, which are traditional for ceramic tools that are subjected to diamond sharpening. Besides, many craters from grains torn out during grinding are found on the ceramic surface previously shown in a series of previous works [17,21,25]. The noted features are defects of abrasive processing in essence, which are inherent in the process physics. The danger lies in the fact that ceramics are very sensitive to structural defects, which serve as stress concentrators and are often sources of tool's working surfaces spalling and chipping. It is possible to minimize or eliminate such defects in additional operations of finishing and polishing, which significantly increases the labor intensity and manufacturing cost of a ceramic product. The topography of the DLC-coated ceramic specimen surface (Figure 13b) demonstrates that the coating significantly modifies a thin surface layer's relief and significantly affects the size and shape of surface microroughness formed during the diamond sharpening stage. The coating fills the micro-grooves on the surface, thereby providing a kind of "healing" of the surface (it should be noted that a similar effect is observed when applying coatings of various compositions). There is reason to expect that the coating will affect the characteristics of ceramic samples taking into account th described effect and the fact that the hardness of the formed DLC coating is not less thar 25 GPa [73–77], which significantly exceeds the initial hardness of the tool ceramics, anc the coating has a lower coefficient of friction and good strength of the adhesive bond wit ceramic basis [78–80].

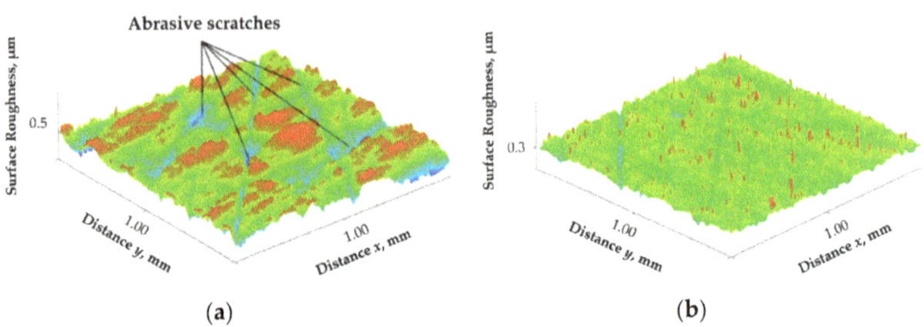

Figure 13. Topography of 80% (90α10β) + 20% TiN ceramic specimens surface areas: (**a**) before DL coating deposition and (**b**) after DLC coating deposition.

Figure 14 shows histograms that illustrate the relationship between the applied critica load, at which destruction occurs, the path of punch movement along the z-axis unti the destruction of the sintered ceramic samples, depending on the version of the powde composition. Data are given for samples before and after DLC coating deposition (quar tification for each option was carried out based on four tests' results). The experimenta data presented show that lower values of forces are required for the fracture of cerami specimens 80% (70α30β) + 20% TiN and 90% (90α10β) + 10% TiN. It indirectly indicate their lower resistance to brittle fracture (this is confirmed by the studies presented i Figure 9). The coating has minimal effect on the increase in mean breaking force for thes specimens. A different picture is observed for a sample of 80% (90α10β) + 20% TiN. Firstl the average critical force is 5.2 kN, which is 10–25% higher than the corresponding indicato for the other two compositions. Secondly, the DLC coating deposition up to 6.8 kN (b 30%) increases the average critical force at which the sintered ceramic specimen break down (an increase in the punch travel path after which fracture occurs is also observed The results obtained allow concluding that the coating can somewhat improve the fractur resistance of sintered ceramics based on SiAlON due to the transformation of the surfac properties. However, it is only observed if the coating is applied to a ceramic base wit a satisfactory combination of crack resistance and strength. It can be assumed that th improvements are a consequence of the effect mentioned above of "healing" of surfac defects to a certain extent.

A series of tribological tests were carried out at a load of 1 N and a sliding speed c 10 cm/s according to the "ball-ceramic disk" scheme (due to the nickel alloy's increase wear counter body, an Al_2O_3 ball was used as a material in these tests) to assess th contribution of the coating to the change in the properties of the ceramics based on SiAlO (80% (90α10β) + 20% TiN) surface layer. Figure 15 systematized the tests' results at a frictio length of 200 m before and after DLC coating deposition during tests without therma exposure and under high-temperature heating conditions. The traditional character of th change in the friction coefficients of ceramic samples over time is observed in the roor temperature results. At the very beginning of the tests, the uncoated ceramic sample ha a friction coefficient of 0.2, which increases rather quickly and reaches 0.8 after 100 m c distance, which remains until the end of the tests. A very stable behavior of the coatin during the entire test cycle, when the friction coefficient was invariably at the level of 0.1, observed after DLC coating deposition. This coating belongs to the anti-friction class and traditionally characterized by a reduced coefficient of friction. Considering that a cerami cutter during the processing of a nickel alloy is subjected to high thermal effects, which ha little in common with operating products at room temperature, the results of tribologic tests when heated to high temperatures are of the most significant interest. Under these conditions, SiAlON (80% (90α10β) + 20% TiN) samples without coating show unstable results, when the friction coefficient changes abruptly. Friction coefficient μ intensively increases after a value of 0.2 at the beginning and reaches a value of more than 0.7 after 25 m, then decreases to 0.45, and then it increases, decreases, and again increases, reaching more than 0.9 by the end of the tests. The authors described similar unstable results of other tool ceramics' behaviors (based on Al_2O_3) in a previously published study [25]. The DLC coating deposition onto ceramic samples strongly changes the conditions of frictional interaction: μ remains at a low level and varies slightly within the range of 0.09–0.15 for a sufficiently long time, and it begins to increase only after passing a distance of 150 m, reaching a value of 0.72 by the end of the tests. It can be assumed that the aforementioned is a consequence of the unique properties that a two-layer coating (CrAlSi)N/DLC possesses upon contact with the counter body under conditions of intense heat exposure. Superficial carbon DLC coating layers show poor results at increased thermal loads and often lose their initial microhardness, known from several authoritative works [12].

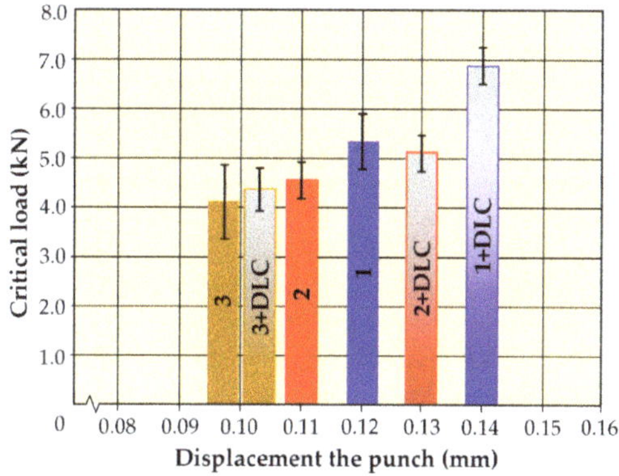

Figure 14. The relationship between the critical (breaking) load, the path of punch movement until the fracture of ceramic samples sintered from various powder compositions, where (1) is 80% (90α10β) + 20% TiN, (2) is for 90% (90α10β) + 10% TiN, and (3) is for 80% (70α30β) + 20% TiN before and after DLC coating deposition.

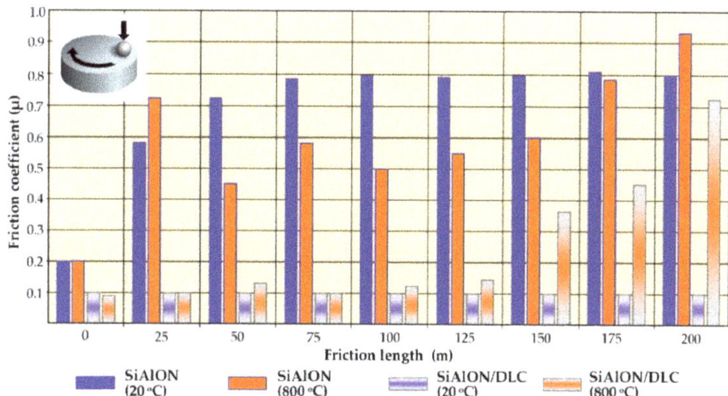

Figure 15. Dependences of the friction coefficient on the surface of experimental specimens made of SiAlON (80% (90α10β) + 20% TiN) ceramics on the friction path before and after DLC coating deposition under various test temperature conditions: without heating (20 °C) and with heating to 800 °C.

Nevertheless, due to the introduction of silicon into its composition during the deposition process, a formed DLC-Si layer makes it possible to significantly increase the DLC layer's thermal stability and expand the field of application of such coatings in this case [75,81,82]. In addition, the presence of a nitride sublayer under the DLC layer based on the Cr-Al-Si system increases the strength of the external DLC layer's adhesive bond with the tool base [83–85] and contributes to the formation of secondary wear-resistant phases during high-temperature heating [86–90]. When heated in oxygen, the coating components can form nonstoichiometric oxide phases, contributing to the change in the contact interaction between the ceramic article and the counter body nature [91–93].

3.3. Performance Testing of Solid Ceramic End Mills

The operational tests' main task was to compare the efficiency of two options of end mills: (1) experimental specimens of SiAlON (80% (90α10β) + 20% TiN) ceramics obtained by spark plasma sintering and (2) industrial designs of commercial ceramic mills based on SiAlON, alloyed with stabilizing additives containing Yb. Another important task was to assess the prospects for using DLC coatings to increase these tools' wear resistance when processing heat-resistant nickel alloys. For the tests, groups of ceramic end mills of identical design were used (Figure 2). Section 2.7 details the performance test procedure, cutting data, and material properties. Figure 16 shows the dependences of flank wear on the operating time of two variants of ceramic end mills before and after DLC coating deposition. It can be seen that the experimental end mills developed within the framework of this work (1) are quite workable, but there is relatively intense wear of their cutting teeth, and they are noticeably inferior in operating time until critical wear (resistance) is reached compared to the reference specimens (2) commercial ceramic mills (5 min versus 7.1 min, respectively). Simultaneously, the spread of wear values for a ceramic end mill's teeth for commercial ceramic mills during operation is noticeably smaller. Such differences in the two types of ceramic end mills' tool life are not unexpected if we consider the experimental data presented in Figure 10. The comparative analysis of the microstructures of two variants of ceramics performed earlier showed that for the experimental samples sintered by the authors, there is a less uniform distribution of components over the volume of the sintered ceramic and a larger average grain size compared with commercial SiAlON ceramic doped with stabilizing additives containing Yb. It is well known that these characteristics are critical for tool ceramics and largely determine the tool's performance, especially under conditions of exposure to cyclic loads typical for milling. The DLC coating deposition makes it possible to reduce the wear rate of the experimental ceramic end mills (3), produced by the authors of the work, and demonstrates an increase in durability by 1.5 times, which is 7.7 min, relative to the uncoated cutting tool; at the same time, the spread of wear values on the ceramic end mill teeth slightly decreased. An important observation that can be made is that after DLC coating, the experimental end mills' durability is comparable to commercial ceramic mills' service lives. It cannot be considered an irregular or random result since the DLC coating deposition on commercial ceramic mills (4) demonstrates the same tendency: the coatings reduce the tool's wear rate and increase its tool life by 1.5, which is 11.2 min. Separately, it should be noted that the phenomena of pronounced chipping of cutting teeth of ceramic end mills were not observed in any of the tool's tested samples when carrying out resistance tests. The increase in the resistance of experimental samples of ceramic mills and commercial ceramic mills after DLC coating deposition established in this work is undoubtedly a consequence of the changes in the properties of the surface and surface layer of ceramics described in detail in Section 3.2.

Additionally, the work evaluated DLC coatings' effects on the surface quality (surface roughness) of machined nickel alloy NiCr20TiAl workpieces (Figure 17). The analysis of the presented curves, which were obtained based on the test results of ceramic end mills without coatings, shows that the height of the formed surface microroughness on the workpiece's surface with the tool operation course has a non-monotonic character and characteristic peaks. Simultaneously, the curves obtained for experimental samples of ceramic end mills (1) and commercial ceramic mills (2) have a similar character. Considering that the level of antifriction properties of the tool surface strongly affects the quality of the workpiece surface layer, the nature of the experimental dependences of the processed workpiece surface roughness on the operating time is explained and in good agreement with the research results presented in Figure 15 (stepwise change with time in the friction coefficient of uncoated SiAlON-based ceramics under heating conditions up to 800 °C). Another regularity in the change in surface roughness can be observed during the operation of experimental (3) and commercial ceramic mills (4) after DLC coating deposition. The coating makes a pronounced contribution to improving the surface layer's quality of the workpiece being processed by changing the conditions of interaction with the cutting tool in the tribocontact zone. In the first minutes of operation, the surface roughness even slightly decreases and then begins to grow steadily. It is explained by the increase in the cutting tool wear zone and the actual area of its contact with the workpiece over time. Under such conditions, the intensity of their frictional and adhesive interaction increases, and the DLC coating contribution is leveled over time. Note that the described regularity correlates with the results of time variation of the friction coefficient SiAlON ceramic with a DLC coating under heating conditions up to 800 °C.

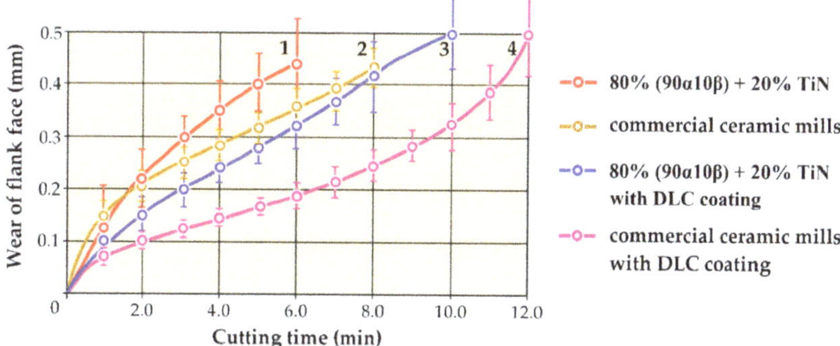

Figure 16. Dependence of flank face wear on the operating time of ceramic end mills before and after DLC coating deposition: (1) 80% (90α10β) + 20% TiN; (2) commercial ceramic mills; (3) 80% (90α10β) + 20% TiN with DLC coating; and (4) commercial ceramic mills with DLC coating.

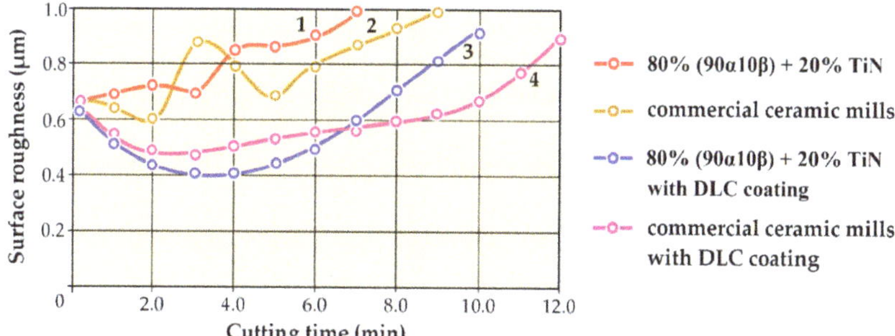

Figure 17. Dependence of the processed surface roughness on the operating time of ceramic end mills before and after DLC coating deposition: (1) 80% (90α10β) + 20% TiN; (2) commercial ceramic mills; (3) 80% (90α10β) + 20% TiN with DLC coating; and (4) commercial ceramic mills with DLC coating.

4. Conclusions

The complex of experimental studies carried out allows obtaining several original results that can be used to develop further the science-intensive direction associated with the creation and maintenance of effective operating conditions for solid end mills made of ceramics based on SiAlON powder compositions for machining heat-resistant nickel alloys:

1. The rational quantitative ratio between the components of the α/β-SiAlON-TiN powder composition for spark plasma sintering ceramics is α/β-SiAlON 80 wt.% (α-SiAlON 90 wt.% + β-SiAlON 10 wt.%) + TiN 20 wt.%. The temperature of 1700 °C is a rational value for the spark plasma sintering powder composition based on SiAlON since the samples made of experimental ceramics exhibit the best combination of theoretical density, crack resistance, and flexural strength at this value. The developed experimental ceramics are characterized by the presence in the structure of equiaxial and non-equiaxial grains with an average grain size of 3–5 μm compared to reference SiAlON-based commercial ceramics (including expensive Yb-containing stabilizing phases) that are used on an industrial scale with grain size of 1.5–3.5 μm. The grains of the commercial ceramics are more evenly distributed over the volume, and the structure is more homogeneous compared to developed ones. Everything indicates that the Yb-containing stabilizing phase is formed along the grain boundaries of the SiAlON-based ceramics (elemental EDX analysis of the fracture of the sample showed Yb of about 7.0%). This ceramic inhibits grain growth during sintering and ensures the formation of a finer-grained and uniform structure compared with samples from experimental α/β-SiAlON-TiN ceramics. Nevertheless, the end mills of experimental ceramics are still efficient in processing the heat-resistant nickel alloy NiCr20TiAl at a cutting speed of 376.8 m/min. However, there is more intense wear of the cutting teeth, and they are noticeably inferior in durability to the reference samples of commercial ceramic end mills (5 and 7.1 min, respectively). Besides, experimental end mills have a wider spread of wear values on the tool's teeth during operation.
2. The deposition of a two-layer coating (CrAlSi)N/DLC with a thickness of about 3.6 μm significantly affects the properties of the surface and surface layer of sintered ceramics. The coating significantly modifies the relief of a thin surface layer and significantly affects the size and shape of surface microroughness and some defects formed at the stage of diamond sharpening. It somehow fills the micro-grooves on the surface, providing a kind of surface layer "healing." The experiments also show that the coating can somewhat improve the fracture resistance of sintered ceramics based on SiAlON due to the transformation of the surface properties, which manifests itself as an increase of up to 30% in the average critical force at which the fracture of the sintered ceramics occurs (5.2 kN for uncoated ceramics and 6.8 kN for ceramic coated (CrAlSi)N/DLC). However, the effect mentioned above is possible when the coating is applied to a ceramic base with a satisfactory combination of fracture toughness and strength.
3. Tribological tests have shown that (CrAlSi)N/DLC coating noticeably changes the conditions of frictional interaction of SiAlON-based ceramics with a counter body and provides a more stable ceramic sample surface layer operation under intense temperature exposure compared to ceramics without coatings. The friction coefficient for uncoated specimens changes abruptly and varies in the range from 0.2 to 0.9 showing instability throughout the entire friction distance. For the (CrAlSi)N/DLC coating, the friction coefficient remains at a low level for a relatively long time and varies slightly within the range of 0.09–0.15, and begins to increase only at the end of the tests, reaching 0.72.
4. Performance tests in laboratory conditions have shown that the application of (CrAlSi)N/DLC coating significantly affects the service life (durability) of solid ceramic end mills based on SiAlON powder compositions (both experimental end mill samples and commercial ceramic end mills) during processing heat-resistant nickel alloys such as NiCr20TiAl. The coating reduces the wear rate of experimental ceramic end mills relative to uncoated tools by a factor of 1.5 at a cutting speed of 376.8 m/min and provides durability of 7.7 min, which is comparable to the service life of commercial ceramic end mills. Furthermore, the application of a diamond-like carbon coating

on commercial ceramic mills demonstrates a similar result when the tool's wear rate decreases and its tool service life increases by 1.5 times and is 11.2 min. It is important to note that the coating contributes to improving the quality of the surface layer of the workpiece being processed by changing the conditions of interaction with the cutting tool in the tribocontact zone. In the first minutes of operation, the workpiece's surface roughness even slightly decreases and then begins to grow steadily. It is explained by the increase over time in the cutting tool wear area (land) and the actual area of its contact with the workpiece being processed. At the moment of failure, it is about 0.7 μm.

5. Deposition of (CrAlSi)N/DLC coating can be considered a promising way to improve the performance of ceramic end mills. Nevertheless, it is impossible to cover all issues and study the behavior of solid end mills made of ceramics in a wide range of changes in operating conditions within the framework of one work. First of all, research is needed at increased cutting speeds, which should be the next step in developing this scientific direction since it is possible to test the efficiency of (CrAlSi)N/DLC coating under high-speed milling conditions of heat-resistant nickel alloys only experimentally. Besides, it is necessary to search and test the effectiveness of new options for coatings and other surface finishing methods of sintered ceramics, which could significantly improve the properties of the surface and surface layer of ceramic end mills and thereby positively affect the performance of the tool.

Author Contributions: Conceptualization, S.N.G.; methodology, M.A.V.; validation, S.V.F. and A.A.O.; formal analysis, P.M.P., P.Y.P. and A.E.; investigation, P.M.P., P.Y.P. and A.E.; resources, P.M.P. and P.Y.P.; data curation, A.A.O. and A.E.; writing—original draft preparation, M.A.V.; writing—review and editing, S.N.G.; visualization, A.A.O.; supervision, M.A.V.; project administration, S.N.G. All authors have read and agreed to the published version of the manuscript.

Funding: The study was supported by a grant of the Russian Science Foundation (project No. 21-79-30058).

Institutional Review Board Statement: Not applicable.

Informed Consent Statement: Not applicable.

Data Availability Statement: The data presented in this study are openly available in Figure 1a at (https://www.mmsonline.com/blog/post/applications-advance-for-solid-ceramic-end-mills, (accessed on 29 April 2021); https://www.cnctimes.com/editorial/solid-ceramic-endmills-from-kennametal-help-meet-critical-delivery-date, (accessed on 29 April 2021)), reference number [7,8]).

Acknowledgments: The work was carried out in the High-efficiency processing department of Moscow State University of Technology "STANKIN".

Conflicts of Interest: The authors declare no conflict of interest.

References

1. Available online: https://www.mitsubishicarbide.com/en/technical_information/tec_rotating_tools/tec_solid_end_mills/tec_solid_end_mills_guide/tec_solid_end_mills_precautions_ceramic (accessed on 29 April 2021).
2. Molaiekiya, F.; Stolf, P.; Paiva, J.M.; Bose, B.; Goldsmith, J.; Gey, C.; Engin, S.; Fox-Rabinovich, G.; Veldhuis, S.C. Influence of process parameters on the cutting performance of SiAlON ceramic tools during high-speed dry face milling of hardened Inconel 718. *Int. J. Adv. Manuf. Technol.* **2019**, *105*, 1083–1098. [CrossRef]
3. Zheng, G.; Zhao, J.; Zhou, Y.; Li, A.; Cui, X.; Tian, X. Performance of graded nano-composite ceramic tools in ultra-high-speed milling of Inconel 718. *Int. J. Adv. Manuf. Technol.* **2013**, *67*, 2799–2810. [CrossRef]
4. Fadok, J. Advanced Gas Turbine Materials, Design and Technology. In *Advanced Power Plant Materials, Design and Technology*; Roddy, D., Ed.; Woodhead Publishing Series in Energy: Cambridge, UK, 2010; pp. 3–31.
5. Grigoriev, S.N.; Grechishnikov, V.A.; Volosova, M.A.; Kozochkin, M.P.; Pivkin, P.; Isaev, A.V.; Peretyagin, P.Y.; Seleznev, A.E.; Minin, I.V. Machining high-temperature alloys by means of solid ceramic end mills. *Russ. Eng. Res.* **2020**, *40*, 79–82. [CrossRef]
6. Uhlmann, E.; Hübert, C. Tool grinding of end mill cutting tools made from high performance ceramics and cemented carbides. *CIRP Ann.* **2011**, *60*, 359–362. [CrossRef]
7. Available online: https://www.mmsonline.com/blog/post/applications-advance-for-solid-ceramic-end-mills (accessed on 29 April 2021).

8. Available online: https://www.cnctimes.com/editorial/solid-ceramic-endmills-from-kennametal-help-meet-critical-delivery-date (accessed on 29 April 2021).
9. Available online: https://www.iscar.ru/Products.aspx/countryid/27/ProductId/12291 (accessed on 29 April 2021).
10. Available online: https://www.sandvik.coromant.com/en-gb/news/press_releases/pages/ceramic-end-mills-take-the-heat-in-hrsa-machining.aspx (accessed on 29 April 2021).
11. Grguras, D.; Kern, M.; Pusavec, F. Suitability of the full body ceramic end milling tools for high speed machining of nickel based alloy Inconel 718. *Procedia CIRP* **2018**, *77*, 630–633. [CrossRef]
12. Bitterlich, B.; Bitsch, S.; Friederich, K. SiAlON based ceramic cutting tools. *J. Eur. Ceram. Soc.* **2008**, *28*, 989–994. [CrossRef]
13. Zheng, G.; Zhao, J.; Gao, Z.; Cao, Q. Cutting performance and wear mechanisms of Sialon–Si3N4 graded nano-composite ceramic cutting tools. *Int. J. Adv. Manuf. Technol.* **2011**, *58*, 19–28. [CrossRef]
14. Eser, O.; Kurama, S. The effect of the wet-milling process on sintering temperature and the amount of additive of SiAlON ceramics. *Ceram. Int.* **2010**, *36*, 1283–1288. [CrossRef]
15. Smirnov, K.L. Sintering of SiAlON ceramics under high-speed thermal treatment. *Powder Metall. Met. Ceram.* **2012**, *51*, 76–82. [CrossRef]
16. Herrmann, M.; Höhn, S.; Bales, A. Kinetics of rare earth incorporation and its role in densification and microstructure for-mation of α-SiAlON. *J. Eur. Ceram. Soc.* **2012**, *32*, 1313–1319. [CrossRef]
17. Volosova, M.A.; Grigor'ev, S.N.; Kuzin, V.V. Effect of titanium nitride coating on stress structural inhomogeneity in ox-ide-carbide ceramic. Part 4. Action of Heat Flow. *Refract. Ind. Ceram.* **2015**, *56*, 91–96. [CrossRef]
18. Smirnov, K.L.; Grigoryev, E.G.; Nefedova, E.V. Current-assisted sintering of combustion-synthesized β-SiAlON ceramics. *Int. J. Self-Propagating High-Temp. Synth.* **2019**, *28*, 28–33. [CrossRef]
19. Kuzin, V.V.; Grigoriev, S.N.; Fedorov, M.Y. Role of the thermal factor in the wear mechanism of ceramic tools. Part 2: Microlevel. *J. Frict. Wear* **2015**, *36*, 40–44. [CrossRef]
20. Chi, I.S.; Bux, S.K.; Bridgewater, M.M.; Star, K.E.; Firdosy, S.; Ravi, V.; Fleurial, J.-P. Mechanically robust SiAlON ceramics with engineered porosity via two-step sintering for applications in extreme environments. *MRS Adv.* **2016**, *1*, 1169–1175. [CrossRef]
21. Kuzin, V.; Grigoriev, S.N.; Volosova, M.A. The role of the thermal factor in the wear mechanism of ceramic tools: Part 1. Macrolevel. *J. Frict. Wear* **2014**, *35*, 505–510. [CrossRef]
22. Grigoriev, S.; Pristinskiy, Y.; Volosova, M.; Fedorov, S.; Okunkova, A.; Peretyagin, P.; Smirnov, A. Wire electrical discharge machining, mechanical and tribological performance of TiN reinforced multiscale SiAlON ceramic composites fabricated by spark plasma sintering. *Appl. Sci.* **2021**, *11*, 657. [CrossRef]
23. Lukianova, O.A.; Novikov, V.Y.; Parkhomenko, A.A.; Sirota, V.V.; Krasilnikov, V.V. Microstructure of spark plasma-sintered silicon nitride ceramics. *Nanoscale Res. Lett.* **2017**, *12*, 293. [CrossRef]
24. Kaidash, O.N.; Fesenko, I.P.; Kryl', Y.A. Heat conductivity, physico-mechanical properties and interrelations of them and structures of pressureless sintered composites produced of Si3N4-Al2O3-Y2O3(-ZrO2) nanodispersed system. *J. Superhard Mater.* **2014**, *36*, 96–104. [CrossRef]
25. Volosova, M.; Grigoriev, S.; Metel, A.; Shein, A. The role of thin-film vacuum-plasma coatings and their influence on the efficiency of ceramic cutting inserts. *Coatings* **2018**, *8*, 287. [CrossRef]
26. Grigoriev, S.; Volosova, M.; Fyodorov, S.; Lyakhovetskiy, M.; Seleznev, A. DLC-coating application to improve the durability of ceramic tools. *J. Mater. Eng. Perform.* **2019**, *28*, 4415–4426. [CrossRef]
27. Kuzin, V.V.; Grigor'ev, S.N.; Volosova, M.A. Effect of a TiC coating on the stress-strain state of a plate of a high-density nitride ceramic under nonsteady thermoelastic conditions. *Refract. Ind. Ceram.* **2014**, *54*, 376–380. [CrossRef]
28. Xu, X.; Nishimura, T.; Hirosaki, N.; Xie, R.-J.; Yamamoto, Y.; Tanaka, H. Fabrication of β-sialon nanoceramics by high-energy mechanical milling and spark plasma sintering. *Nanotechnology* **2005**, *16*, 1569–1573. [CrossRef]
29. Szutkowska, M.; Cygan, S.; Podsiadło, M.; Laszkiewicz-Łukasik, J.; Cyboroń, J.; Kalinka, A. Properties of TiC and TiN rein-forced alumina–zirconia composites sintered with spark plasma technique. *Metals* **2019**, *9*, 1220. [CrossRef]
30. Jojo, N.; Shongwe, M.B.; Tshabalala, L.C.; Olubambi, P.A. Effect of sintering temperature and yttrium composition on the densification, microstructure and mechanical properties of spark plasma sintered silicon nitride ceramics with Al_2O_3 and Y_2O_3 additives. *Silicon* **2019**, *11*, 2689–2699. [CrossRef]
31. Lee, C.H.; Lu, H.H.; Wang, C.A.; Nayak, P.K.; Huang, J.L. Microstructure and mechanical properties of TiN/Si_3N_4 nanocomposites by spark plasma sintering (SPS). *J. Alloys Compd.* **2010**, *508*, 540–545. [CrossRef]
32. Ayas, E.; Kara, A.; Kara, F. A novel approach for preparing electrically conductive α/β SiAlON-TiN composites by spark plasma sintering. *J. Ceram. Soc. Jpn.* **2008**, *116*, 812–814. [CrossRef]
33. Kandemir, A.; Sevik, C.; Yurdakul, H.; Turana, S. First-principles investigation of titanium doping into β-SiAlON crystal in TiN-SiAlON composites for EDM applications. *Mater. Chem. Phys.* **2015**, *162*, 781–786. [CrossRef]
34. Calloch, P.; Trompetter, W.J.; Brown, I.W.M.; MacKenzie, K.J. Oxidation resistance of β-Sialon/TiN composites: An ion beam analysis (IBA) study. *J. Mater. Sci.* **2018**, *53*, 15348–15361. [CrossRef]
35. Canarslan Özgür, S.; Rosa, R.; Köroğlu, L.; Ayas, E.; Kara, A.; Veronesi, P. Microwave sintering of SiAlON ceramics with TiN addition. *Materials* **2019**, *12*, 1345. [CrossRef]
36. Wang, L.; Wu, T.; Jiang, W.; Li, J.; Chen, L. Novel fabrication route to Al_2O_3-TiN nanocomposites via spark plasma sintering. *J. Am. Ceram. Soc.* **2006**, *89*, 1540–1543. [CrossRef]

7. Nordberg, L.-O.; Nygren, M.; Käll, P.-O.; Shen, Z. Stability and oxidation properties of RE-α-Sialon ceramics (RE = Y, Nd, Sm, Yb). *J. Am. Ceram. Soc.* **2005**, *81*, 1461–1470. [CrossRef]
8. Nekouee, K.A.; Khosroshahi, R.A. Synthesis and properties evaluation of β-SiAlON prepared by mechanical alloying followed by different sintering technique. *Int. J. Mater. Res.* **2016**, *107*, 1129–1135. [CrossRef]
9. Lu, H.H.; Huang, J.L. Effect of Y_2O_3 and Yb_2O_3 on the microstructure and mechanical properties of silicon nitride. *Ceram. Int.* **2001**, *27*, 621–628. [CrossRef]
0. Kim, M.S.; Go, S.I.; Kim, J.M.; Park, Y.J.; Kim, H.N.; Ko, J.W.; Yun, J.D. Microstructure and mechanical properties of β-SiAlON ceramics fabricated using self-propagating high-temperature synthesized β-SiAlON powder. *J. Korean Ceram. Soc.* **2017**, *54*, 292–297. [CrossRef]
1. Sobol', O.V.; Andreev, A.A.; Grigoriev, S.N.; Gorban', V.F.; Volosova, M.A.; Aleshin, S.V.; Stolbovoy, V.A. Physical charac-teristics, structure and stress state of vacuum-arc TiN coating, deposition on the substrate when applying high-voltage pulse during the deposition. *Probl. At. Sci. Technol.* **2011**, *4*, 174–177.
2. Grigoriev, S.; Volosova, M.; Vereschaka, A.; Sitnikov, N.; Milovich, F.; Bublikov, J.; Fyodorov, S.; Seleznev, A. Properties of (Cr,Al,Si)N-(DLC-Si) composite coatings deposited on a cutting ceramic substrate. *Ceram. Int.* **2020**, *46*, 18241–18255. [CrossRef]
3. Sousa, V.F.C.; Silva, F.J.G. Recent advances on coated milling tool technology—A comprehensive review. *Coatings* **2020**, *10*, 235. [CrossRef]
4. Sobol, O.V.; Andreev, A.A.; Grigoriev, S.N.; Volosova, M.A.; Gorban, V.F. Vacuum-arc multilayer nanostructured TiN/Ti coatings: Structure, stress state, properties. *Met. Sci. Heat Treat.* **2012**, *54*, 28–33. [CrossRef]
5. Volosova, M.A.; Grigor'ev, S.N.; Kuzin, V.V. Effect of tinaium nitride coatings on stress structural inhomogeneity in oxide-carbide ceramic. Part 2. Concentrated force action. *Refract. Ind. Ceram.* **2015**, *55*, 487–491. [CrossRef]
6. Vereschaka, A.S.; Grigoriev, S.N.; Tabakov, V.P.; Sotova, E.S.; Vereschaka, A.A.; Kulikov, M.Y. Improving the efficiency of the cutting tool made of ceramic when machining hardened steel by applying nano-dispersed multi-layered coatings. *Key Eng. Mater.* **2014**, *581*, 68–73. [CrossRef]
7. Uhlmann, E.; Hühns, T.; Richarz, S.; Reimers, W.; Grigoriev, S. Development and application of coated ceramic cutting tools. *Adv. Sci. Technol.* **2006**, *45*, 1155–1162. [CrossRef]
8. ISO 3310-1:2016. *Test Sieves—Technical Requirements and Testing—Part 1: Test Sieves of Metal Wire Cloth*; ISO: Geneva, Switzerland, 2016.
9. Grigoriev, S.N.; Volosova, M.A.; Peretyagin, P.Y.; Seleznev, A.E.; Okunkova, A.A.; Smirnov, A. The effect of TiC additive on mechanical and electrical properties of Al_2O_3 ceramic. *Appl. Sci.* **2018**, *8*, 2385. [CrossRef]
0. Li, Y.; Zheng, G.; Cheng, X.; Yang, X.; Xu, R.; Zhang, H. Cutting performance evaluation of the coated tools in high-speed milling of AISI 4340 steel. *Materials* **2019**, *12*, 3266. [CrossRef] [PubMed]
1. Metel, A.; Bolbukov, V.; Volosova, M.; Grigoriev, S.; Melnik, Y. Source of metal atoms and fast gas molecules for coating deposition on complex shaped dielectric products. *Surf. Coat. Technol.* **2013**, *225*, 34–39. [CrossRef]
2. Vereschaka, A.; Grigoriev, S.; Popov, A.; Batako, A. Nano-scale multilayered composite coatings for cutting tools operating under heavy cutting conditions. *Procedia CIRP* **2014**, *14*, 239–244. [CrossRef]
3. Santecchia, E.; Hamouda, A.; Musharavati, F.; Zalnezhad, E.; Cabibbo, M.; Spigarelli, S. Wear resistance investigation of titanium nitride-based coatings. *Ceram. Int.* **2015**, *41*, 10349–10379. [CrossRef]
4. Vereschaka, A.; Volosova, M.; Grigoriev, S. Development of wear-resistant complex for high-speed steel tool when using process of combined cathodic vacuum arc deposition. *Procedia CIRP* **2013**, *9*, 8–12. [CrossRef]
5. Metel, A.; Bolbukov, V.; Volosova, M.; Grigoriev, S.; Melnik, Y. Equipment for deposition of thin metallic films bombarded by fast argon atoms. *Instrum. Exp. Tech.* **2014**, *57*, 345–351. [CrossRef]
6. Kohlscheen, J.; Bareiss, C. Effect of hexagonal phase content on wear behaviour of AlTiN arc PVD coatings. *Coatings* **2018**, *8*, 72. [CrossRef]
7. Vereschaka, A.; Tabakov, V.; Grigoriev, S.; Aksenenko, A.; Sitnikov, N.; Oganyan, G.; Seleznev, A.; Shevchenko, S. Effect of adhesion and the wear-resistant layer thickness ratio on mechanical and performance properties of ZrN—(Zr,Al,Si)N coatings. *Surf. Coat. Technol.* **2019**, *357*, 218–234. [CrossRef]
8. Hao, G.; Liu, Z. Experimental study on the formation of TCR and thermal behavior of hard machining using TiAlN coated tools. *Int. J. Heat Mass Transf.* **2019**, *140*, 1–11. [CrossRef]
9. Grigoriev, S.N.; Melnik, Y.A.; Metel, A.S.; Panin, V.V.; Prudnikov, V.V. A compact vapor source of conductive target material sputtered by 3-keV ions at 0.05-Pa pressure. *Instrum. Exp. Tech.* **2009**, *52*, 731–737. [CrossRef]
0. Metel, A.S.; Grigoriev, S.N.; Melnik, Y.A.; Bolbukov, V.P. Characteristics of a fast neutral atom source with electrons inject-ed into the source through its emissive grid from the vacuum chamber. *Instrum. Exp. Tech.* **2012**, *55*, 288–293. [CrossRef]
1. Sobol', O.V.; Andreev, A.A.; Grigoriev, S.N.; Gorban', V.F.; Volosova, M.A.; Aleshin, S.V.; Stolbovoi, V.A. Effect of high-voltage pulses on the structure and properties of titanium nitride vacuum-arc coatings. *Met. Sci. Heat Treat.* **2012**, *54*, 195–203. [CrossRef]
2. Anders, S.; Callahan, D.L.; Pharr, G.M.; Tsui, T.Y.; Bhatia, C.S. Multilayers of amorphous carbon prepared by cathodic arc deposition. *Surf. Coat. Technol.* **1997**, *94*, 189–194. [CrossRef]
3. Durmaz, Y.M.; Yildiz, F. The wear performance of carbide tools coated with TiAlSiN, AlCrN and TiAlN ceramic films in intelligent machining process. *Ceram. Int.* **2019**, *45*, 3839–3848. [CrossRef]

64. Fominski, V.Y.; Grigoriev, S.N.; Gnedovets, A.G.; Romanov, R.I. Pulsed laser deposition of composite Mo–Se–Ni–C coatings using standard and shadow mask configuration. *Surf. Coat. Technol.* **2012**, *206*, 5046–5054. [CrossRef]
65. Grigoriev, S.N.; Sobol, O.V.; Beresnev, V.M.; Serdyuk, I.V.; Pogrebnyak, A.D.; Kolesnikov, D.A.; Nemchenko, U.S. Tribological characteristics of (TiZrHfVNbTa)N coatings applied using the vacuum arc deposition method. *J. Frict. Wear.* **2014**, *35*, 359–364. [CrossRef]
66. Danek, M.; Fernandes, F.; Cavaleiro, A.; Polcar, T. Influence of Cr additions on the structure and oxidation resistance of multilayered TiAlCrN films. *Surf. Coat. Technol.* **2017**, *313*, 158–167. [CrossRef]
67. Volosova, M.A.; Grigoriev, S.N.; Ostrikov, E.A. Use of laser ablation for formation of discontinuous (discrete) wear-resistant coatings formed on solid carbide cutting tool by electron beam alloying and vacuum-arc deposition. *Mech. Ind.* **2016**, *17*, 720. [CrossRef]
68. Siwawut, S.; Saikaew, C.; Wisitsoraat, A.; Surinphong, S. Cutting performances and wear characteristics of WC inserts coated with TiAlSiN and CrTiAlSiN by filtered cathodic arc in dry face milling of cast iron. *Int. J. Adv. Manuf. Technol.* **2018**, *97*, 3883–3892. [CrossRef]
69. Grigoriev, S.N.; Volosova, M.A.; Fedorov, S.V.; Mosyanov, M. Influence of DLC coatings deposited by PECVD technology on the wear resistance of carbide end mills and surface roughness of $AlCuMg_2$ and $41Cr_4$ workpieces. *Coatings* **2020**, *10*, 1038. [CrossRef]
70. Derakhshandeh, M.R.; Eshraghi, M.J.; Hadavi, M.M.; Javaheri, M.; Khamseh, S.; Sari, M.G.; Zarrintaj, P.; Saeb, M.R.; Mozafari, M. Diamond-like carbon thin films prepared by pulsed-DC PE-CVD for biomedical applications. *Surf. Innov.* **2018**, *6*, 167–175. [CrossRef]
71. ASTM. *ASTM G99—17 Standard Test Method for Wear Testing with a Pin-on-Disk Apparatus*; ASTM: West Conshohocken, PA, USA, 2017.
72. British Standards Institution. *BS EN 10269:2013 Steels and Nickel Alloys for Fasteners with Specified Elevated and/or Low Temperature Properties*; British Standards Institution: London, UK, 2013.
73. Robertson, J. Diamond-like amorphous carbon. *Mater. Sci. Eng. R Rep.* **2002**, *37*, 129–281. [CrossRef]
74. Wei, C.; Yang, J.-F. A finite element analysis of the effects of residual stress, substrate roughness and non-uniform stress distribution on the mechanical properties of diamond-like carbon films. *Diam. Relat. Mater.* **2011**, *20*, 839–844. [CrossRef]
75. Hainsworth, S.V.; Uhure, N.J. Diamond like carbon coatings for tribology: Production techniques, characterisation methods and applications. *Int. Mater. Rev.* **2007**, *52*, 153–174. [CrossRef]
76. Zou, C.W.; Wang, H.J.; Feng, L.; Xue, S.W. Effects of Cr concentrations on the microstructure, hardness, and tempera-ture dependent tribological properties of Cr-DLC coatings. *Appl. Surf. Sci.* **2013**, *286*, 137–141. [CrossRef]
77. Jeon, Y.; Park, Y.S.; Kim, H.J.; Hong, B.; Choi, W.S. Tribological properties of ultrathin DLC films with and without metal interlayers. *J. Korean Phys. Soc.* **2007**, *51*. [CrossRef]
78. Martinez-Martinez, D.; De Hosson, J. On the deposition and properties of DLC protective coatings on elastomers: A critical review. *Surf. Coat. Technol.* **2014**, *258*, 677–690. [CrossRef]
79. Aijaz, A.; Ferreira, F.; Oliveira, J.; Kubart, T. Mechanical properties of hydrogen free diamond-like carbon thin films deposited by high power impulse magnetron sputtering with Ne. *Coatings* **2018**, *8*, 385. [CrossRef]
80. Dolgov, N.A. Analytical methods to determine the stress state in the substrate–coating system under mechanical loads. *Strength Mater.* **2016**, *48*, 658–667. [CrossRef]
81. Wu, Y.; Li, H.; Ji, L.; Ye, Y.; Chen, J.; Zhou, H. Vacuum tribological properties of a-C:H film in relation to internal stress and applied load. *Tribol. Int.* **2014**, *71*, 82–87. [CrossRef]
82. Nakazawa, H.; Kamata, R.; Miura, S.; Okuno, S. Effects of frequency of pulsed substrate bias on structure and properties of silicon-doped diamond-like carbon films by plasma deposition. *Thin Solid Film.* **2015**, *574*, 93–98. [CrossRef]
83. Liu, X.; Yang, J.; Hao, J.; Zheng, J.; Gong, Q.; Liu, W. A near-frictionless and extremely elastic hydrogenated amorphous carbon film with self-assembled dual nanostructure. *Adv. Mater.* **2012**, *24*, 4614–4617. [CrossRef] [PubMed]
84. Lubwama, M.; Corcoran, B.; McDonnell, K.; Dowling, D.; Kirabira, J.; Sebbit, A.; Sayers, K. Flexibility and frictional behaviour of DLC and Si-DLC films deposited on nitrile rubber. *Surf. Coat. Technol.* **2014**, *239*, 84–94. [CrossRef]
85. Zhang, T.; Pu, J.; Xia, Q.; Son, M.; Kim, K. Microstructure and nano-wear property of si-doped diamond-like carbon films deposited by a hybrid sputtering system. *Mater. Today Proc.* **2016**, *3*, S190–S196. [CrossRef]
86. Vopát, T.; Sahul, M.; Haršáni, M.; Vortel, O.; Zlámal, T. The tool life and coating-substrate adhesion of AlCrSiN-coated carbide cutting tools prepared by LARC with respect to the edge preparation and surface finishing. *Micromachines* **2020**, *11*, 166. [CrossRef] [PubMed]
87. Bobzin, K.; Brögelmann, T.; Kruppe, N.; Carlet, M. Nanocomposite (Ti,Al,Cr,Si)N HPPMS coatings for high performance cutting tools. *Surf. Coat. Technol.* **2019**, *378*, 124857. [CrossRef]
88. Vereshchaka, A. Improvement of working efficiency of cutting tools by modifying its surface properties by application of wear-resistant complexes. *Adv. Mat. Res.* **2013**, *712*, 347–351.
89. Georgiadis, A.; Fuentes, G.G.; Almandoz, E.; Medrano, A.; Palacio, J.F.; Miguel, A. Characterisation of cathodic arc evaporated CrTiAlN coatings: Tribological response at room temperature and at 400 °C. *Mater. Chem. Phys.* **2017**, *190*, 194–201. [CrossRef]
90. Endrino, J.; Fox-Rabinovich, G.; Gey, C. Hard AlTiN, AlCrN PVD coatings for machining of austenitic stainless steel. *Surf. Coat. Technol.* **2006**, *200*, 6840–6845. [CrossRef]

91. Shuai, J.; Zuo, X.; Wang, Z.; Guo, P.; Xu, B.; Zhou, J.; Wang, A.; Ke, P. Comparative study on crack resistance of TiAlN monolithic and Ti/TiAlN multilayer coatings. *Ceram. Int.* **2020**, *46*, 6672–6681. [CrossRef]
92. Chang, Y.-Y.; Chuang, C.-C. Deposition of multicomponent AlTiCrMoN protective coatings for metal cutting applications. *Coatings* **2020**, *10*, 605. [CrossRef]
93. Li, G.; Li, L.; Han, M.; Luo, S.; Jin, J.; Wang, L.; Gu, J.; Miao, H. The performance of TiAlSiN coated cemented carbide tools enhanced by inserting Ti interlayers. *Metals* **2019**, *9*, 918. [CrossRef]

Review

Quo Vadis: AlCr-Based Coatings in Industrial Applications

Joerg Vetter [1,*], Anders O. Eriksson [2], Andreas Reiter [2], Volker Derflinger [2] and Wolfgang Kalss [2]

1 Oerlikon Balzers Coating Germany GmbH, Am Boettcherberg 30–38, 51427 Bergisch Gladbach, Germany
2 Oerlikon Balzers, Oerlikon Surface Solutions AG, Iramali 18, LI-9496 Balzers, Liechtenstein; anders.o.eriksson@oerlikon.com (A.O.E.); andreas.reiter@oerlikon.com (A.R.); Volker.Derflinger@oerlikon.com (V.D.); wolfgang.kalss@oerlikon.com (W.K.)
* Correspondence: joerg.vetter@oerlikon.com

Abstract: AlCr-based hard nitride coatings with different chemical compositions and architectures have been successfully developed and applied over the last few decades. Coating properties are mainly influenced by deposition conditions and the Al/Cr content. The fcc structure is dominant for an Al-content up to $Al_{0.7}Cr_{0.3}N$ and is preferred for most cutting applications. Different (AlCrX)N alloying concepts, including X = Si, W, B, V, have been investigated in order to enhance oxidation resistance and wear behaviour and to provide tribological properties. AlCr-based oxynitrides and even pure oxides $(Al_{1-x}Cr_x)_2O_3$ with different crystalline structures have been explored. Multi- and nanolayered coatings within the AlCr materials system, as well as in combination with (TiSi)N, for example, have also been implemented industrially. The dominant deposition technology is the vacuum arc process. Recently, advanced high-power impulse magnetron sputtering (HiPIMS) processes have also been successfully applied on an industrial scale. This paper describes basic coating properties and briefly addresses the main aspects of the coating processes as well as selected industrial applications.

Keywords: AlCr-based; CrAl-based; (AlCrX)N; $(Al_{1-x}Cr_x)_2O_3$; arc; HiPIMS; nanolayers; nanocomposite; structure; properties

Citation: Vetter, J.; Eriksson, A.O.; Reiter, A.; Derflinger, V.; Kalss, W. Quo Vadis: AlCr-Based Coatings in Industrial Applications. *Coatings* 2021, *11*, 344. https://doi.org/10.3390/coatings11030344

Academic Editor: Armando Yáñez-Casal

Received: 12 February 2021
Accepted: 15 March 2021
Published: 18 March 2021

1. Introduction

Surface solutions using PVD (Physical Vapor Deposition) coatings for the improvement of the wear resistance and tribological properties of tools and components started on an industrial scale around the middle of the 1970s in the Soviet Union. The first component coating was MoN [1]. The first successful industrial coating for tooling applications in Western Europe was the "golden" TiN coating applied by low-voltage arc evaporation developed at Oerlikon Balzers [2]. A modification of this coating with carbon followed, and the TiNC coating types were implemented in mass production at the end of the 1980s. In parallel, Cr-based coatings were also introduced, including modifications using oxygen and carbon [3]. Besides nitride, carbonitride and oxynitride coatings, hard amorphous carbon coatings, the DLC (Diamond Like Carbon) coatings, were developed and applied as well [4,5]. These coatings were deposited using several coating architectures [6]. Many coating developments focused on alloying TiN using different additional elements [5].

The most important industrial development after TiN, CrN and DLC was achieved in the Ti-Al-N system [7]. Around the year 2000, this coating family was the dominant coating, especially for cutting tools [8,9]. However, in the Ti-Al-N system, only minor improvements of coating properties and the performance of cutting tool applications were possible, e.g., by alloying with Si, Y, Cr, C, O [10–13]. A big step forward was achieved through the investigation of totally new coating compositions. This led to the development of a new coating generation based on Al-Cr-N. Coatings in the AlCrN system span the compositional range from Al-rich to Cr-rich. In this paper, the terminology CrAl-based is used for coatings with a Cr content that is higher than that of Al (Cr/Al > 1 in at.%), and conversely, AlCr-based is used for coatings in which Al/Cr > 1.

Results dealing with coatings with a low Al content (Al/(Al + Cr) ratio < 0.5) and (CrAl)N coatings, deposited by sputtering, were published for the first time in 1990 [14,15]. In the same year, coatings with an Al/(Al + Cr) composition of about 0.5 deposited using a modified hollow cathode discharge process were also presented [16]. Arc evaporation was used to deposit (AlCr)N coatings with a high Al content with an (Al/(Al + Cr) ratio of >0.5 [17,18]. The first multilayer coating with an Al content greater than the Cr content was a multilayer architecture with CrN [19]. Another big step was the development of (TiSi)N coatings [20]. Coating solutions were also developed which combined the two coating types in multilayer structures [21].

A challenge for turning tools was, and still is in part, the development of alpha alumina using PVD. A PVD solution to add metallic doping elements to aluminium was patented in 1992 [22]. Processes to deposit oxides based on arc evaporation were successfully developed on the basis of Al-Cr cathodes [23].

This paper will review the main industrial-oriented coating developments using CrAl- and AlCr-based coatings, covering a wide variety of chemical and structural fields as well as architectures. The main deposition methods used will be described briefly. The review will concentrate on arc deposition processes, although sputtering techniques including high-power impulse magnetron sputtering (HiPIMS), will be discussed as well. The main focus is on arc deposition processes and nitride coatings with an Al content greater than the Cr content and on (AlCr)ON coatings. Coatings with a cubic structure and various multilayer coatings are used most commonly in industrial applications. Selected application examples will be presented in the last section.

2. Deposition Technologies and Coating Systems

Various deposition technologies have been used to synthesise (CrAl)N and (AlCr)N coatings over a period of about 30 years. Basic information about the deposition technologies is beyond the scope of this review. The interested reader is invited to look at available publications [24–31]. Some more exotic evaporation methods are hollow cathode arc evaporation [16] and activated reactive evaporation [32]. The first application of AlCr-based coatings for cutting tools deposited by hollow cathode arc was mentioned very early [16,33]. However, primarily sputtering processes and arc processes were used both for basic investigations as well as for applications.

The first reported (CrAl)N coatings were DC (Direct Current) magnetron sputtered [14,15]. The first arc coatings were deposited using elemental cathodes of pure Al and pure Cr [17]. The first multilayer coating (AlCr)N was (AlCr)N/CrN deposited by arc [19]. Sometimes even hybrid methods, such as arc plus hollow cathode discharge [34] or, more commonly arc plus sputtering [35], have been used. Alloyed CrAl-based and AlCr-based coatings are most commonly deposited either by magnetron sputtering or by arc evaporation. For sputtered coatings, the preferred methods are various HiPIMS processes in industrial applications, e.g., the S3p process [30]. The deposition of alpha alumina coating is based on a dedicated pulsed arc process, the P3e process [23].

It should be mentioned that the main differences between arc deposition processes and sputtering are:

Growth defects: Arc-deposited coatings have more growth defects in the coating than sputtered coatings due to macroparticle generation in the arc evaporation process. Sputtered coatings are thus preferred in applications requiring particularly low surface roughness.

Growth rate: The arc process allows a higher growth rate to be achieved than sputtering, depending on the PVD system set-up.

Degree of ionisation and ion energies: Arc evaporation is known to produce plasmas with a high degree of ionisation. Multiple charge states are formed as well in arc evaporation [36]. Sputtering processes are characterised by lower degrees of ionisation [26].

Energy consumption: The specific energy consumption of the arc process is lower than that of sputtered coatings for the same coating thickness [37].

Process stability: Arc processes tend to exhibit higher process stability than sputtering processes.

Larger coating volumes: Arc processes can be more easily upscaled for coating long parts (several metres).

The different basic methods used for CrAl-based and AlCr-based coatings are listed here briefly.

2.1. *Arc Deposition Methods*

CrAl-based and AlCr-based coatings are deposited by reactive direct vacuum arc using either elemental cathodes or composite cathodes. The first (CrAl)N and (AlCr)N coatings were deposited by the reactive co-evaporation of elemental chromium and aluminium cathodes available at that time [17–19].

The first composite cathodes suitable for arc evaporation manufactured by a powder metallurgical process were produced by Plansee AG, Austria, and were used successfully for tool coatings in 1995 [38]. Later, composite cathodes became the standard, especially in industrial applications [39,40]. A special pulsed process was developed for alpha $(Al_{1-x}Cr_x)_2O_3$ coatings, the P3e process [23]. The main goal of the pulsing is to overcome process difficulties generated by the oxide process (cathode, anode reactions).

2.2. *Sputtering Methods*

2.2.1. DC and RF Sputtering

Several sputtering methods as well as combinations of different sputtering methods have been used to deposit both CrAl-based and AlCr-based coatings from elemental or composite cathodes. Even special processes based on the combination of composite targets combined with elemental targets have been realised [41]. The first CrAl-based coatings were deposited by reactive DC magnetron sputtering of a composite target $(Cr_{75}Al_{75})$ [14,15]. Additionally, newer research on a laboratory scale as well as industrial-scale PVD systems are using DC magnetron sputtering [41–46]. It should be noted that special magnetic field configurations can be achieved within the PVD systems by means of the magnetron set-ups, e.g., the CFUBM (Closed Field Unbalanced Magnetron Sputtering) [47,48]. RF (Radio Frequency) sputtering has been used for basic investigation [49,50]. Even the combination of RF sputtering plus DC sputtering has been applied [51,52], though RF sputtering is typically not used for industrial applications.

2.2.2. Standard Pulsed Sputtering and HiPIMS

Besides standard pulsed magnetron sputtering [41], the mode of dual pulsed magnetrons has also been used [48]. The most advanced pulsed sputtering method is high-power impulse magnetron sputtering. Pulses with peak power densities in the range of several hundred W/cm^2 up to several thousand W/cm^2 are achieved in classic high-power impulse magnetron sputtering (HiPIMS), also referred to as HPPMS, MPP or HIPAC. The pulse duration is usually in the range of 25 to 250 μs [41,53–56]. The longest pulse lengths of around 1000 μs are used in the case of MPP [57].

A dedicated HiPIMS process, known as the S3p process, is a process that runs with a constant pulse current and can operate using longer pulses than the classic HiPIMS process. The pulse duration often lies in the range of several milliseconds [30].

2.3. *Industrial Coating Systems*

A brief description of selected aspects of industrial systems will be provided here. In-depth descriptions of deposition systems are available elsewhere [27,30]. Both system configurations, set-ups with magnetrons and set-ups with arc evaporators, use particle sources with either a circular, rectangular or tubular geometry for the active evaporation surfaces, as shown in Figure 1. The circular form dominates for arc evaporation, whereas the rectangular and the cylindrical forms are used most commonly for sputtering. In addition to arc and sputtering systems, hybrid systems, e.g., arc plus HiPIMS, are also in use [29].

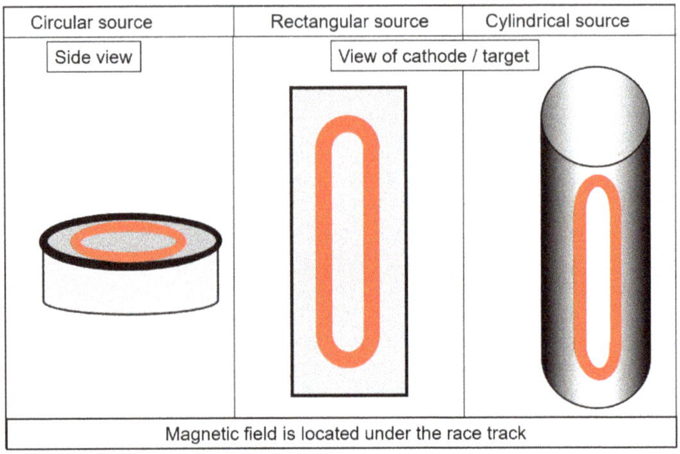

Figure 1. Schematic drawing of the circular, rectangular or tubular active evaporation surfaces, redrawn after [27], original © Vulkan-Verlag, Germany.

For the deposition of AlCr-based coatings, powder metallurgically manufactured composite targets are used most commonly. Sputtering has the advantage that composite targets in the form of segmented targets or targets with plugs of a second type of material (e.g., Cr plugs in Al target plates) can also be used [48].

Modern coating systems run in fully automatic operation. They must fulfil the following criteria with respect to high productivity and quality: high production reliability, short cycle times, high flexibility in coating types and substrate holders, easy maintenance. Additional aspects are CE (Conformité Européenne), conformity and high occupational safety standards. Environmental sustainability is also gaining importance, including the influence of factors such as energy consumption.

Of course, the basic components of the coating systems must also be optimised, e.g., vacuum pumps, power supplies, particle sources, heaters, substrate holders. In addition, all process steps including loading, pumping, heating, ion cleaning, coating, cooling, unloading and maintenance must be optimised for short cycle times and efficient operation. The choice between batch systems or inline systems depends on the use and the required flexibility. Batch systems are predominant for industrial coating applications. The system size is selected based on the expected batch size as determined by the dimensions and number of parts to be coated, but also by operating economy.

The usable volume of a batch system is defined by the interior size of the deposition chamber. Usually, coating systems are designed with a circular geometry along a central axis, allowing the rotation of substrates. The maximum useful volume therefore constitutes a cylindrical body. Small coating systems typically have a diameter of <0.5 m and a height of <0.5 m, which is adequate for small-scale series or research facilities. Medium size systems, with roughly a diameter of 1 m and a height of 1 m, are predominant in industrial production. Figure 2 shows an arc system equipped with circular arc evaporators, and a coatable diameter of 0.7 m and a height of 0.9 m. Special-purpose systems are available for large-scale manufacturing, with roughly a diameter of 1.5 m and a height of 1.5 m. Systems for oversized parts, such as broaches or plastic extrusion screws, have a coating height of up to 4.5 m.

Figure 2. Medium-sized arc system equipped with circular arc evaporators, diameter 0.7 m, height 0.9 m, courtesy of Oerlikon Balzers.

3. Basic Properties of CrAlN and AlCrN Coatings

CrAl-based and AlCr-based coatings deposited by arc evaporation processes or by magnetron sputtering are widely used for cutting tools, moulds, dies and for various components. The success of these coatings with their predominantly fcc structure is due to their outstanding mechanical and tribological properties (such as high hot hardness, good abrasive and sliding wear resistance) combined with high oxidation and corrosion resistance. The present chapter highlights selected basic coating properties.

To begin with, an important remark must be made on the way the coating composition is reported in papers. A full compositional characterisation, including metallic and non-metallic elements as well as impurities such as oxygen, is the most complete, but is not always reported. Several publications neglect the stoichiometric aspects, the deviation of coating composition from the cathode/target composition, and the incorporation of residual gas components. Sometimes, only the cathode/target composition or only one of the values Al/Cr or Al/(Al + Cr) are given, presumably because methods such as EDS are most suited for the characterisation of metallic and heavy elements. Many papers state that the coatings consist stoichiometrically of 50 at.% metallic and 50 at.% non-metallic elements, and they are simply described as $(Al_{1-x}Cr_x)N$.

One positive example is described for arc-deposited coatings using cathodes of $Al_{70}Cr_{30}$. The coating was characterised by XPS as having Al 33.1 at.%, Cr 15.8 at.%, N 48.1 at.%, and O 3.0 at.%. The metallic content is thus 48.9 at.% and the non-metallic content is 51.1 at.%. The ratio Al/(Al + Cr) is 0.68, meaning 68 at.% of the metallic content is Al. This corresponds to a deviation of 2 at.% from the cathode material. The Al/Cr ratio in the coating is 2.09. The coating contains oxygen from the residual gas. The coating is slightly over-stoichiometric [58]. The total formula has to be $(Al_{1-x}Cr_x)(N_{1-w}O_w)$ plus the stochiometric ratio $(Al_{1-x}Cr_x)/(N_{1-w}O_w)$. Unfortunately, however, detailed compositional data are not reported in many publications.

As a guide for the reading of coating compositions, the following terminology is used throughout this paper.

1. If only the cathode composition is given and the stoichiometry in the coating is assumed to be equivalent to the cathode composition, the coating is described as $(Al_{(100-x)}Cr_x)N$ with 100 in at.%, e.g., (Al70Cr30)N.
2. If the metallic elements (and metalloids) were measured, but only a general statement about the stoichiometry is made, the coatings are described as $(Al_{(100-x)}C_x)N$ with 100 in at.%, e.g., $(Al_{70}Cr_{30})N$.
3. If both metallic elements (and metalloids) and the N, O (and C) content were measured, all values are given as $Al_xCr_yN_uO_w$, where $x + y + u + w = 100$ at.%, e.g., $Al_{33}Cr_{16}N_{48}O_3$.

3.1. The Influence of Al Content on Lattice Parameters, Phases and Microhardness

A schematic diagram of different basic crystal structures as a function of the Al portion of the coating's metallic content in at.% is shown in Figure 3. It should be mentioned that in publications, the formula $Cr_{1-x}Al_xN$ or $Al_{1-x}Cr_xN$ is sometimes used for the same coating in dependence on the Al content. Nowadays, the most commonly used terminology in industrial applications is the short name (AlCr)N for Al-rich coatings. The authors suggest that coatings with a chemical composition in at.% of Cr > Al should be named (CrAl)N. If Al > Cr, then the coating should be named (AlCr)N. (CrAl)N coatings always have an fcc crystal structure. This structure type is also referred to as B1, a NaCl structure or c in certain publications, depending on the convention. With increasing Al content, a phase evolution to an hcp crystal structure takes place, which is also referred to as B4, a ZnS-type structure, wurtzite, w, h or hcp in publications. In the following, fcc and hcp will be used

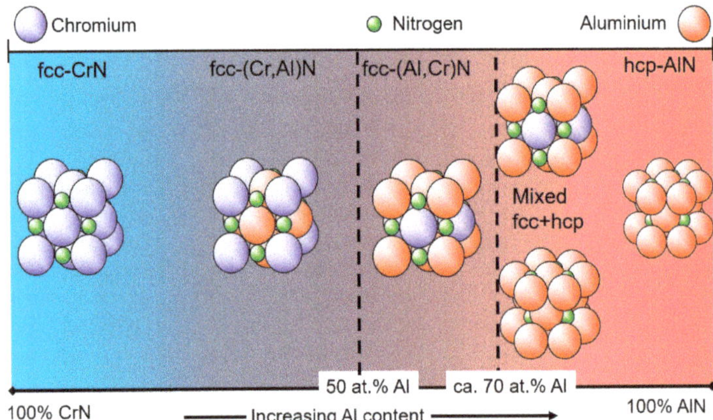

Figure 3. Schematic diagram showing the basic crystal structure of (CrAl)N and (AlCr)N coatings as a function of the Al content.

XRD investigations of rf-sputtered coatings have shown that at 57 at.% Al, a pure fcc structure was obtained, while at 75 at.% Al, the hcp structure was observed. The transition range between fcc and fcc + hcp was in the range of 57 at.% < Al_{max} < 67 at.% [49]. With pulsed closed-field magnetron sputtering, the formation of hcp phases was observed at an Al content of 64 at.%, whereas fcc phases were detected at up to 60.9 at.%, giving a transition range of 60.9 at.% < Al_{max} < 64 at.% [59]. A systematic experimental investigation of (CrAl)N and (AlCr)N coatings synthesised using the cathodic arc method showed that the crystal structure changed from a pure fcc structure to a mixed-phase structure of fcc and hcp at an Al content of about 60–70 at.% of the metal content in the cathodes [60], whereas an fcc structure was observed at up to 71 at.% in the coating by the authors of [61].

The reported different maximum Al contents for the X-ray-diffraction-measured pure fcc phase differ over a range of about 60–70 at.%, as shown in Table 1.

Table 1. Maximum Al content for pure fcc phase generation measured by X-ray diffraction for arc-deposited and sputtered coatings.

Deposition Method	Source Composition (Targets/Cathodes)	Pure fcc Phases by X-Ray at Al [at.%] Coating or Source	Fcc + hcp Phases by X-Ray at Al [at.%] Coating or Source	Range of Transition Al_{max} [at.%]	Reference
RF sputtering	Al and Cr targets	Coating 57 at RT 56 at 300 °C	Coating 75 at RT 67 at 300 °C	Coating $57 < Al_{max} < 67$	[49]
RF/DC sputtering	Al and Cr targets	Coating 63 at 300 °C	-	$63 < Al_{max}$	[51]
Pulsed CFUBM sputtering	Al and Cr targets	Coating 60.9 at 175 °C	Coating 64 at 175 °C	Coating $60.9 < Al_{max} < 64$	[59]
Arc	Alloyed cathodes	Cathodes 60 at 600 °C	Cathodes 70 at 600 °C	Cathodes $60 < Al_{max} < 70$	[60]
Arc	Alloyed cathodes	Coating 70 at 500 °C TEM traces hcp	-	Coating $70 < Al_{max}$	[62]
Arc	Alloyed cathodes	Coating 71 at 450 °C	Coating 75 at 450 °C	Coating $71 < Al_{max} < 83$	[61]
Arc	Alloyed cathodes	Coating 71 at 450 °C	Coating 75 at 450 °C	Coating $71 < Al_{max} < 75$	[63]
DC sputtering	Segmented target	Coating 70 at 400 °C	-	Coating $70 < Al_{max}$	[64]
DC sputtering	Alloyed targets	Targets [1] 70 at 500 °C	-	Targets $70 < Al_{max}$	[65]
Arc	Alloyed cathodes	Coating 70 at 500 °C	Coating 82 at 500 °C	Coating $70 < Al_{max} < 82$	[66]

[1] source: no measurement of chemical composition in the coating, values from cathodes/targets.

The deposition process itself (source properties, parameters) has an influence at the maximum Al concentration on whether a pure fcc phase is obtained using X-ray diffraction, as will be shown for arc evaporation in Section 3.4.

Taking measurement uncertainties into account, a well-accepted maximum critical value of the transition is about 70 at.% Al [61–66]. It should be mentioned that coatings with Al contents of 65–70 at.% might contain some traces of hcp phases, which may, however, be difficult to detect using XRD. For example, this effect was observed with SAED (Selected Area Electron Diffraction) measurement of (AlCr)N coatings deposited using arc evaporation from cathodes of $Al_{70}Cr_{30}$ on sapphire [62]. The critical value of the transition from fcc to the mixed fcc plus hcp structure also depends slightly on the deposition conditions, influenced, for example, by the evaporator magnetic field set-up analogously to AlTiN [9], but bias and deposition pressure also have an influence on the "fine" structure in the area of the transition [45].

Figure 4 shows experimental results for the hardness and lattice parameters of arc-deposited (CrAl)N and (AlCr)N coatings [60,61]. Both studies show a hardness increase of about 60% compared to CrN for fcc-structured coatings in the range of Al 60–70 at.%. The differences in the absolute values are likely related to different deposition techniques and hardness measurement conditions. Furthermore, in the region of the mixed-phase structure of fcc + hcp at high Al-content, the hardness is equivalent to CrN. The lattice parameters decrease from 0.415–0.416 nm (CrN)) to 0.413 nm in the region of Al 60–70 at.%.

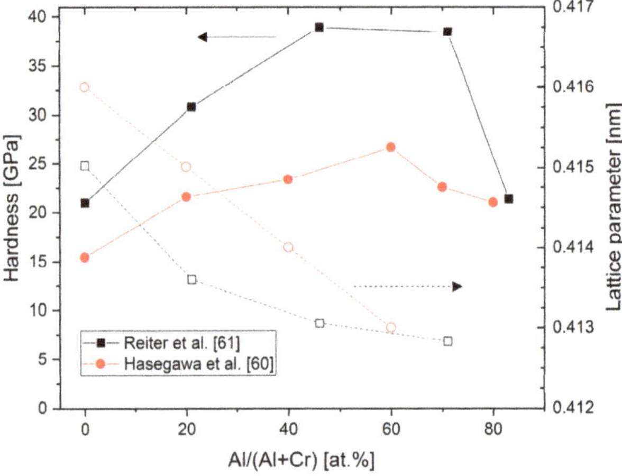

Figure 4. Hardness and lattice parameters of (CrAl)N and (AlCr)N coatings versus Al content deposited by cathodic vacuum arc, redrawn after [60,61], original © Elsevier.

Lattice parameters for (CrAl)N, (AlCr)N, CrN and AlN coatings calculated ab initio are shown in Figure 5 and are compared with experimental XRD results [63]. The good fit of the calculated and the experimental data shows how well the basic properties of AlCr-based coatings can be calculated.

3.2. Mechanical and Physical Properties

3.2.1. Thermal Expansion

The thermal expansion coefficients (TECs) of fcc-structured (Cr,Al)N and (Al,Cr)N coatings deposited by sputtering were investigated using synchrotron X-ray diffraction at up to 600 °C [65]. It was shown that the thermal expansion coefficient increases with an increasing Al content, from about 7×10^{-6}/K at room temperature to 10×10^{-6}/K at 600 °C, see Figure 6. Higher mean values of 14.5×10^{-6}/K were reported both for sputtered and arc-deposited $(Al_{70}Cr_{30})N$ coatings at room temperature [43]. A fair general estimate is thus $(10.5 \pm 3.5) \times 10^{-6}$/K.

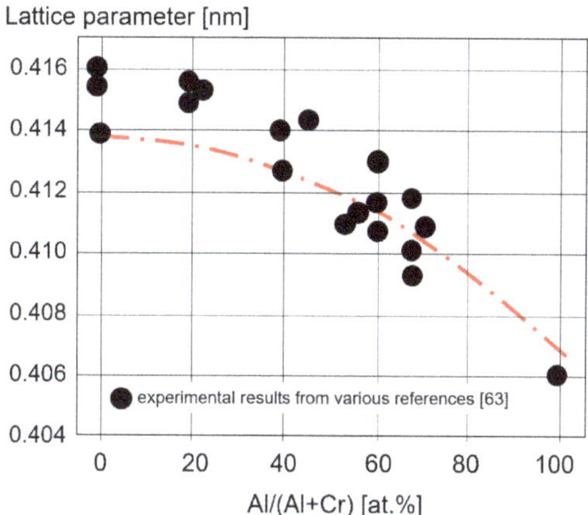

Figure 5. Lattice parameter calculated ab initio for CrN, (CrAl)N, (AlCr)N, and AlN coatings and experimentally measured XRD values, redrawn after [63], original © Elsevier.

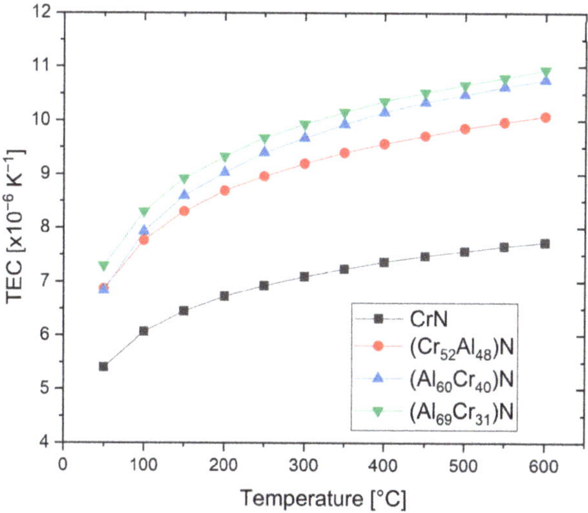

Figure 6. Temperature-dependent thermal expansion coefficients (TECs) of (CrAl)N and (AlCr)N coatings derived from synchrotron experiments, redrawn after [65], original © Elsevier.

3.2.2. Thermal Conductivity

Figure 7 shows the thermal conductivity of $(Ti_{50}Al_{50})N$, $(Al_{66}Ti_{34})N$, and $(Al_{70}Cr_{30})N$ coatings [67]. The thermal conductivity is temperature-dependent and it is interesting to note that the thermal conductivity of $(Al_{70}Cr_{30})N$ drops at about 200 °C and, in the temperature range of 250–450 °C, is significantly lower than that of $(Al_{66}Ti_{34})N$. The thermal properties of $Cr_{25}Al_{20.5}Si_{4.5}N_{50}$ coatings were measured using pulsed photothermal radiometry. A very low thermal conductivity of ca. 2.75 W/mK at room temperature and 3.5 W/mK at 400 °C was found [68]. PVD coatings exhibit a certain anisotropy of thermal conductivity perpendicular and parallel to the direction of growth. This effect is particularly pronounced for multilayer coatings. When engineering the thermal properties of a coated part, not only the intrinsic thermal conductivity of the coatings, but also the concentration and dimension of different growth defects, e.g., holes and droplets, must be taken into account [69].

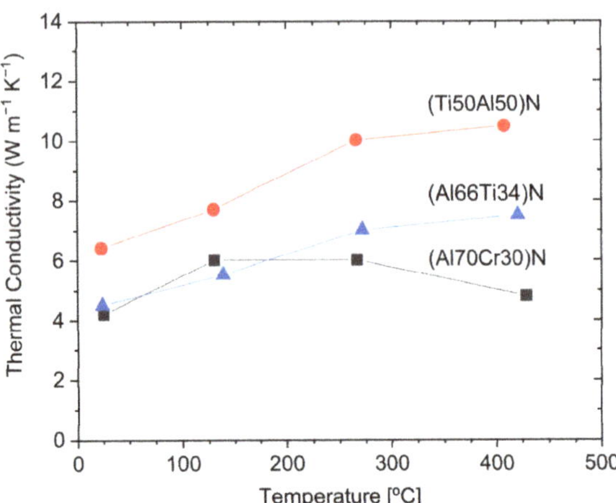

Figure 7. The temperature dependence of thermal conductivity for arc-deposited ($Ti_{50}Al_{50}$)N, ($Al_{66}Ti_{34}$)N and ($Al_{70}Cr_{30}$)N coatings, measured using the picosecond thermal reflection method, redrawn after [67], original © Elsevier.

3.2.3. Electrical Resistivity

A two-probe measurement method was used to estimate the surface resistance as a function of the Al/(Al + Cr) ratio for arc-deposited coatings. It can be seen in Figure 8 that the electrical resistivity increases sharply with the Al portion of the metallic content in at.% [19].

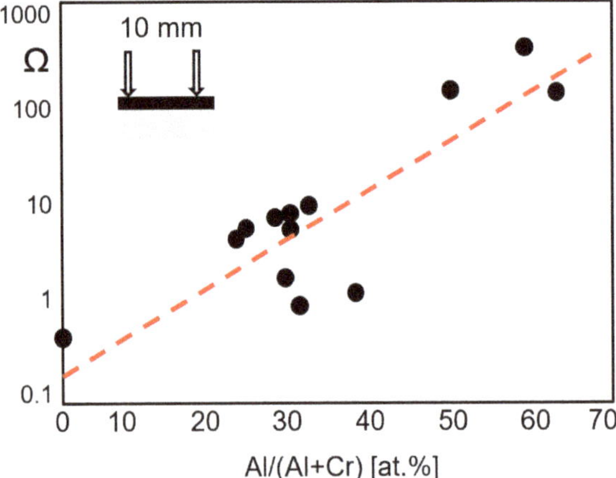

Figure 8. Surface resistance as a function of the Al content of fcc-(Cr,Al)N and fcc-(Al,Cr)N arc deposited coatings, redrawn after [19], original © Elsevier.

3.2.4. Poisson's Ratio, Young's Moduli, Fracture Toughness

Ab initio calculations have shown that the Poisson's ratio drops from about 0.27 for low Al contents to about 0.2 for an Al content of 70 at.% [70]. Figure 9 shows the ab initio calculated Young's moduli (E) as a function of the Al content for fcc (Al,Cr)N [70,71]. Selected Young's moduli measured by nanoindentation from [64,71] have been added.

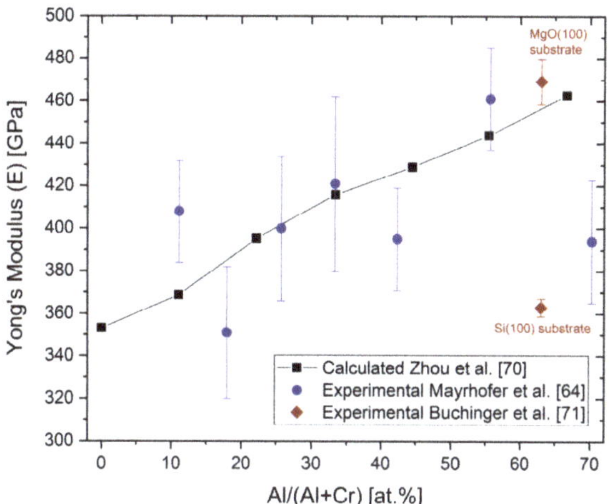

Figure 9. Ab initio calculated Young's moduli E and experimentally obtained indentation moduli (black symbols) for different Al concentrations in fcc-(Cr,Al)N and fcc-(Al,Cr)N coatings, redrawn after [70], original © AIP Publishing, data from [64,71].

The Young's modulus increases with an increasing Al content. The same relative tendency, but with lower absolute values, was experimentally shown in [43]. The calculated values are in good agreement with measured nanoindentation moduli [64,72]. The Young's modulus of (AlCr)N coatings drops significantly when the mixed-phase structure fcc + hcp is reached [43], as well as for Si-alloyed (AlCr)N coatings of, for example, $Al_{33.4}Cr_{18.3}Si_{2.3}N_{46}O_{0.7}$ [73]. It should be noted that large variations of measured E values for fcc-(Al,Cr)N coatings have been published for coatings of nearly the same composition, for example, from 300 GPa (Al/(Al + Cr) content of 68 at.%) [73] to 469 GPa (Al/(Al + Cr) content of 63 at.%) [71]. Different sample conditions (thickness and substrate) and measurement systems themselves will influence the measured value.

In addition to the chemical composition, the Young's modulus is also dependent on the grain size (grain boundary fraction), the compressive stress state, and the texture. A variation from 363 to 469 GPa was measured for $Al_{28.4}Cr_{18.6}N_{55}$ in different coating states. The high value was measured for a highly 100-oriented coating deposited on MgO (100) substrates [71]. Thus, a relatively wide variation of experimental results, besides the measuring conditions, must be anticipated.

The fracture toughness K_{IC} of sputtered $Al_{28.4}Cr_{18.6}N_{55}$ was measured by micromechanical bending tests at 1.3 ± 0.1 MPa $\sqrt{m}$, which is lower than that of TiN at 2.0 ± 0.1 MPa $\sqrt{m}$ related to domain size effects, but it is possible to raise this in a superlattice combination of the two materials [71].

3.2.5. Thermal Phase Stability and Hardness after Annealing

The thermal stability of arc-deposited (CrAl)N and (AlCr)N coatings has been investigated through heat treatment in an Ar atmosphere with a hold time of 2 h at temperatures between 600 and 1300 °C in combination with XRD analysis [61]. The fcc structure was found to be stable up to 800 °C. Three reactions were observed sequentially for the (AlCr)N-system:

1. Transformation of fcc (Al,Cr)N to hcp (Al,Cr)N at the grain boundaries first [74],
2. Segregation of Cr_2N,
3. Segregation of pure chromium.

Above 900 °C, the entire coating transforms to a mixed-phase structure. No phase transformation occurred in the hcp-AlN coatings, as they were already in their thermodynamically stable state. This is shown schematically in Figure 10.

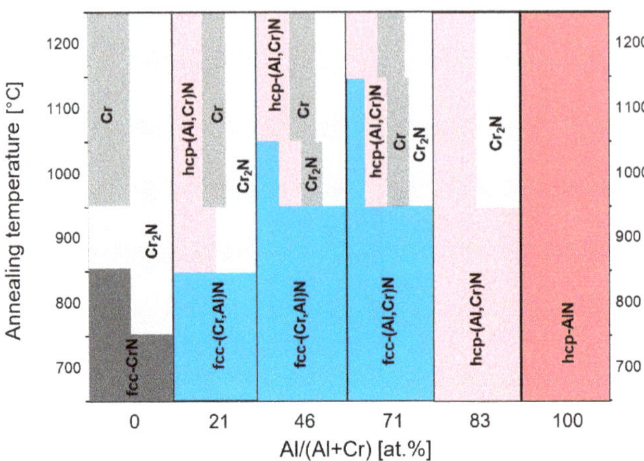

Figure 10. Schematic diagram of phase stability for (CrAl)N and (AlCr)N as determined by XRD spectra after annealing at different temperatures, redrawn after [61], original © Elsevier.

The effect of the annealing on the hardness of the same coatings was characterised as well, see Figure 11.

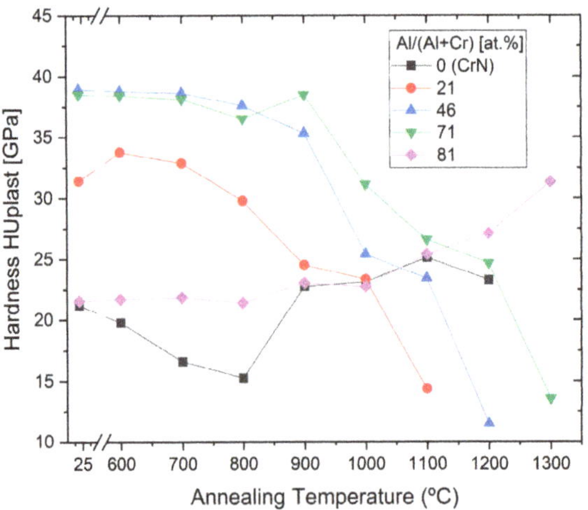

Figure 11. Microhardness for (CrAl)N, (AlCr)N, CrN and AlN coatings after annealing at different temperatures, redrawn after [61], original © Elsevier.

The highest hardness was shown for ($Cr_{54}Al_{46}$)N and ($Al_{71}Cr_{29}$)N, and was retained up to 800 and 900 °C, respectively. The subsequent drop in hardness at higher annealing temperatures is correlated with the phase decomposition [61].

The influence of the coating stress at the decomposition temperature was investigated by varying the deposition temperature. The coating deposition was performed by arc using $Al_{70}Cr_{30}$ cathodes. It was shown that the decomposition temperatures of the metastable fcc-($Al_{63.5}Cr_{36.5}$)N phase depends significantly on the stress level in the coatings. The decomposition process starts at the same compressive stress level of around 4.3 GPa for all coatings that were investigated [75].

Figure 12 shows a comparison of the hardness after annealing at high temperatures for different commercial hard coatings, showing that the ($Al_{70}Cr_{30}$)N coating has the highest hot hardness for temperatures exceeding 950 °C.

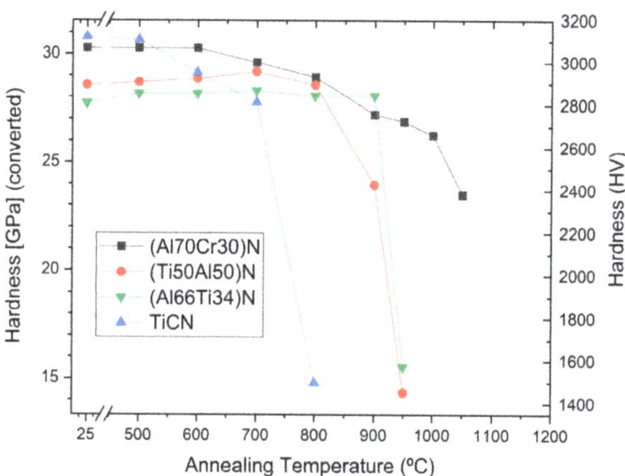

Figure 12. Hardness after annealing of commercial TiCN, $(Ti_{50}Al_{50})N$, $(Al_{66}Ti_{34})N$ and $(Al_{70}Cr_{30})N$ coatings on cemented carbide.

3.2.6. Oxidation Behaviour

The oxidation behaviour of PVD hard coatings is an important property for applications such as dry high speed cutting and high temperature stressed components such as turbochargers. In a study of Cr-rich (CrAl)N sputtered coatings (Cr/Al = 2.94), a relatively small Al content ($Cr_{44}Al_{15}N_{40}$ at.%) improved the oxidation stability compared to CrN [15]. The first publication about the oxidation characteristics of arc-deposited coatings (AlCr)N with a Al/Cr ratio > 0.5 showed a lower oxidation rate in comparison with CrN [19], as has been confirmed in further studies [51,61,76,77]. Oxidation tests of (CrAl)N and (AlCr)N coatings deposited by activated reactive evaporation showed excellent behaviour [32]. The authors stated that the investigated (CrAl)N and (AlCr)N coatings exhibited significantly better oxidation behaviour than TiAlN coatings, which has been confirmed in further studies [40,78]. A systematic comparison of the oxidation behaviour of arc-deposited (AlTi)N and (AlCr)N coatings, including long time exposure, was performed in [40]. It was shown that the oxidation resistance of both coatings improved with an increase in the Al content. The oxidation resistance of (AlCr)N coatings was significantly superior to that of (AlTi)N coatings. Figure 13 shows the oxide layer thickness of CrN, (CrAl)N and (AlCr)N coatings after annealing in the temperature range of 800 to 1000 °C in an ambient atmosphere for 30 min in 50 and 100 °C intervals. The addition of 20 at.% of Al to CrN deferred the start of oxidation by 100 °C and reduced the oxidation rate. The onset temperature for oxidation increased and the oxidation rate decreased with a further increase in the Al-content up to 71 at.% Al. At a higher Al content, significantly decreased oxidation resistance was detected for a coating with 83 at.% Al and for AlN. Oxidation then starts already at 800 °C. These coatings already have an hcp structure. The negative influence of the hcp structure on the oxidation resistance was confirmed by [79].

A model of the oxidation behaviour for both the onset temperature and the oxidation rate depends on the coating composition and the related phase structure, as described above. However, it was observed in general that Cr and Al ions diffuse to the surface, forming a dense oxide layer acting as a diffusion barrier, thereby limiting the inward diffusion of oxygen [32,76,80]. The onset of oxidation of fcc (Cr,Al)N and fcc (Al,Cr)N always starts with the dissociation to h-Cr_2N and nitrogen in the coating. The presence of thermally stable Al–N bonding in the fcc-(Cr,Al)N structure can suppress the reduction of nitrogen in the coating. A dense $(CrAl)_2O_3$ or $(AlCr)_2O_3$ oxide layer (either amorphous or crystalline) is formed at an early stage of oxidation [76]. This can act as an effective diffusion barrier hindering the inward diffusion of the oxygen. All further reactions are influenced by the Al content at a given temperature.

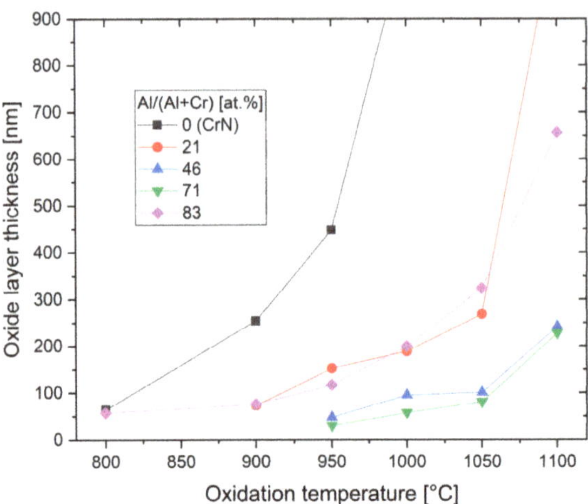

Figure 13. Thickness of oxide layer for CrN, (CrAl)N and (AlCr)N coatings after annealing at different temperatures in an air atmosphere, redrawn after [61], original © Elsevier.

It should be mentioned that there are several ways to increase the oxidation stability of (CrAl)N and (AlCr)N coatings, e.g., by means of synergistic alloying of the coatings with small amounts of Si [81], see also Section 4.4. By way of illustration, Figure 14 shows the excellent oxidation resistance of $(Al_{70}Cr_{30})N$ coatings in comparison to industrial-standard PVD coatings.

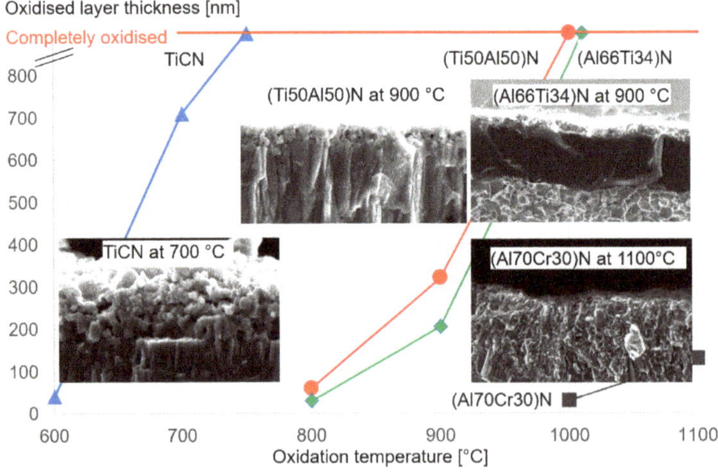

Figure 14. Thickness of the oxide layer as a function of the oxidation temperature for different industrial-standard PVD coatings.

3.3. Selected Tribological Properties

The tribological properties of (CrAl)N and (AlCr)N depend not only on the coating properties, but also on the tribosystem itself (e.g., counterpart, lubrication, temperature, loads).

3.3.1. Dry Friction against Steel

The dry friction value of (AlCr)N coatings against steel is slightly higher than that of CrN at ca. 0.65 [19], as measured using the pin–on-disc method against hardened bearing steel, 100Cr6. A friction value for (AlCr)N of about 0.5 was reported for AISI4340 as a pin [82]. The friction value at low load (5 N) was similar to that of (TiAl)N, whereas at high load conditions (20 N), (AlCr)N displayed a significantly lower friction coefficient. The superior tribological behaviour, namely a low wear rate and low friction, of an (AlCr)N coating against AISI 4340 steel is associated with the higher proportion of distinct (dispersive-polar) interactions, as evidenced by the measured surface energies. (AlCr)N and (TiAl)N coatings have a high polar component of 19.8 and 15.4 mJ/m^2, and a low dispersive component of 2.9 and 6.5 mJ/m^2, respectively, while the AISI 4340 steel presented a similar proportion between the polar, 14.3 mJ/m^2 and dispersive 13.5 mJ/m^2 components [82]. Measurement of the dry friction of (AlCr)N coatings by pin-on-disc against austenitic stainless steel (DIN1.4301, AISI 304, hardness 274 HV1) showed a friction value of 0.85 at room temperature, which dropped to about 0.6 at higher temperatures [83].

3.3.2. Dry Friction and Wear against Ceramics

The coefficient of friction measured by pin-on-disc tests against Al_2O_3 balls for (CrAl)N and (AlCr)N coatings of different Al contents showed a friction coefficient of about 0.6 at room temperature, independent of the Al content [83]. The friction value increased for 500 °C and dropped to 0.6 again at 700 °C, and was also nearly independent of the Al content. Figure 15 shows the decreasing abrasive wear rate with increasing Al content for all investigated temperatures [83].

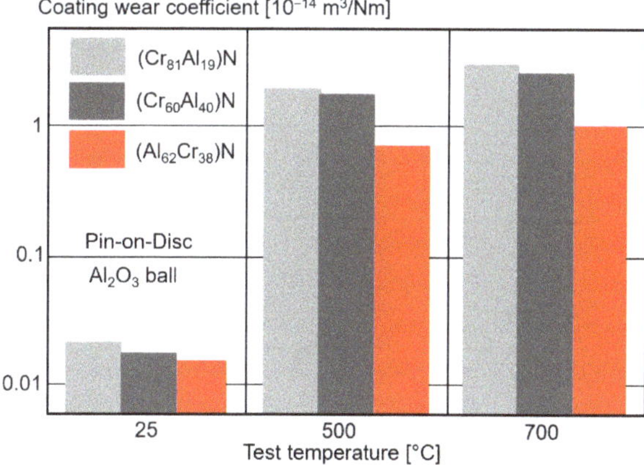

Figure 15. Wear coefficient of (CrAl)N and (AlCr)N coatings with different Al contents by pin-on-disc tests compared to Al_2O_3 balls for different test temperatures, redrawn after [83], original © Springer Nature.

The dry tribological properties of arc-deposited ($Al_{70}Cr_{30}$)N, ($Al_{67}Ti_{33}$)N, and CrN coatings against a Si_3N_4 ball as the counterpart showed friction values of 0.73, 0.79, and 0.77 at the end of the test, respectively [84]. A special effect related to the wear rate was observed. The highest wear rate was measured for ($Al_{67}Ti_{33}$)N, but CrN had a slightly lower wear rate than ($Al_{70}Cr_{30}$)N. The authors claim this is caused by the tribological oxidation behaviour, which has a great influence on the wear mechanism and the debris removal behaviour of the coatings.

3.4. Control of Coating Morphology, Stress and Texture

(CrAl)N and (AlCr)N coatings synthesised using PVD processes display a rich variety of microstructures, from fine-grained morphology up to coarse columnar structures, which make different (AlCr)N coatings suitable for many diverse application areas. This section focuses on (AlCr)N coatings, though similar effects are also valid for (CrAl)N coatings. The relationships between the deposition parameters and the coating morphology are summarised in generic structure-zone diagrams [85] to provide guidelines that can be applied for (AlCr)N coatings in general. The deposition conditions can also be used to control the stress state of the coatings [86], which often is an important part of engineering coatings and in adapting them to different use cases and requirements. For example, thick coating layers are at risk of peeling off if the stress state is too high, while high compressive stresses can be advantageous in interrupted cutting operations. The stress state has also been pointed out as a key determining factor for thermal stability and decomposition pathways [75].

The effect of bias voltage on the stress state and coating properties of (AlCr)N coatings deposited using $Al_{60}Cr_{40}$ targets has been investigated systematically [87]. Increasing the bias voltage led to higher levels of compressive residual stresses as a function of the increased ion bombardment, up to a threshold in the range of 100 V, see Figure 16. Further higher bias values reduced the overall stress state, which was attributed to the annihilation of defects and stress relaxation. A concurrent reduction in grain size, the lattice parameter and a modified preferential orientation in XRD from (200) to a mixture of (200) and (111) were also observed as an effect of increasing the bias [87].

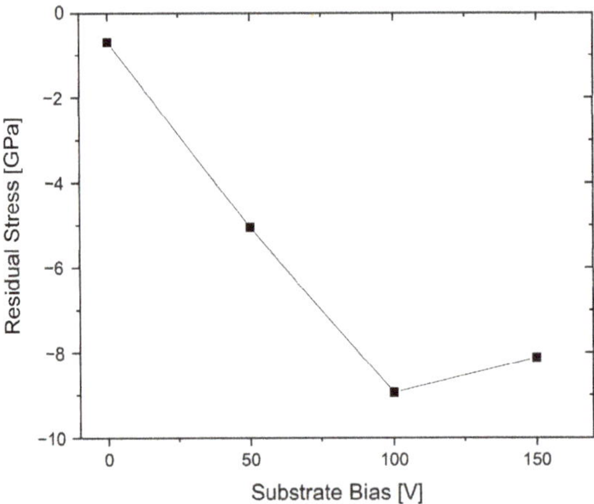

Figure 16. Compressive residual stress versus the bias voltage for arc-deposited $(Al_{60}Cr_{40})N$ coatings redrawn after [87], original © Elsevier.

Similar trends for texture and increased hardness with bias were also reported for coatings deposited using $Al_{70}Cr_{30}$ and $Al_{75}Cr_{25}$ cathodes. Coatings with a higher Al content, deposited using $Al_{85}Cr_{15}$ and $Al_{90}Cr_{10}$, formed dual-phase structures at bias voltages higher than 40 V [66]. The influence of phase structure on the stress state has been exemplified in the direct comparison of $(Al_{80}Cr_{20})N$ coatings, which had lower compressive stress values relative to $(Al_{70}Cr_{30})N$. This effect was attributed to stress relief through the presence of hexagonal phases in the former case [88].

The stress state of (AlCr)N coatings can also be influenced by alloying elements. For example, B-alloyed coatings with 2.3–9.1 at.% of B have been demonstrated to have lower compressive state levels compared to unalloyed coatings. The B-alloying also caused grain size refinement and an increase in hardness attributed to a combination of solid solution hardening and Hall-Petch hardening [89]. Further aspects of alloying will be discussed later in this review.

Surface morphology, and in particular the density of macroparticles in arc deposition processes, can be influenced by the deposition pressure where fewer macroparticles are generated and smoother coatings are obtained at higher pressure [90–92], see also the example in Figure 17. This effect can be rationalised through a higher degree of poisoning on the target surface at an elevated pressure. Columnar morphology is typically achieved over a large process window of deposition pressures, though with variation in the column width. This is illustrated in Figure 17a–d. Further reported effects of deposition pressure include increased hardness and an influence on the phase structure [91,92].

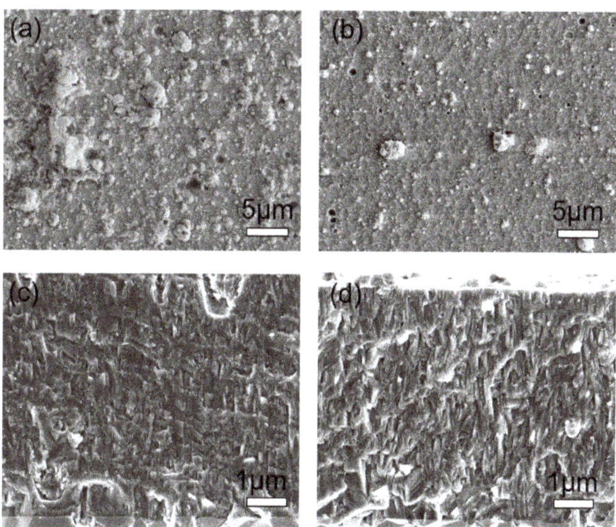

Figure 17. Surface and cross-sectional morphology of (AlCr)N monolayer coatings deposited for (**a**,**c**) at 2.5 Pa N_2 and for (**b**,**d**) at 7.5 Pa N_2, for illustration of relative trends.

The structure of arc-deposited AlCr-based coatings depends not only on the cathode composition, process temperature, substrate bias potential, arc current and reactive gas pressure, but also on the arc source design, in a manner that is similar to the case of AlTi-based coatings [9]. The influence of the magnetic set-up was investigated in one coating process using two arc sources, both equipped with $Al_{70}Cr_{30}$ cathodes, strong and weak magnetic fields, respectively [93]. Figure 18 shows the differences between the two deposited coatings by means of XRD investigation, SEM cross-section and hardness measurement. The coating properties vary significantly. Although EPMA (Electron Probe Micro Analysis) measurements showed a lower Al/(Al + Cr) content, a minor hcp face is visible in the coating deposited by the weak field, which explains the finer growth structure (smaller grains) of the coating. Furthermore, the coating deposited with the weak field contains less nitrogen and the hardness is about 30% lower compared to the coating deposited with strong magnetic fields.

3.5. Features of (AlCr)N Coatings with Mixed fcc Plus hcp Structure

Dual-phase coatings with Al concentrations of more than 70 at.% have been reported as monolayer coatings [88,90,94,95], as well as multilayers [66,96]. The coatings are characterised by relatively low hardness compared with fcc (Al,Cr)N, lower oxidation resistance than fcc (Al,Cr)N with a high Al content, but also with the positive effect of exhibiting lower stress.

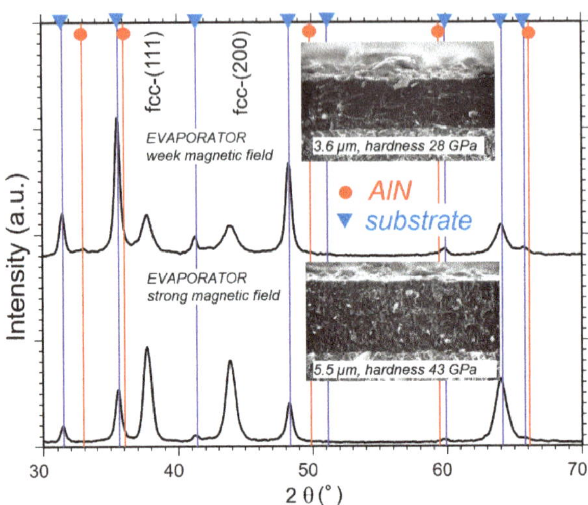

Figure 18. Phase formation of coatings deposited with pm cathodes $Al_{70}Cr_{30}$ for two evaporator with different magnetic field set-ups [93].

4. Alloying of AlCrN

The alloying (doping) of (AlCr)N is amongst the most important approaches used t develop and adapt these coatings. The addition of at least one alloying element (such as S B, C, O, W, Y and others) into (AlCr)N coatings is used to further optimise properties an performance. Through alloying, the base coating may be improved in terms of increase hardness, higher oxidation resistance, extended thermal stability, higher wear resistanc and lower friction values. The main directions in the development of alloyed (AlCr)N coatings will be reviewed in this section. Figure 19 shows an overview of alloying concept Metals, metalloids, and oxygen are added as single elements or in combination with other

Aspects of environmental sustainability have not yet been discussed in the scientifi and engineering community until now. These important ecological aspects, such as selec ing specific alloying elements to minimise environmental impacts in the production of th coating, the use of the coating and the recycling, were overlooked. In general, the additio of O, C and Si seems to offer the most sustainable solutions. An overview of differen alloying directions that have been reported is given in Table 2.

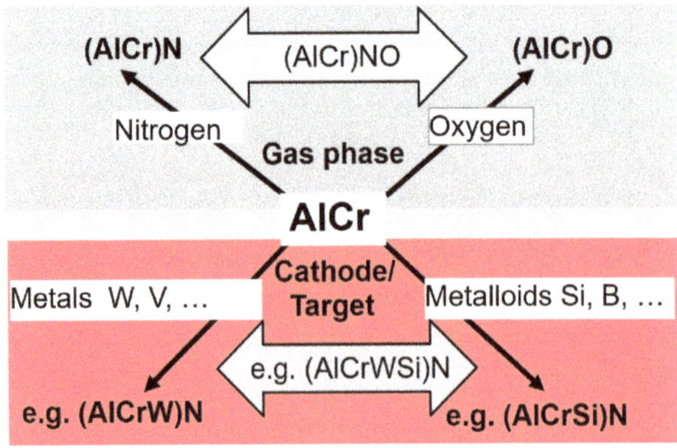

Figure 19. Directions of alloying for (CrAl)N and (AlCr)N designed to tune selected properties.

Table 2. Selected coatings alloyed with one metal and metalloid component.

Alloying Element	Coating Composition	Cathodes/Targets	Deposition Method	Results	Reference
Ti	e.g., $Al_{30.5}Cr_{14}Ti_{5.5}N_{50}$	$Al_{66}Ti_{29}Cr_5$, $Al_{67}Cr_{33}$, $Al_{34}Cr_{33}$, $Al_{67}Ti_{33}$	Arc/ different AlCrTi composition in one batch	Age hardening	[97]
	$(Al_{66}Cr_{24}Ti_{10})N$	$Al_{70}Cr_{20}Ti_{10}$	Arc	Age hardening	[98]
	e.g., $(Al_{69}Cr_{20}Ti_{11})N$	$Al_{65}Cr_{21}Ti_{14}$, $Al_{71}Cr_{19}Ti_{10}$, $Al_{74}Cr_{18}Ti_8$	Arc	Oxidation onset 1000 °C	[99]
V	$Al_{25}Cr_{22}V_3N_{50}$	$Al_{55}Cr_{45}$, $Al_{55}Cr_{40}V_5$	Arc	Higher hardness, reduced oxidation onset	[100]
	e.g., $Al_{34}Cr_{10.5}V_{5.5}N_{50}$	$Al_{70}Cr_{30-x}V_x$; x = 0, 10, 15, 20, 25, and 30	Arc	Metastable solubility limit decreases at low bias	[101]
	$Cr_{29.5}Al_{10.5}V_{10}N_{50}$	$Al_{80}Cr_{20}$ $Cr_{80}Al$ insert targets, and Cr, V	DC sputter Plus HiPIMS	Lower friction (Magnéli phases)	[102]
	(AlCrV)N	$Al_{70}Cr_{30}$, $Cr_{70}V_{30}$	Arc	Cutting test hardness, structure	[103]
Zr	$Cr_{38.8}Al_{4.5}V_2N_{54.8}$	$Al_{90}Cr_{10}$, $Al_{80}Cr_{20}$, and Cr, Al, Zr	DC sputter	Lower friction and wear	[104]
Mo	e.g., $Cr_{20}Al_{18.5}Mo_7N_{54.5}$	$Al_{60}Cr_{40}$, $Al_{70}Cr_{30}$, $Al_{55}Mo_{45}$, with inserts Mo,Cr,$Al_{60}Cr_{40}$, $Al_{85}Cr_{15}$	DC sputter	H/E optimisation	[105]
	e.g., $(Al_{50}Cr_{37}Mo_{13})N$	$Cr_{35}Al_{65}$, and Cr, Mo	Arc	Triboactive, element Mo	[106]
	(AlCrMo)N	$Al_{70}Cr_{30}$, $Cr_{70}Mo_{30}$	Arc	Cutting test hardness, structure	[103]

Table 2. *Cont.*

Alloying Element	Coating Composition	Cathodes/Targets	Deposition Method	Results	Reference
Hf	e.g., $Al_{37}Cr_{11}Hf_2N_{50}$	$AlCr_{20}$ (Cr in Al) inserted, and Hf	DC sputter	Higher oxidation resistance	[107]
Nb	e.g., $Al_{27.5}Cr_{17.5}Nb_5N_{50}$	$Al_{55}Cr_{45}$, $Al_{60}Cr_{35}Nb_5$, $Al_{60}Cr_{30}Nb_{10}$	Arc	Higher hardness, lower oxidation resistance	[108]
W	e.g., $(Al_{64}Cr_{31}W_5)N$	$Al_{70}Cr_{30}$, $Al_{70}Cr_{25}W_5$	Arc	Improved oxidation resistance at 1100 °C	[81]
Cu	e.g., $(Al_{47}Cr_{34}Cu_{19})N$	$Cr_{35}Al_{65}$, and Cr, Cu	DC sputter	Triboactive, hardness decrease	[106]
Y	e.g., $Al_{26.5}Cr_{22.5}Y_1N_{50}$	Al/Cr = 1.5 with 0, 2, 4, 8 at.% Y	DC sputter	Structure, improved oxidation resistance	[109,110]
	e.g., $Al_{23.8}Cr_{23.1}Y_{0.7}N_{50}O_{2.4}$	$Al_{50}Cr_{50}$, Y	DC sputter	Improved oxidation	[111]
	e.g., $Cr_{25.8}Al_{15.3}Y_{3.4}N_{55.5}$	$Al_{90}Cr_{10}$, $Al_{80}Cr_{20}$, and Y	DC sputter	Structure, improved oxidation	[112]
La	e.g., $Al_{28.1}Cr_{12.5}La_{1.4}N_{58}$	$Al_{70}Cr_{30}$, and La	DC sputter	Wear, friction, structure	[113]
C	e.g., $Cr_{44.4}Al_{12}C_{27.6}N_{16}$	Cr_2AlC	DC sputter	Wear, friction, structure	[114]
	e.g., $Cr_{24}Al_{24}N_{48}C_4$	$AlCr_{20}$, $AlCr_{24}$, Cr inserts in Al, and Cr	DC sputter HiPIMS	Structure, hardness	[115]

Table 2. Cont.

Alloying Element	Coating Composition	Cathodes/Targets	Deposition Method	Results	Reference
B	e.g., $Al_{26.7}Cr_{21.7}B_{2.3}N_{49.3}$	$Al_{60}Cr_{40}$, $Al_{70}Cr_{30}$, B alloyed Al/Cr =1.8, with 10, 20, 30 at.%	Arc	Structure, phases, hardness, nano composite	[89]
	e.g., $Al_{30.4}Cr_{16.3}B_{2.1}N_{48.1}O_{3.1}$	$Al_{55}Cr_{35}B_{10}$, $Al_{70}Cr_{30}$	Arc	Structure, wear, tribology	[58]
	e.g., $(Al_{64.5}Cr_{32.9}B_{2.6})N$	$Al_{70}Cr_{30}$, $Al_{70}Cr_{25}B_5$,	DC sputter	Oxidation, structure	[81]
Si	See Section 4.4 paragraph Cr-Al-Si N and Al-Cr-Si-N				

4.1. Additional Metallic Elements

Modifications of the coating properties are possible by the addition of one or mor metallic elements, e.g., Ti, V, Mo, Hf, Nb, W, Cu.

4.1.1. Coatings with the Addition of Ti, V, Zr

Al-Cr-Ti-N

Various quaternary cubic (AlCrTi)N coatings with a composition of about $(Al_{60}(Cr,Ti)_{40})N$ and varying Ti content (0, 1, 2, 11 at.%) were deposited by cathodic arc evaporation. Th hardness of the $(Al_{30.5}Cr_{14}\ Ti_{5.5}N_{50})$ coating was retained after annealing up to 1100 °C The coating also showed an age hardening process caused by spinodal decompositio similar to (AlTi)N [97]. However, the oxidation resistance was found to be slightly reduce compared to unalloyed (AlCr)N coatings. The age hardening effect was also discusse for AlCrTiN coatings with a higher Al content $(Al_{66}Cr_{24}Ti_{10})N$ [98]. It was shown tha the oxidation resistance of $(Al_{69}Cr_{20}Ti_{11})N$ was significantly increased as compared t $(Ti_{50}Al_{50})N$ [99].

Al-Cr-V-N

$Al_{26}Cr_{24}N_{50}$ and $Al_{25}Cr_{22}V_3N_{50}$ coatings were deposited by arc evaporation. It wa shown that V doping increased the hardness from ca. 31 to 33 GPa, but the oxidatio resistance was decreased [100]. Systematic investigations of arc-deposited coatings usin $(Al_{70}Cr_{30})_{1-x}V_x$ cathodes with compositions of x = 0, 10, 5,20, 25, and 30 at.% showed tha the incorporation of V into AlCrN coatings triggers a phase separation into fcc-CrN an hcp-AlN structured grains, even at a low content [101]. Higher bias voltages stabilise th fcc structure.

$Cr_{29.5}Al_{10.5}V_{10}N_{50}$ sputtered coatings showed a decrease in the friction coefficient a higher temperatures (600 and 800 °C) due to phase transformation towards the Magnél phase V_3O_7 in pin-on-disc tribometer testing [102].

(AlCrV)N coatings were deposited by arc using $Al_{70}Cr_{30}$ and $Cr_{70}V_{30}$ cathodes [103 The (AlCrV)N coating had a higher hardness and a lower friction value and wear rate tha the $(Al_{70}Cr_{30})N$ coating. The coating containing V showed a higher service life in a millin cutting test. A reason might be the formation of lubricious VO_x phases.

Cr-Al-Zr-N

A sputtered Cr-rich (CrAlZr)N coating with the composition of $Cr_{38.8}Al_{4.5}V_2N_{54.8}$ wa shown to form a Cr_2O_3–rich tribolayer that decreases friction and wear [104].

4.1.2. Coatings with the Addition of Refractory Metals

Al-Cr-Mo-N

Basic investigations of sputtered coatings showed that the addition of Mo significantl increases the hardness in a comparison, for example, of $Cr_{20}Al_{18.5}\ Mo_7N_{54.5}$ at 37 GPa t unalloyed $Al_{23}Cr_{20}N_{57}$ at 29 GPa [105]. Other (AlCrMo)N coatings (e.g., $(Al_{50}Cr_{37}Mo_{13})N$ were deposited by means of cathodic arc evaporation. The addition of Mo was shown t have a high potential to act as a triboactive coating when lubricated with lubricants co taining sulphur [106]. (AlCrMo)N coatings have also been deposited by arc using $Al_{70}Cr_{3}$ and $Cr_{70}Mo_{30}$ cathodes [103]. The (AlCrMo)N coating had slightly higher hardness and lower friction value and wear rate than the $(Al_{70}Cr_{30})N$ coating. The coating containin Mo showed a higher service life in a milling cutting test. A reason might be the formatio of lubricious MoO_x phases.

Al-Cr-Hf-N

AlCrN (Al/Cr = 3.2) and AlCrHfN coatings with an Al/Cr ratio > 3 and varying H content between 0 and 11.6 at.% have been synthetised by DC magnetron sputtering. Th hardness was found to decrease slightly with an increasing Hf content, from 22 GPa t 19 GPa. The oxidation resistance of AlCrN increased by alloying it with a small amour of Hf of about 2 at.% (e.g., $Al_{37}Cr_{11}Hf_2N_{50}$); however, it decreases again at the highest Hf content [107].

Al-Cr-Nb-N

Fcc-structured $Al_{26}Cr_{24}N_{50}$, $Al_{23.5}Cr_{19}Nb_{2.5}N_{50}$ and $Al_{27.5}Cr_{17.5}Nb_5N_{50}$ coatings were deposited by cathodic arc evaporation. The hardness increased slightly with the addition of Nb from 31 to 32 GPa at room temperature, but the increase in hardness is more pronounced after annealing at a temperature of around 1000 °C (plus 3 GPa). The phase stability was shifted to include higher temperatures with the addition of Nb, but the oxidation resistance decreases [108].

Al-Cr-W-N

(AlCr)N and (AlCrW)N coatings have been deposited by the cathodic vacuum arc method using cathodes of $Al_{70}Cr_{25}W_5$ and $Al_{70}Cr_{30}$. The coating had a W content that was about the same as in the cathode and an Al/Cr ratio of 2, ($Al_{64}Cr_{31}W_5$)N. Oxidation tests showed that the (AlCrW)N coating had higher oxidation stability than the (AlCr)N coating at 1100 °C; however, it was worse than (AlCr)N at 900 °C [81].

4.1.3. Coatings with the Addition of Non-Ferrous Metals

Al-Cr-Cu-N

($Al_{53}Cr_{47}$)N and ($Al_{47}Cr_{34}Cu_{19}$)N were deposited by means of cathodic arc evaporation. It was shown that the Cu is present mainly in metallic form. The alloying with Cu was found to decrease the hardness from 30 to 25 GPa [106].

4.2. Coatings with the Addition of Rare Earth Elements

The effect of Y and La has been studied in AlCrN-based coatings.

Al-Cr-Y-N

The oxidation behaviour of $Al_{27}Cr_{46}N_{50}$ and Y-doped coatings was investigated. $Al_{26.5}Cr_{22.5}Y_1N_{50}$ and $Al_{26.5}Cr_{21.5}Y_2N_{50}$ coatings showed improved oxidation resistance, whereas a further increase (4 at.% in the coating) in Y had a negative influence. Experimental and computational studies on the effect of yttrium on the phase formation of sputtered (AlCrY)N showed a decrease in the maximum Al content with retained fcc structure to 68 at.% at a Y content of 2 at.% of the total metal content. The authors concluded that, theoretically, the $Al_{34}Cr_{15}Y_1N_{50}$ coatings exhibit the most promising oxidation resistance within the group of coatings containing Y as compared to (AlCr)N coatings [109,110]. (CrAlY)N coatings with a Y content up to 2.3 at.% were co-sputtered using $Cr_{50}Al_{50}$-composite and pure Y targets. The hardness increased from ca. 16 to 24 GPa with increasing Y content. However, oxidation experiments (1100 °C) demonstrated a lower Y content of 0.3 to 0.7 at.% (e.g., $Al_{23.8}Cr_{23.1}Y_{0.7}N_{50}O_{2.4}$) to be beneficial for the oxidation resistance. In excess of 1.3 at.% Y, the oxidation resistance deteriorated as the result of the formation of porous and non-protective oxide scales [111]. A further study of co-sputtered coatings containing Y showed that excellent oxidation behaviour was achieved with a Y content of 3.4 at.% for Cr-rich $Cr_{25.8}Al_{15.3}Y_{3.4}N_{55.5}$ coatings, whereas for Al-rich coatings, a lower Y content of 2.6 at.% was the best with $Al_{24.9}Cr_{18.1}Y_{2.6}N_{54.4}$. It was also found that higher Y contents promote the hcp phase formation [112].

It can thus be concluded that a low Y content in the range of about 0.7 to 3.4 at.% of the total elemental composition of the coating can have a positive effect on the oxidation. However, the absolute value seems to be dependent on the Al/Cr ratio in the coating.

Al-Cr-La-N

$Al_{29.1}Cr_{13.5}N_{57.4}$ and (AlCrLa)N coatings were deposited by sputtering using $Al_{70}Cr_{30}$ and La targets. The concentration of La was varied in the range of 1.39 to 7.73 at.% by adjusting the sputtering power for the La target. A small amount of La, e.g., $Al_{28.1}Cr_{12.5}La_{1.4}N_{58}$, formed a solid solution in the (AlCr)N lattice, resulting in grain refinement. The hardness of (AlCrLa)N containing 1.4 at.% La was significantly higher than that of the undoped

(AlCr)N. This coating showed the lowest wear rate. An increased amount of La resulted in a grain boundary segregation, thus an nc-(AlCrLa)N coating consisting of an (AlCrLa)N matrix combined with a-La_2O_3/a-LaN segregations. The hardness decreased with an increase in La content at the same time as the friction value (against Si_3N_4) was reduced [113].

4.3. Coatings with the Addition of Carbon

Al-Cr-N-C

$Cr_{44.4}Al_{12}C_{27.6}N_{16}$ coatings have been deposited by reactive magnetron sputtering using a Cr_2AlC target. The coatings showed a phase mixture of c-CrC, h-AlN and Cr with 23 GPa hardness. In addition, the structure started to change at 800 °C (h-Cr_2AlC, Cr_3C_2, Cr_7C_3, AlN) [114]. Better coating properties than those of the (CrAlC)N coating were described for the more complex (CrAlTi)CN coating.

DC/HiPIMS sputtered $Cr_{25}Al_{24}N_{51}$ and (CrAl)NC coatings with a Cr/Al of ca. 1 and N > C were investigated with a carbon content in the range of 1–18 at.% [115]. Nanocomposite structured (CrAl)NC/a-C coatings are formed when a critical carbon content is exceeded (a few at.%). Crystalline (CrAl)NC phases are surrounded by an amorphous carbon phase. This effect was also detected for carbon-doped AlTi-based coatings, both for sputtering [116] and arc deposition [13].

It was shown that first the hardness significantly increases from 22 to 30 GPa for a carbon content of up to about 4 at.%, e.g., $Cr_{24}Al_{24}N_{48}C_4$, then the hardness drops to about 15 GPa at a carbon content of 18 at.%. The hardness drop is enforced by two effects; by the continually growing amorphous phase—containing amorphous carbon and amorphous CrC_x—and by the formation of hcp AlN. The carbon-doped (CrAl)NC coatings showed improved frictional characteristics in metal forming.

4.4. Coatings with the Addition of Metalloids

4.4.1. Coatings with the Addition of B

Al-Cr-B-N

(AlCr)N and (AlCrB)N coatings were arc deposited using $Al_{70}Cr_{30}$, $Al_{60}Cr_{40}$ and B-doped cathodes with a constant Al/Cr atomic ratio of 1.8 and varying B content of 10, 20 and 30 at.% [89]. The resulting coatings had a B content between 2.3, $Al_{26.7}Cr_{21.7}B_{2.3}N_{49.3}$, and 9.1 at.%, $Al_{22.7}Cr_{20.5}B_{9.1}N_{47.7}$, corresponding to (Al + Cr)/B ratios in the range 0.04–0.21. The authors stated that even at low B contents, B segregates to the grain boundaries where (AlCrB)N crystallites are at least partially covered by an a-BN_x tissue phase. A nanocomposite structure was formed at compositions of ≥5.7 at.% B, consisting of fcc-(Al,Cr,B)N grains surrounded by an amorphous BN_x phase dependent on the B content The hardness was significantly increased from ca. 30 GPa for coatings containing no B to a maximum of about 40 GPa for a B content of 5.7 at.%. The addition of B acts as a grain refiner where the highest B content of 9.1 at.% reduced the grain size down to 5 nm. It was also shown that the addition of B decreases the coating stress.

A similar investigation of arc deposited coatings was carried out using cathodes of $Al_{70}Cr_{30}$ and $Al_{55}Cr_{35}B_{10}$ [58]. The (AlCrB)N coating contained 2.1 at.% of B for $Al_{30.4}Cr_{16.3}B_{2.1}N_{48.1}O_{3.1}$, with (Al+Cr)/B = 0.045 and Al/Cr = 1.86, and was reported to be composed of an fcc solid solution when measured by XRD. XPS measurements showed peaks of B-N and Cr-B bondings. Thus, a BN_x tissue phase formation is highly probable The B alloying significantly increased the hardness from the B-free $Al_{33.1}Cr_{15.8}N_{48.1}O_3$ at 31 to 37 GPa. The grain size concurrently decreases from 16 to 5 nm. The friction value and the wear rate measured by pin-on-disc were lower than that of the coating with no B.

(AlCrB)N coatings have been deposited by the cathodic vacuum arc method using cathodes of $Al_{70}Cr_{25}B_5$ and $Al_{70}Cr_{30}$. The (AlCrB)N coating had an Al/Cr ratio of 1.96 and a B/(Al + Cr)/ratio of 0.04. Oxidation tests at 900 °C showed that this particular $(Al_{64.5}Cr_{32.9}B_{2.6})N$ coating had a lower oxidation stability than the AlCrN coating [81].

In general, it can be concluded that the addition of B decreases the grain size in combination with an increase in hardness of around 30%.

4.4.2. Coatings with the Addition of Si

Si addition is the most investigated alloying element, with the goal of achieving an improvement of mechanical properties and oxidation resistance [72,117–121]. The properties are mainly dependent on the Si content at a given Al/Cr ratio, and also on the deposition technology. The Al/Cr ratio and the addition of Si have to be optimised to keep the results primarily in the structural zone of fcc-CrAlSiN or fcc-AlCrSiN so that suitable mechanical properties can be achieved, particularly for applications on cutting tools.

Figure 20 shows a schematic representation of the structural evolution as a function of the Al/Cr ratio and Si content. In a first approximation, the limit of the Al content to obtain a coating dominated by the fcc phase is assumed to be about the same as for Si-free coatings (ca. 70 at.% Al).

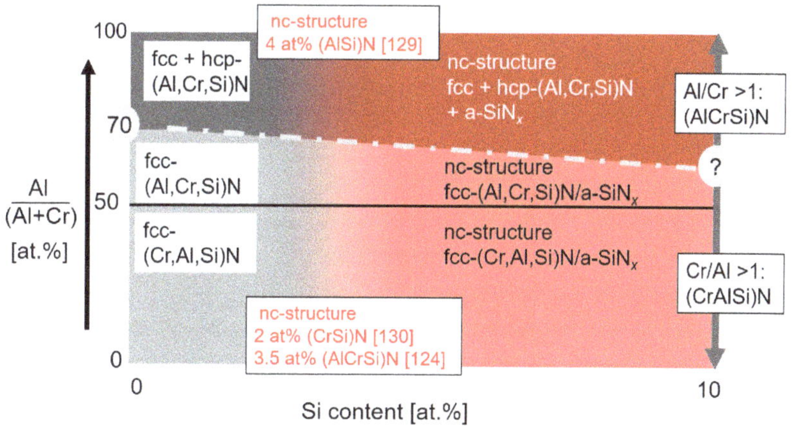

Figure 20. Schematic diagram of the phase evolution for (CrAlSi)N and (AlCrSi)N for different Al and Si contents in the coating, nc = nanocomposite.

However, it was shown that the addition of Si promotes the formation of the hcp phase. Arc-deposited coatings use a constant Al content of 70 at.% in the cathode, but with a different Cr and Si content (0, 1, 2, 5 at.% Si) [72]. For coatings synthesised at low bias voltages (40 V), a phase separation into fcc (Al,Cr,Si)N and hcp (Al,Cr,Si)N for Si contents of around 1 at.% in the coating, e.g., $Al_{33}Cr_{16}Si_1N_{51}$, was observed. This results in a congruent drop in mechanical properties. The hardness decreased from 30 ± 2 GPa for $Al_{32}Cr_{17}N_{51}$ coatings to 21 GPa for the coating $Al_{32}Cr_{14.5}Si_{2.5}N_{51}$ deposited using $Al_{70}Cr_{25}Si$ targets. Higher bias voltage favoured the fcc phase formation. However, it is unclear if this is due to growth effects or due to a decrease in the Al and Si contents as a secondary effect of the bias voltage [72].

AlCrSi cathodes with a constant Cr content of 30 at.%, but with varying Al and Si silicon concentrations (0, 1, 2, 5, 10 at.% Si), were deposited by cathodic arc evaporation [73]. The Al/(Al + Cr) ratio of the coatings decreased from about 0.68 to about 0.64 at the highest Si content. Up to a Si-content of about 1 at.% (corresponding to cathodes with 2 at.% Si), the coatings showed an fcc structure, $Al_{33.9}Cr_{18.1}Si_{1.1}N_{51}O_{0.9}$. The coating with a composition of about $Al_{33.4}Cr_{18.3}Si_{2.3}N_{46}O_{0.7}$ (5 at.% in the cathodes) displayed the formation of the hcp phase at an Al/(Al + Cr) ratio of only about 0.65 and a Si content of about 2 at.%.

Investigations of whether the addition of Si decreases the phase stability of the fcc phase at higher temperatures are lacking. In addition, it is also not clear how much Si can be substitutionally incorporated into (CrAl)N and (AlCr)N at a given Al content, and what the critical Si content is for the segregation of an amorphous SiN_x phase at the grain boundaries.

Above the solid solubility limit of Si, a nanocomposite (nc) structure is formed [35,121–127]. The Si segregates along the grain boundaries, thus an amorphous SiN_x phase grows. This effect has been observed for both (CrSi)N and (AlSi)N. A schematic structural model for (CrSi)N was shown in [128]. It was determined that for (AlSi)N, the tissue phase is generated for (AlSi)N in a coating containing 4 at.% of Si, $Al_{46}Si_4N_{50}$. The authors showed that the nanocomposite was formed at 6 at.% Si in the coating for $Al_{44}Si_6N_{50}$ [129].

A nanocomposite structure was also detectable for (CrSi)N with a Si content of about 2 at.% or more [130]. Analogous phase evolution was shown for coatings with a low aluminium content (Cr/Al ca. 3) for (CrAlSi)N [35]. An (AlCrSi)N coating with a nanocomposite structure was demonstrated at a Si content of 3.5 at.% [124].

Unfortunately, no systematic investigations of the minimum Si content for different Al contents in a (CrAlSi)N or (AlCrSi)N coating are available. Roughly, it can be deduced that a nanocomposite structure might be generated at a Si content of between 2 and 4 at.% of the total chemical composition (Al + Cr + Si + N content in at.% equal to 100 at.%). It was shown that the addition of Si not only has an influence on the thermal properties of (AlCr)N, but also acts as a grain-size refiner. Depending on the amount of Si, the grain size is in the range of less than 10 nm [124,131].

A comparison of the oxidation behaviour of (AlCr)N (Al/Cr = 1.25), (CrAlSi)N (Cr/Al = 1.16) and (AlCrSi)N (Al/Cr = 1.89) coatings revealed that the coating (AlCrSi)N with 3.3 at.% Si, $Al_{32.7}Cr_{17.4}Si_{3.3}N_{44.3}O_{2.3}$, showed the highest oxidation resistance [80]. Additional work was performed to investigate the oxidation of various Si-free and Si-containing coatings [119]. Figure 21 shows the specific weight gain of fcc-structured (CrAl)N, (CrAlSi)N and (AlCr)N, (AlCrSi)N coatings measured using TGA in synthetic air. The authors showed that the formation of a crystalline corundum type mixed and/or layered $(Al_xCr_{1-x})_2O_3$ oxide scale, composed of Cr-rich and Al-rich areas, is typical for (CrAl)N, (AlCr)N and (CrAlSi)N and (AlCrSi)N coatings for coatings with low Si contents. The oxide layer formed prevents further oxidation. A special case was observed for high Si contents, e.g., a $(Cr_{45}Al_{39}Si_{16})N$ coating, where the highest onset temperature for oxidation was observed accompanied with the lowest weight gain. The reason is that a crystalline SiO_2 phase grows in addition to $(Al_xCr_{1-x})_2O_3$. This effect might be used for some coating solutions requiring a low oxidation rate.

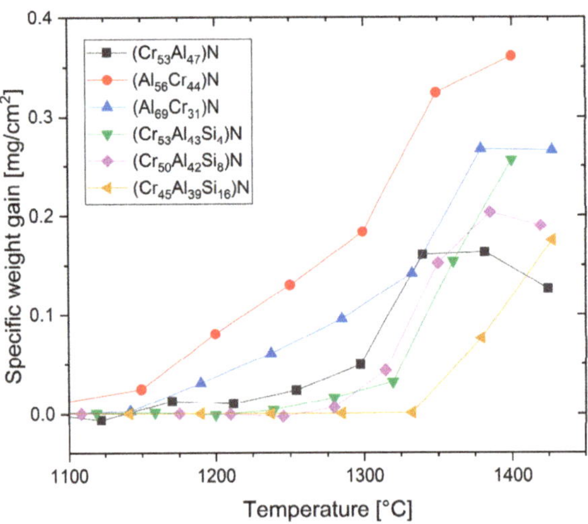

Figure 21. Weight gain measured by dynamic TGA measurements of fcc-structured (AlCr)N, (AlCrSi)N and (CrAl)N, (CrAlSi)N coatings up to 1440 °C in synthetic air, redrawn after [119], original American Vacuum Society.

The (CrAlSi)N and (AlCrSi)N coatings can also be doped with one additional element to address modifications of various properties, e.g., O [132], Y [133] B [134], W [135,136], Ni [137], and others. Additionally, more than one element was added, e.g., Y and O [138]. Multilayer architectures were built using CrAlSiN and AlCrSiN combined, for example, with Si-free AlCrN layers [139], with AlSiN [140], or with coatings containing Si, such as Cr-doped AlSiN [127], with MoN, NbN [139], or with CrN [141], and others.

4.5. Coatings with the Addition of Two or More Elements

Quarternary and higher-order systems have been explored to further optimise coating properties. However, this has also been pursued from an industrial perspective dealing with patent issues to obtain unique selling points (USPs) or to supersede patents from competitors. However, the complexity quickly increases, and there is the potential for cross-reactions between the alloying elements. Even if the results achieved do not generally outperform the simpler coatings, a study of that research will reveal indicators of potential directions for coating development. A detailed discussion is beyond the scope of this paper; however, selected publications on Al-Cr based multinary coatings are listed below.

Si-Mo sputtered [142], Si-Ni sputtered [143,144], Si-Ti arc [145,146], Si-W arc [78,135], Si-W sputtered [136], Si-Ni sputtered [137], Si-C sputtered [133,147], Si-B sputtered [134], Si-Mg [148], Ti-C sputtered [114], MoCu sputtered [106], Si-Y-O sputtered [138], Ti-Si-Y arc [149], and Nb-Ti-Si sputtered [150].

In addition, even high-entropy coating modifications are the focus of R&D activities [150,151].

5. Complex Coating Architectures

Coatings used in industrial applications are seldom homogeneous monolayers, but often comprise several coating layers or complex architectures. This allows several functional requirements to be met through coating layers serving different functions, for example, good adhesion to the substrate material through a bonding layer, mechanical stability through a main layer, and tribological properties through a top layer. Furthermore, in nanocomposite or multilayered designs, the additional interfaces created can help to enhance performance, for example, by providing increased fracture resistance, beyond the properties of the individual layers. A multitude of approaches have been demonstrated for AlCr*X*-based coatings.

A schematic drawing of the main coating architectures is shown in Figure 22. It must be noted that the bond coat which is often applied essentially produces a two-layer coating that has not been considered explicitly in this overview.

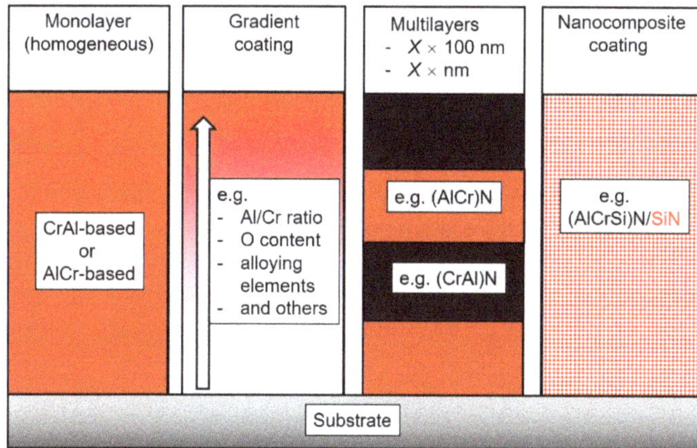

Figure 22. Main types of coating architectures.

It should be mentioned that in practical coating designs, different architecture types are often mixed in one and the same coating, e.g., gradient coatings are predominantly realised more or less stepwise. The nanocomposite coating shown might be a part of the multilayer coating or part of the gradient coating. A primary example of a nanocomposite structure is the system (AlCrSi)N/a-SiN_x, as discussed in Section 4.4.

5.1. Multilayer and Nanomultilayer Coatings

Multilayered coatings comprise at least two different layers, and most commonly this is a periodic repetition of two or more layers. The most frequently used is the classic … A/B/A/B/A … type of layering, where the same two layers are sequentially repeated. When the periodicity is within the size scale of up to a few tenths of nanometers, the terms nanomultilayers or nanolayered coatings are often used. The sublayers can be synthesised using different strategies. Keeping the same (nominal) coating chemistry, a variation of the process parameters, such as bias or deposition pressure, can be used to create layers of different stress states, for example. Changing the reactive gas composition is a further approach to alternate nitride, oxynitride and oxide layers on the same metal base, for example. Perhaps most common though is the combination of heterogeneous coating materials obtained through the inclusion of targets of different composition in the PVD deposition system. Furthermore, particularly in industrial-scale deposition systems, co-deposition from spatially separated deposition sources with different target materials may be combined with substrate rotation to create multilayered coatings. There are a variety of different approaches for the materials used in the second single layers. Some examples are shown below.

- *CrAl- or AlCr-based structural multilayer*

The combination of fcc (Al,Cr)N with hcp (Al,Cr)N was investigated to address the potential for adapting stress, hardness and wear properties [88,96].

- *CrAl- or AlCr-based combined with a binary nitride coating*

The first publication dealt with (AlCr)N and CrN. One aim was to have a "ductile' coating similar to CrN combined with a high oxidation resistance. [19]. A combination with VN was utilised to improve the tribological behaviour at elevated temperatures [152–154] TiN/(AlCr)N superlattices were deposited by sputtering. The coatings showed both an improved fracture toughness and hardness as compared to the monolayers [71].

- *CrAl- or AlCr-based combined with a ternary nitride coating*

The coating properties of (AlCrX)N can be combined with AlTiN [155,156]. There are some coating systems that contain V in a nitride coating to tailor tribological properties (formation of lubricant oxides), such as (TiV)N [157], (CrV)N and (CrMo)N [103]. The combination with (TiSi)N is focused on high hardness and oxidation stability [158].

- *CrAl- or AlCr-based combined with a quaternary nitride coating*

The comination with quarternary nitride layers is a very complex approach. Many coating properties are influenced in parallel. The following systems have been explored (AlTiSi)N [151,159], (AlCrSi)N [139,160], (TiTaAl)N [161].

- *CrAl- or AlCr-based combined with an oxynitride or oxide coating*

The arc deposition of an fcc-(Al,Cr)ON layer on top of an fcc-(Al,Cr)N-based coating resulted in improved wear resistance [162]. The combination of alternating single layers consisting of a nitride fcc-(Al_xCr_{1-x})N layer followed by an oxide coating $(Al_xCr_{1-x})_2O_3$ with a hardness of up to 22 GPa was demonstrated [163].

5.2. Gradient Coatings

Coating designs with gradient layers imply that coating properties change sequentially in the direction of the coating thickness. Frequently, the primary factor that is varied is the composition of the coating. Both the sputtering and arc deposition methods are used to generate compositional gradients continuously or stepwise. Two examples of coatings with a Si gradient in which the layer containing less silicon is at the bottom and the layer with more silicon is at the top are shown to demonstrate this.

In the first example, CrAlSiN monolayers (0, 4, 1.8, 5.9, 8.5 at.% Si) and CrAlSiN coatings with a continuous Si gradient (Cr/Al ca. 1.6) were deposited by co-sputtering using elemental targets of Cr, Si, and Al with a Si content of 5 and 8 at.% in the top layer and were compared with (CrAlSi)N monolayers [122]. The authors described the formation of a nanocomposite (CrAlSi)N/a-SiN_x structure for monolayer coatings with a Si content of 5.9, $Cr_{23.4}Alr_{15.6}Si_{5.9}N_{53.8}O_{1.3}$, or higher by means of an X-ray diffraction pattern and for 8.5 at.% Si as well using TEM examinations. The gradient coatings exhibited higher toughness in terms of scratch/crack propagation resistance relative to monolayers with comparable Si content.

In the second example, (AlCrSi)N coatings were arc deposited using cathodes of $Al_{70}Cr_{30}$ and $Al_{60}Cr_{30}Si_{10}$ as a gradient coating for the Si content [121,125]. Different levels of Si content were achieved by adjusting the number of cathodes per coating deposition step. The coatings containing Si showed a nanocomposite structure (AlCrSi)N/a-SiN_x in all compositional regions. Cutting tests showed that the gradient coating provided better performance than the pure $(Al_{70}Cr_{30})N$ coating [121,125].

Coating designs with a gradient on the O content were realised by arc deposition in the top layer region of a part for solar energy thermal conversion in the form of (AlCr)N/(AlCr)NO/(AlCr)O. The functional coating properties indicate that this coating with its characteristic of selective absorption is a candidate for photo-thermal conversion at high temperatures [164].

6. From (CrAl)N or (AlCr)N to Oxynitride and Oxide Coatings

The special features of the $MeN_{1-x}O_x$ coating arise from the addition of oxygen to MeN, which results in coating properties between those of metal nitrides, MeN, and those of the insulating oxides, MeO [165]. Tuning the oxygen/nitrogen ratio allows the physical, mechanical and tribological characteristics to be tailored. For AlCr-based coatings, both oxynitrides as well as pure oxide coatings of AlCr are of interest. It was shown that oxynitride and oxide coatings (alpha and cubic crystal modifications) have a potential for use in various tribological applications and for oxidation protection [166–170], and moreover, even for use in solar absorbers [171]. Both vacuum arc deposition and various sputtering methods, including HiPIMS processes and hybrid processes (HiPIMS plus DC magnetron sputtering), were applied to investigate coatings of the types (AlCr)ON and (AlCr)O [23,172–175].

In the context of oxynitrides, it should be noted that in many industrialised nitride deposition processes, residual amounts of oxygen (up to several at.%) are incorporated [111, 122,146,172,175,176]. The effects of substrate rotation additionally influence details of the structure due to the changing position in the chamber relative to the source [132]. These effects occur in industrial coating systems and are frequently not discussed. However, for example, it was observed that a small amount of oxygen can decrease the hardness of nanocomposites significantly [177]. A detailed discussion of these effects is beyond the scope of this paper. However, actively controlling oxygen content in the coatings is a potent modifier of structure and properties, as will be exemplified in this section.

Oxynitride coatings and oxide coatings have often been investigated in one and the same paper. Figure 23 shows a schematic representation of the phase evolution in (CrAl)NO and (AlCr)NO systems dependent on Cr/(Al + Cr) and O/(O + N) at a typical deposition temperature of about 500–600 °C based on results from [166,167,172,173,178]. The differentiation between Cr-rich and Al-rich coating types is ignored here in the interest of facilitating visual perception.

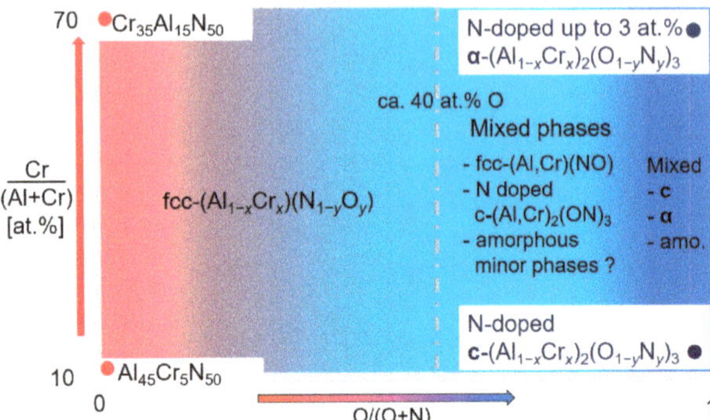

Figure 23. Schematic representation of phase evolution in the (AlCr)NO system dependent on Cr/(A + Cr) and O/(O + N) at a deposition temperature in the range of about 500–600 °C, c = cubic structur amo. = amorphous structure.

Oxynitrides are formed up to an estimated oxygen content of about 40 at.%. Abov this threshold, a mixed phase structure is formed, consisting of oxynitrides and oxid phases as well as sometimes a minor fraction of amorphous phases as well. At hig oxygen contents, oxides are formed. A cubic oxide structure (gamma or fcc) grows at low Cr/(Cr + Al) content, whereas a hexagonal oxide structure (alpha) is predominant at higher Cr/(Cr + Al) content. The formation of an oxide coating structure with a dominatin fcc phase has also been reported a number of times [166,179,180]. The lowest reporte value to obtain a pure corundum phase was Cr/(Cr + Al) = 0.3 [178]. There are also severa reports on the existence of minor amorphous phases [175,178,179].

Selected results will be presented in more detail for the two deposition methods, ar and sputtering.

6.1. Arc Evaporation: Oxynitride and Oxide Coatings

One of the earliest approaches to synthesising oxides by arc evaporation is reporte in a patent application (1992) from Schulz and Bergmann for the deposition of α-Al_2O type coatings by DC arc evaporation at a lower temperature than in CVD (Chemical Vapc Deposition) processes, when the Al is doped using Cr or other selected elements [22]. Th field, however, remained relatively unexplored well into the 2000s.

Investigations of coatings containing Al, Cr, O and N deposited by the vacuum ar method using different cathode compositions (Al/Cr) and different reactive gas compos tions (N_2/O_2) are shown in Figure 24.

Upon increasing the oxygen fraction in the gas flow, first an oxynitride is forme followed by an N-doped oxide and then a pure oxide [166]. Different phases are forme depending on the oxygen content in the reactive gas and the metal composition of th cathodes. In oxynitride coatings, the phase composition is first predominantly cubic, then mixed zone is observable, comprising the solid solutions fcc (Cr,Al)N, and nitrogen-dope cubic $(Cr,Al)_2O_3$ (gamma or fcc) or, finally, α $(Cr,Al)_2O_3$ at a pure oxygen flow. The oxid structures formed were either cubic or α $(Cr,Al)_2O_3$. Figure 25 shows the tendencies for th formation of either cubic or α $(Cr,Al)_2O_3$.

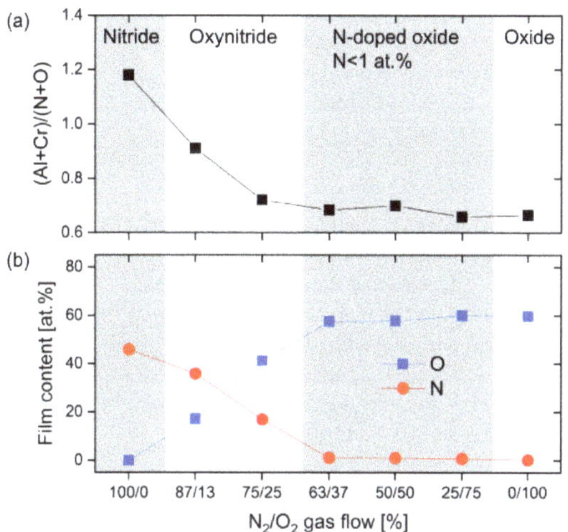

Figure 24. (**a**) Al + Cr to N + O ratio (**b**) N and O content in Al-Cr-O-N coatings for different N_2/O_2 reactive gas mixtures, redrawn after [166], original © Elsevier.

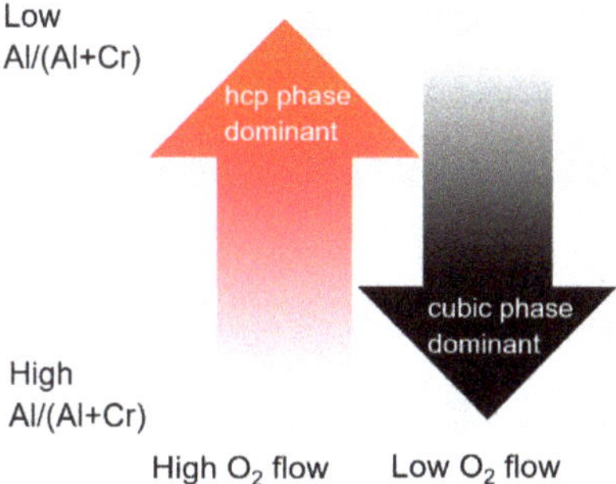

Figure 25. Schematic representation of the influence the coating composition has on the trend of phase formation in $(AlCr)_2O_3$, redrawn after [166], original © Elsevier.

A low level of O and a high Al content (e.g., using cathodes of $Al_{66}Cr_{34}$) in the coatings results in the formation of cubic $(Cr,Al)_2O_3$ layers. Higher levels of O and Cr (e.g., using cathodes of $Cr_{75}Al_{25}$) in the coatings increase the probability of the deposition of α $(Cr,Al)_2O_3$ layers.

The Cr-rich (CrAl)O coatings show the highest hardness values of 31–34 GPa, while the Al-rich coatings have lower values of 24–28 GPa. Metal cutting tests using cemented carbide inserts in turning operations showed good wear properties for mainly oxygen-rich coatings. These results were better than with the presence of the corundum phase of the oxide. Fcc-$(Cr,Al)_2O_3$ dominated coatings have also been shown to have wear properties similar to those of α-$(Cr,Al)_2O_3$ coatings [166].

In conclusion, the ideal structure is a corundum-type solid solution, preferably with a high Al-content (>60 at.% on the metal sublattice). However, at a high Al-content, the tendency towards the formation of dual-phase compositions containing metastable phase fractions has been observed, which may negatively influence the mechanical properties and performance. The authors concluded that the cathodic arc evaporation of $(Al,Cr)_2O_3$-based coatings is very complex when the microstructure of the arc cathode used is included as a factor [174].

In combination with (AlCr)N, a small addition of O using $Al_{70}Cr_{30}$ in the top layer of a double layer coating ((AlCr)N/(AlCr)NO) has been reported to increase the service life of tools in milling applications [181]. The effect of the oxygen content in doped AlCr based coatings, e.g., (AlCrSi)N, has also been investigated [132].

It should be mentioned that a special pulsed arc process, the P3e^{TM}, was developed to prevent oxide contaminations of the cathodes and increase the reactivity of the metal vapour and reactive gas [23].

6.2. Sputtering: Oxynitride and Oxide Coatings

The deposition of AlCr-based oxynitride and oxide coatings through sputtering has been approached using both RF-sputtering [173,179,182,183] as well as HiPIMS techniques [170,175]. Systematic investigations of RF-sputtered Al-Cr-O-N coatings have shown that, depending on the gas composition (Ar, O_2, N_2), the total gas pressure and the Al/Cr ratio, coatings with a single phase or mixed phase are grown. These are a corundum-type α-$(Al_{1-x},Cr_x)_2(O_{1-y},N_y)_3$ structure and a CrN-type fcc-$(Al_{1-x},Cr_x)(O_{1-y},N_y)$ structure, as well as a mixed-phase composition of both single phases [173,183]. The corundum phase can contain up to 3 at.% nitrogen [138,183].

The coating composition and structure with a given sputtering target composition are strongly dependent on the sputtering mode. Different Al/Cr ratios, deposition rates, and oxygen contents were reported for the comparison of DC magnetron, HiPIMS and a hybrid mode employing both [175]. Thick oxynitride coatings, thicker than 35 µm, were deposited by gas-flow sputtering [170].

7. Selected Industrial Coating Types and Main Applications

A large number of commercial coating solutions involving AlCr-based coatings are offered by PVD system manufacturers, job coating service centres and by tool and part manufacturers worldwide. The chemical compositions and the coating architectures vary, as do the deposition technologies and coating processes used. In the dominant arc source technology, the magnetic field set-up is particularly important as it influences the plasma conditions and coating properties. Within the group of sputtering technologies, a a great many more parameters can be varied, particularly in HiPIMS processes, e.g., pulse form and length.

The functionality of the coating-substrate system results from the quality and reproducibility of the whole process chain. Proper selection of the pre-treatment processing of the parts, for example, the cutting edges of cutting tools, is essential and must be carried out outside the coating chamber prior to deposition. Ion etching processing of the parts is performed to remove surface contamination and obtain sufficient coating adhesion Process parameters are selected to manage stress in the coating. For commercial reasons, the process details are typically protected by the companies offering CrAl- and AlCr-based coatings.

It must be pointed out that a new challenge, that of sustainable surface engineering, e.g., the minimisation of energy consumption, the selection of the coating material, reducing waste, and facilitating recycling, is attracting increasing interest. A comprehensive overview of all types of commercial AlCr-based coatings is beyond the scope of this review. However, a brief introduction to the application field of AlCr-based coatings will be provided through a characterisation of three different coating types.

7.1. Examples for Industrially Applied Coatings Types

Within the authors' company, more than 10 variations of AlCr-based coatings have been or are in use (different coating thicknesses and other minor variations are not counted) As an example of the technology status, three coating types will be roughly characterised to highlight features of the deposition process and the coating property ranges. Selected coating properties are shown in Table 3. Two deposition methods are applied: arc and HiPIMS (S3p). The coatings have different alloying elements and various architectures.

Table 3. Brief characteristics of selected industrially applied AlCrN-based coating types.

Feature	Monolayer Arc (e.g., BALINIT® ALCRONA PRO)	Multilayer (e.g., BALINIT® HELICA)	Monolayer Sputtering (e.g., BALIQ® ALCRONOS)
Source type	Magnetic steered arc	Magnetic steered arc	Magnetron optimised for S3p (HiPIMS)
Cathode type/form	Powder metallurgical/circular	Powder metallurgical and vacuum melting/circular	Powder metallurgical/circular
Coating temp. [°C]	400–500 °C	400–500 °C	400–500 °C
Chemical composition	Al > Cr content at.% - stoichiometric N content	- Layer A: like monolayer - Layer B: e.g., TiSi - stoichiometric N content	Al > Cr content at.% - stoichiometric N content
Alloying types	- metals, e.g., W, Mg - metalloids, e.g., Si, B - metal + metalloid	-	- metals, e.g., W, Mg - metalloids, e.g., Si, B - metal + metalloid
Coating hardness/Indentation E-modul range [GPa]	29–37 390–420	32–39 380–410	31–38 370–430
Thickness [um] Speciality	0.5–6 Universal use	1–4 High hardness	0.5–5 Low defect density, smooth

The first coating type is a monolayer deposited by cathodic vacuum arc (e.g., BALINIT® ALCRONA PRO). The second coating type is a multilayer coating employing the layer sequence (AlCr)N/(TiSi)N (e.g., BALINIT® HELICA), also applied by arc deposition. The third coating type is a monolayer sputter coating deposited using an advanced HiPIMS process (S3p), which can also be employed for multilayers.

The coating morphologies are shown in the cross-sectional images obtained by SEM together with a ball crater. The fine columnar structure of the arc monolayer (AlCr)N is shown in Figure 26.

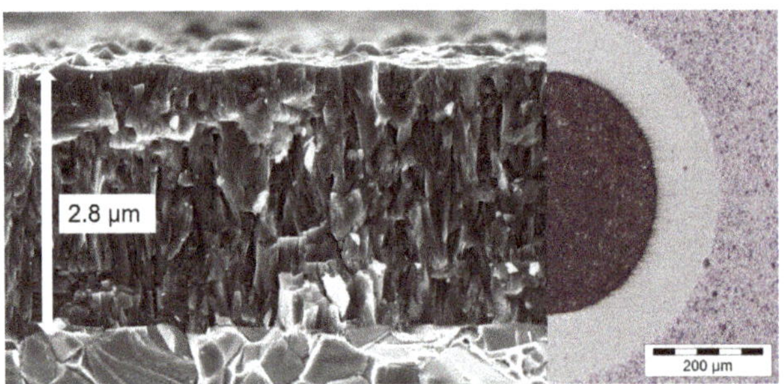

Figure 26. Cross-section and calotte of an industrially applied arc coating, BALINIT® ALCRONA PRO, courtesy of Oerlikon Balzers.

The fine-grained morphology of an arc multilayer (AlCr)N/(TiSi)N is demonstrated in Figure 27. This is a classical multilayer deposited by alternated switching on the sources with the different cathode materials. The added (TiSi)N sublayers increase the hardness of the coating.

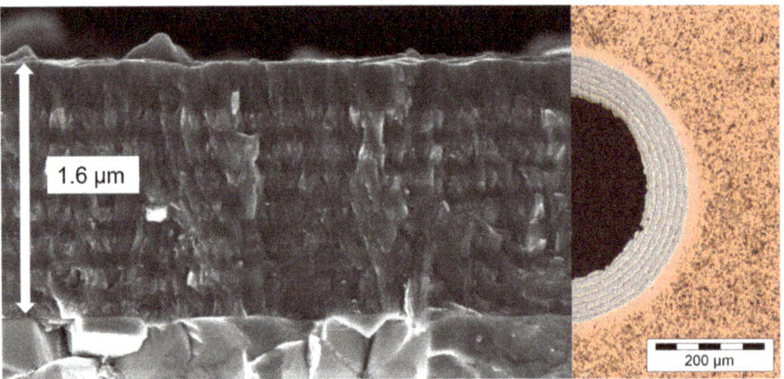

Figure 27. Cross section and calotte of an industrially applied arc multilayer coating (AlCr)N/(TiSi)N with top layer of (TiSi)N, BALINIT® HELICA, courtesy of Oerlikon Balzers.

The dense columnar structure with a low defect density and high smoothness of HiPIMS (AlCr)N monolayers is clearly seen in Figure 28. This coating is particularly advantageous for micro tools and other small parts that are sensitive to droplets in arc deposited coatings.

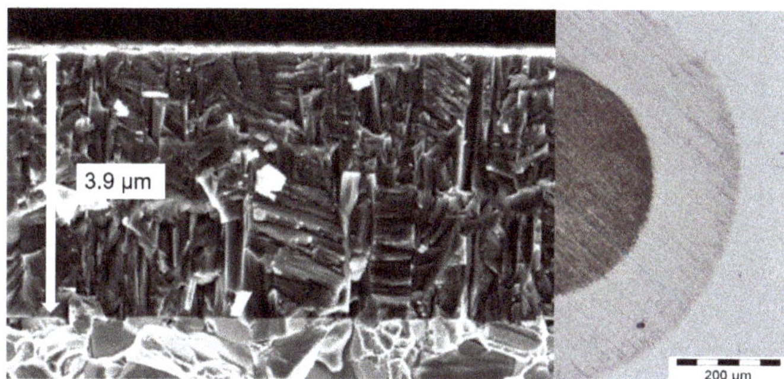

Figure 28. Cross-section and calotte of an industrially applied HiPIMS (S3p) (AlCr)N coating, BALIQ® ALCRONOS, courtesy of Oerlikon Balzers.

7.2. Typical Application Fields of AlCr-Based Coatings

Applications for AlCr-based coatings cover a wide range of cutting tools and forming tools, but also components. The dominant application is for tools [67,149,184–188]. Figure 29 shows examples of tool applications. Besides the standard applications, additional application potentials must be considered as well. Examples of this are: extremely thick coatings of more than 30 µm for turbine blades [170], or multilayer coatings with an oxygen gradient for solar absorbers [164].

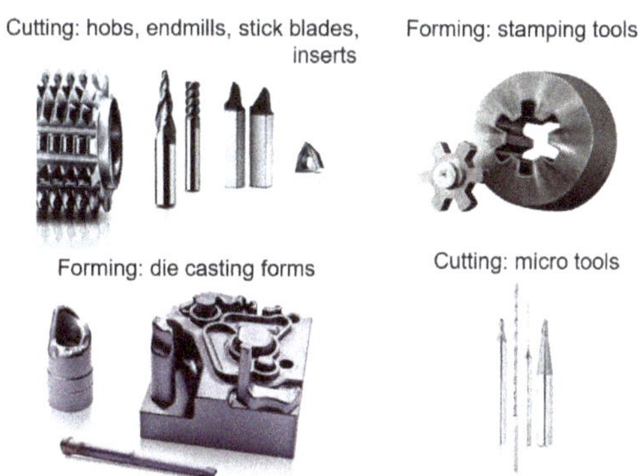

Figure 29. Selected tool applications for AlCr-based coatings, courtesy of Oerlikon Balzers.

8. Summary and Outlook

(Cr,Al)N and (Al,Cr)N coatings are amongst the main PVD coatings in commercial applications and offer a high degree of freedom to tailor the coating structure and properties by controlling the chemical composition and tuning the deposition process parameters. The coating architectures which can be achieved range from simple monolayers to heterogenous multilayers combining AlCr-based layers with other material systems. The present status and current trends can be summarised briefly as follows:

1. Besides arc, HiPIMS deposition methods will also be used more and more in addition to arc evaporation.
2. Alloying to adapt coating properties makes use of at least one element from the metals, metalloids or rare earth elements.
3. The main alloying element presently in use is silicon.
4. The variety in alloying will continue to increase to optimise coatings for dedicated applications.

5. The addition of oxygen offers a possibility for tuning properties, e.g., tribological and optical.
6. Multilayer architectures, including nano multilayers, are increasingly being applied in combination with binary, ternary and quaternary hard coatings.
7. Besides AlTi-based coatings, CrAl- and AlCr-based coatings are the predominant coating type applied to tools.
8. More and more general engineering parts will be coated in addition to tools, also with thicker coatings.
9. Sustainability aspects (sustainable surface engineering) will be taken into account, for example, selecting a specific alloying element, minimising the environmental impact in the production of the coating, the use of the coating, and recycling.

Funding: This research received no external funding.

Institutional Review Board Statement: Not applicable.

Informed Consent Statement: Not applicable.

Acknowledgments: The authors are thankful for the input received and the fruitful discussions enjoyed with all of our Oerlikon Balzers colleagues, including Jürgen Ramm, Arnd Müller, Ali Khatibi, Siegfried Krassnitzer, Denis Kurapov, and Hossein Najafi (Oerlikon Metco). In addition, the authors also wish to thank Christian Mitterer (Montanuniversität Leoben, Austria), Paul Mayrhofer (TU Vienna, Christian Doppler Laboratories, Austria) and Michael Stüber (KIT, Karlsruhe, Germany) for their constructive input in the form of discussions.

Conflicts of Interest: The authors declare no conflict of interest.

References

1. Aksenov, I.I.; Andreev, A.A. Vacuum arc coating technologies at NSC KIPT. *Probl. At. Sci. Technol. Ser. Plasma Phys.* **1999**, *3*, 242–246.
2. Vogel, J. *Harte Schichten, Goldene Zeiten (Hard Coatings, Golden Times)*; Informationsmappe; Oerlikon-Bührle Holding AG: Zurich Switzerland, 1982; Volume 5004.
3. Sue, J.A.; Perry, A.J.; Vetter, J. Young's modulus and stress of CrN deposited by cathodic vacuum arc evaporation. *Surf. Coat Technol.* **1994**, *68*, 126–130. [CrossRef]
4. Vetter, J. Vacuum arc coatings for tools: Potential and application. *Surf. Coat. Technol.* **1995**, *76*, 719–724. [CrossRef]
5. Vetter, J. 60 years of DLC coatings: Historical highlights and technical review of cathodic arc processes to synthesize various DLC types, and their evolution for industrial applications. *Surf. Coat. Technol.* **2014**, *257*, 213–240. [CrossRef]
6. Vetter, J.; Burgmer, W.; Dederichs, H.G.; Perry, A.J. The architecture and performance of multilayer and compositionally gradient coatings made by cathodic arc evaporation. *Surf. Coat. Technol.* **1993**, *61*, 209–214. [CrossRef]
7. PalDey, S.; Deevi, S.C. Single layer and multilayer wear resistant coatings of (Ti,Al)N: A review. *Mater. Sci. Eng. A* **2003**, *342* 58–79. [CrossRef]
8. Vetter, J. (Al_x:Ti_y)N coatings deposited by cathodic vacuum arc evaporation. *J. Adv. Mater.* **1999**, *31*, 41–47.
9. Andersson, J.M.; Vetter, J.; Müller, J.; Sjölén, J. Structural effects of energy input during growth of $Ti_{1-x}Al_xN$ ($0.55 \leq x \leq 0.66$) coatings by cathodic arc evaporation. *Surf. Coat. Technol.* **2014**, *240*, 211–220. [CrossRef]
10. Tanaka, Y.; Ichimiya, N.; Onishi, Y.; Yamada, Y. Structure and properties of Al-Ti-Si-N coatings prepared by the cathodic arc ion plating method for high speed cutting applications. *Surf. Coat. Technol.* **2001**, *146*, 215–221. [CrossRef]
11. Donohue, L.A.; Smith, I.J.; Münz, W.-D.; Petrov, I.; Greene, J.E. Microstructure and oxidation resistance of $Ti_{1-x-y-z}Al_xCr_yY_zN$ layers grown by combined steered arc/unbalanced magnetron- sputter deposition. *Surf. Coat. Technol.* **1997**, *94*, 226–231 [CrossRef]
12. Vetter, J.; Krug, T.; von der Heide, V. AlTiCrNO coatings for dry cutting deposited by reactive cathodic vacuum arc evaporation. *Surf. Coat. Technol.* **2003**, *174*, 615–619. [CrossRef]
13. Vetter, J.; Ishikawa, T.; Shima, N. Nanocomposite AlTiNCO coatings deposited by reactive cathodic arc evaporation. *Plasma Process. Polym.* **2007**, *4*, S668–S672. [CrossRef]
14. Knotek, O.; Atzor, M.; Barimani, C.; Jungblut, F. Development of low temperature ternary coatings for high wear resistance. *Surf. Coat. Technol.* **1990**, *42*, 21–28. [CrossRef]
15. Hoffmann, S.; Jehn, H.A. Oxidation behaviour of CrN_x and (Cr,Al)N hard coatings. *Werkst. Korros.* **1990**, *47*, 756–760. [CrossRef]
16. Schulze, D.; Wilberg, R.; Fleischer, W.; Lunow, T. Multicomponent Hard Thin Films Deposited by Hollow Cathode Arc Evaporator (HCA). In Proceedings of the International Conference on Metallurgical Coatings and Thin Films, San Diego, CA, USA, 2–6 April 1990.
17. Knotek, O.; Löffler, F.; Scholl, H.J. Properties of arc-evaporated CrN and (Cr,Al)N. *Surf. Coat. Technol.* **1991**, *45*, 53–58. [CrossRef]

8. Knotek, O.; Löffler, F.; Scholl, H.J.; Barimani, C. The multisource arc process for depositing ternary Cr- and Ti-based Coatings. *Surf. Coat. Technol.* **1994**, *68*, 309–313. [CrossRef]
9. Vetter, J.; Lugscheider, E.; Guerreiro, S.S. (Cr:Al)N coatings deposited by the cathodic vacuum are evaporation. *Surf. Coat. Technol.* **1998**, *98*, 1233–1239. [CrossRef]
0. Ishikawa, T.; Fuji, F. Improvement of the Cutting Performance of TiSiN Coated Cemented Carbide-Tools. In Proceedings of the 34th International Conference and Metallurgical Coatings and Thin Films ICMCTF, San Diego, CA, USA, 23–27 April 2007; p. B6-1-8.
1. Li, W.; Liu, P.; Meng, J.; Zhang, K.; Ma, F.; Liu, X.; Chen, X.; He, D. Microstructure and mechanical prop erty of TiSiN nanocomposite film with inserted CrAlN nanomultilayers. *Surf. Coat. Technol.* **2016**, *286*, 313–318. [CrossRef]
2. Schulz, H.; Bergmann, E. Hard Layer, Work Piece Coated with Such a Layer and Process for Coating with the Layer. EP 513,662. 11 November 1992.
3. Ramm, J.; Ante, M.; Bachmann, T.; Widrig, B.; Brändle, H.; Döbeli, M. Pulse enhanced electron emission (P3e^{TM}) arc evaporation and the synthesis of wear resistant Al-Cr-O coatings in corundum structure. *Surf. Coat. Technol.* **2007**, *202*, 876–883. [CrossRef]
4. Boxman, R.L.; Philip, J.M.; David, M.S. *Handbook of Vacuum Arc Science and Technology: Fundamentals and Applications*, 1st ed.; Noyes Publications: Park Ridge, NJ, USA, 1995.
5. Anders, A. *Cathodic Arcs: From Fractal Spots to Energetic Condensation*; Springer: New York, NY, USA, 2008.
6. Anders, A. A review comparing cathodic arcs and high power impulse magnetron sputtering (HiPIMS). *Surf. Coat. Technol.* **2014**, *257*, 308–325. [CrossRef]
7. Vetter, J. PVD processes for depositing wear- and friction- reducing and decorative coatings. In *Handbook of Thermoprocessing Technologies*; Beneke, F., Nacke, B., Pfeifer, H., Eds.; Vulkan-Verlag Eifel: Essen, Germany, 2015; pp. 661–692.
8. Vetter, J. Surface treatments for automotive applications. In *Coating Technology for Vehicle Applications*; Cha, S.C., Erdemir, A., Eds.; Springer: Cham, Switzerland, 2015; pp. 91–132.
9. Vetter, J.; Kubota, K.; Isaka, M.; Mueller, J.; Krienke, T.; Rudigier, H. Characterization of advanced coating architectures deposited by an arc-HiPIMS hybrid process. *Surf. Coat. Technol.* **2018**, *350*, 154–160. [CrossRef]
0. Vetter, J.; Berger, M.; Derflinger, V.; Krassnitzer, S. Plasma-Assisted Coating Processes. Available online: https://www.oerlikon.com/balzers/com/en/portfolio/surface-technologies/plasma-assisted-coating-processes-new-pvd-and-cvd-book/ (accessed on 12 February 2020).
1. Lundin, D.; Minea, T.; Gudmundsson, T. *High Power Impulse Magnetron Sputtering: Fundamentals, Technologies, Challenges and Applications*; Elsevier Science: Amsterdam, The Netherlands, 2019.
2. Ide, Y.; Inada, K.; Nakamura, T. Formation of Al-Cr-N films by an activated reactive evaporation (ARE) method. *High Temp. Mater. Process.* **2000**, *19*, 265–274. [CrossRef]
3. Schulze, D.; Wilberg, R. Multicomponent hard thin films deposited by hollow cathode arc evaporator (HCA). In Proceedings of the 4th joint International Symposium on Trends and New Applications in Thin Films TATF '94 and the 11th Conference om High Vacuum, Interfaces and Thin Films HVITF '94; DGM-Informationsges: Chemnitz, Germany, 1994.
4. Holzherr, M.; Falz, M.; Schmidt, T. Influence of hollow cathode plasma on AlCrN-thin film deposition with vacuum arc evaporation sources. *Surf. Coat. Technol.* **2008**, *203*, 505–509. [CrossRef]
5. Park, I.W.; Kang, D.S.; Moore, J.J.; Kwon, S.C.; Rha, J.J.; Kim, K.H. Microstructures, mechanical properties, and tribological behaviors of Cr-Al-N, Cr-Si-N, and Cr-Al-Si-N coatings by a hybrid coating system. *Surf. Coat. Technol.* **2007**, *201*, 5223–5227. [CrossRef]
6. Anders, A. Ion charge state distributions of vacuum arc plasmas: The origin of species. *Phys. Rev. E* **1997**, *55*, 969–981. [CrossRef]
7. Vetter, J.; Müller, J.; Erkens, G. Domino platform: PVD coaters for arc evaporation and high current pulsed magnetron sputtering. *IOP Conf. Ser. Mater. Sci. Eng.* **2012**, *39*, 012004. [CrossRef]
8. Vetter, J. *Entwicklung und Erprobung von Beschichtungen für Stanz-, Zieh- und Prägewerkzeuge für die Schmiermittelarme und Schmiermittelfreie Fertigung*; Final Report of Joint Project 13N6271; Metaplas Ionon GmbH: Bergisch Gladbach, Germany, 1996.
9. Kawate, M.; Kimura, A.; Suzuki, T. Microhardness and lattice parameter of $Cr_{1-x}Al_xN$ films. *J. Vac. Sci. Technol. A* **2002**, *20*, 569–571. [CrossRef]
0. Kawate, M.; Hashimoto, A.K.; Suzuki, T. Oxidation resistance of $Cr_{1-x}Al_xN$ and $Ti_{1-x}Al_xN$ films. *Surf. Coat. Technol.* **2003**, *165*, 163–167. [CrossRef]
1. Bagcivan, N.; Bobzin, K.; Theiß, S. $(Cr_{1-x}Al_x)N$: A comparison of direct current, middle frequency pulsed and high power pulsed magnetron sputtering for injection, molding components. *Thin Solid Film.* **2013**, *528*, 180–186. [CrossRef]
2. Wuhrer, R.; Yeung, W.Y. A comparative study of magnetron co-sputtered nanocrystalline titanium aluminium and chromium aluminium nitride coatings. *Scr. Mater.* **2004**, *50*, 1461–1466. [CrossRef]
3. Tritremmel, C. Comparison of Magnetron Sputtering and Arc Evaporation by Al-Cr-N Hard Coating. Diploma Thesis, Montanuniversität Leoben, Leoben, Austria, 2007.
4. Bobzin, K.; Lugscheider, E.; Nickel, R.; Bagcivan, N.; Kramer, A. Wear behavior of $Cr_{1-x}Al_xN$ PVD-coatings in dry running conditions. *Wear* **2007**, *263*, 1274–1280. [CrossRef]
5. Wang, Y.X.; Zhang, S.; Lee, J.W.; Lew, W.S.; Li, B. Influence of bias voltage on the hardness and toughness of CrAlN coatings via magnetron sputtering. *Surf. Coat. Technol.* **2012**, *206*, 5103–5107. [CrossRef]

46. Drnovšek, A.; Rebelo de Figueiredo, M.; Vo, H.; Xia, A.; Vachhani, S.J.; Kolozsvár, S.; Hosemann, P.; Franz, R. Correlating high temperature mechanical and tribological properties of CrAlN and CrAlSiN hard coatings. *Surf. Coat. Technol.* **2019**, *372*, 361–368. [CrossRef]
47. Kim, G.S.; Lee, S.Y. Microstructure and mechanical properties of AlCrN films deposited by CFUBMS. *Surf. Coat. Technol.* **2006**, *201*, 4361–4366. [CrossRef]
48. Weirather, T.; Czettl, C.; Polcik, P.; Kathrein, M.; Mitterer, C. Industrial-scale sputter deposition of $Cr_{1-x}Al_x$ coatings with $0.21 \leq x \leq 0.74$ from segmented targets. *Surf. Coat. Technol.* **2013**, *232*, 303–310. [CrossRef]
49. Sugishima, A.; Kajioka, H.; Makino, Y. Phase transition of pseudobinary Cr-Al-N films deposited by magnetron sputtering method. *Surf. Coat. Technol.* **1997**, *97*, 590–594. [CrossRef]
50. Makino, Y.; Nogi, K. Synthesis of pseudobinary Cr-Al-N films with B1 structure by rf-assisted magnetron sputtering method. *Surf. Coat. Technol.* **1998**, *98*, 1008–1012. [CrossRef]
51. Banakh, O.; Schmid, P.E.; Sanjinés, R.; Lévy, F. High-temperature oxidation resistance of $Cr_{1-x}Al_xN$ thin films deposited by reactive magnetron sputtering. *Surf. Coat. Technol.* **2003**, *163*, 57–61. [CrossRef]
52. Shah, H.N.; Jayaganthan, R. Influence of Al contents on the microstructure, mechanical, and wear properties of magnetron sputtered CrAlN coatings. *J. Mater. Eng. Perform.* **2012**, *21*, 2002–2009. [CrossRef]
53. Anders, A. Tutorial: Reactive high power impulse magnetron sputtering (R-HiPIMS). *J. Appl. Phys.* **2017**, *121*, 171101. [CrossRef]
54. Hsiao, Y.-C.; Lee, J.-W.; Yang, Y.-C.; Lou, B.-S. Effects of duty cycle and pulse frequency on the fabrication of AlCrN thin films deposited by high power impulse magnetron sputtering. *Thin Solid Film.* **2013**, *549*, 281–291. [CrossRef]
55. Avila, P.R.T.; da Silva, E.P.; Rodrigues, A.M.; Aristizabal, K.; Pineda, F.; Coelho, R.S.; Garcia, G.L.; Soldera, F.; Walczak, M.; Pinto, H.C. On manufacturing multilayer-like nanostructures using misorientation gradients in PVD films. *Sci. Rep.* **2019**, *9*, 15898. [CrossRef]
56. Tang, J.F.; Lin, C.Y.; Yang, F.C.; Chang, C.L. Influence of nitrogen content and bias voltage on residual stress and the tribological and mechanical properties of CrAlN films. *Coatings* **2020**, *10*, 546. [CrossRef]
57. Zheng, J.; Zhou, H.; Gui, B.; Luo, Q.; Li, H.; Wang, Q. Influence of power pulse parameters on the microstructure and properties of the AlCrN coatings by a modulated pulsed power magnetron sputtering. *Coatings* **2017**, *7*, 216. [CrossRef]
58. Chen, W.; Hu, T.; Hong, Y.; Zhang, D.; Meng, X. Comparison of microstructures, mechanical and tribological properties of arc-deposited AlCrN, AlCrBN and CrBN coatings on Ti-6Al-4V alloy. *Surf. Coat. Technol.* **2020**, *404*, 126429. [CrossRef]
59. Lin, J.; Mishra, B.; Moore, J.J.; Sproul, W.D. Microstructure, mechanical and tribological properties of $Cr_{1-x}Al_xN$ films deposited by pulsed-closed field unbalanced magnetron sputtering (P-CFUBMS). *Surf. Coat. Technol.* **2006**, *201*, 4329–4334. [CrossRef]
60. Hasegawa, H.; Masahiro Kawate, M.; Suzuki, T. Effects of Al contents on microstructures of $Cr_{1-x}Al_xN$ and $(Zr_{1-x}Al_x)N$ films synthesized by cathodic arc method. *Surf. Coat. Technol.* **2005**, *200*, 2409–2413. [CrossRef]
61. Reiter, A.E.; Derflinger, V.H.; Hanselmann, B.; Bachmann, T.; Sartory, B. Investigation of the properties of $Al_{1-x}Cr_x$ coatings prepared by cathodic arc evaporation. *Surf. Coat. Technol.* **2005**, *200*, 2114–2122. [CrossRef]
62. Willmann, H.; Mayrhofer, P.H.; Persson, P.O.A.; Reiter, A.E.; Hultman, L.; Mitterer, C. Thermal stability of Al-Cr-N hard coatings. *Scr. Mater.* **2006**, *54*, 1847–1851. [CrossRef]
63. Mayrhofer, P.H.; Willmann, H.; Reiter, A.E. Structure and phase evolution of Cr-Al-N coatings during annealing. *Surf. Coat. Technol.* **2008**, *202*, 4935–4938. [CrossRef]
64. Mayrhofer, P.H.; Music, D.; Reeswinkel, T.; Fuß, H.-G.; Schneider, J.M. Structure, elastic properties and phase stability of $Cr_{1-x}Al_xN$. *Acta Mater.* **2008**, *56*, 2469–2475. [CrossRef]
65. Bartosik, M.; Holec, D.; Apel, D.; Klaus, M.; Genzel, C.; Keckes, J.; Arndt, M.; Polcik, P.; Koller, C.M.; Mayrhofer, P.H. Thermal expansion of Ti-Al-N and Cr-Al-N coatings. *Scripta Mater.* **2017**, *127*, 182–185. [CrossRef]
66. Sabitzer, C.; Paulitsch, J.; Kolozsvári, S.; Rachbauer, R.; Mayrhofer, P.H. Influence of bias potential and layer arrangement on structure and mechanical properties of arc evaporated Al-Cr-N coatings. *Vacuum* **2014**, *106*, 49–52. [CrossRef]
67. Kalss, W.; Reiter, A.; Derflinger, V.; Gey, C.; Endrino, J.L. Modern coatings in high performance cutting applications. *Int. J. Refrac. Met. Hard Mater.* **2006**, *24*, 399–404. [CrossRef]
68. Martan, J.; Benes, P. Thermal properties of cutting tool coatings at high temperatures. *Thermochim. Acta* **2012**, *539*, 51–5 [CrossRef]
69. Böttger, P.H.M.; Gusarov, A.V.; Shklover, V.; Patscheider, J.; Sobiech, M. Anisotropic layered media with microinclusions: Thermal properties of arc-evaporation multilayer metal nitrides. *Int. J. Therm. Sci.* **2014**, *77*, 75–83. [CrossRef]
70. Zhou, L.; Holec, D.; Mayrhofer, P.H. First-principles study of elastic properties of cubic $Cr_{1-x}Al_xN$ alloys. *J. Appl. Phys.* **2013**, *11* 043511. [CrossRef]
71. Buchinger, J.; Wagner, A.; Chen, Z.; Zhang, Z.L.; Holec, D.; Mayrhofer, P.H.; Bartosik, M. Fracture toughness trends of modulus matched TiN/(Cr,Al)N thin film superlattices. *Acta Mater.* **2021**, *202*, 376–386. [CrossRef]
72. Soldán, J.; Neidhardt, J.; Sartory, B.; Kaindl, R.; Čerstvý, R.; Mayrhofer, P.H.; Tessadri, R.; Polcik, P.; Lechthaler, M.; Mitterer, C. Structure-property relations of arc-evaporated Al-Cr-Si-N coatings. *Surf. Coat. Technol.* **2008**, *202*, 3555–3562. [CrossRef]
73. Warcholinski, B.; Gilewicz, A.; Myslinski, P.; Dobruchowska, E.; Murzynski, D.; Kuznetsova, T.A. Effect of silicon concentration on the properties of Al-Cr-Si-N coatings deposited using cathodic arc evaporation. *Materials* **2020**, *13*, 4717. [CrossRef] [PubMed]
74. Willman, H.; Mayrhofer, P.H.; Hultman, L.; Mitterer, C. Hardness evolution of Al-Cr-N coatings under thermal load. *Mater. Res.* **2008**, *23*, 2880–2885. [CrossRef]

75. Meindlhumer, M.; Klima, S.; Jäger, N.; Stark, A.; Hruby, H.; Mitterer, C.; Kecks, J.; Daniel, R. Stress-controlled decomposition routes in cubic AlCrN films assessed by in-situ high-temperature high-energy grazing incidence transmission X-ray diffraction. *Sci. Rep.* **2019**, *9*, 1–14. [CrossRef] [PubMed]
76. Lin, J.; Mishra, B.; Moore, J.J.; Sproul, W.D. Study of the oxidation behavior of CrN and CrAlN thin films inir using DSC and TGA analyses. *Surf. Coat. Technol.* **2008**, *202*, 3272–3283. [CrossRef]
77. Chim, Y.C.; Ding, Z.Z.; Zeng, X.T.; Zhang, S. Oxidation resistance of TiN, CrN, TiAlN and CrAlN coatings deposited by lateralrotating cathode arc. *Thin Solid Film.* **2009**, *517*, 4845–4849. [CrossRef]
78. Feng, Y.-P.; Zhang, L.; Ke, R.-X.; Wan, Q.-L.; Wang, Z.; Lu, Z.-H. Thermal stability and oxidation behavior of AlTiN, AlCrN and AlCrSiWN coatings. *Int. J. Refract. Met. Hard Mater.* **2014**, *43*, 241–249. [CrossRef]
79. Khamseh, S.; Nose, M.; Kawabata, T.; Matsuda, K.; Ikeno, S. Oxidation resistance of CrAlN films with different microstructures prepared by pulsed DC balanced magnetron sputtering system. *Mater. Trans.* **2010**, *51*, 271–276. [CrossRef]
80. Polcar, T.; Cavaleiro, A. High temperature properties of CrAlN, CrAlSiN and AlCrSiN coatings—Structure and oxidation. *Mater. Chem. Phys.* **2011**, *129*, 195–201. [CrossRef]
81. Endrino, J.; Fox-Rabinovich, G.; Reiter, A.; Veldhuis, S.; Galindo, R.E.; Albella, J.; Marco, J. Oxidation tuning in AlCrN coatings. *Surf. Coat. Technol.* **2007**, *201*, 4505–4511. [CrossRef]
82. Souza, P.S.; Santos, A.J.; Cotrim, M.A.P.; Abrão, A.M.; Câmara, M.A. Analysis of the surface energy interactions in the tribological behavior of AlCrN and TiAlN coatings. *Tribol. Int.* **2020**, *146*, 106206. [CrossRef]
83. Reiter, A.E.; Mitterer, C.; de Figueiredo, M.R.; Franz, R. Abrasive and adhesive wear behavior of arc-evaporated $Al_{1-x}Cr_xN$ hard coatings. *Tribol. Lett.* **2010**, *37*, 605–611. [CrossRef]
84. Mo, J.L.; Zhu, M.H. Tribological oxidation behaviour of PVD hard coatings. *Tribol. Int.* **2009**, *42*, 1758–1764. [CrossRef]
85. Anders. A structure zone diagram including plasma-based deposition and ion etching. *Thin Solid Film.* **2010**, *518*, 4087–4090. [CrossRef]
86. Abadias, G.; Chason, E.; Keckes, J.; Sebastiani, M.; Thompson, G.B.; Barthel, E.; Doll, G.L.; Murray, C.E.; Stoessel, C.H.; Martinu, L. Review Article: Stress in thin films and coatings: Current status, challenges and prospects. *J. Vac. Sci. Technol. A* **2018**, *36*, 020801. [CrossRef]
87. Lomello, F.; Sanchette, F.; Schuster, F.; Tabarant, M.; Billard, A. Influence of bias voltage on properties of AlCrN coatings prepared by cathodic arc deposition. *Surf. Coat. Technol.* **2013**, *224*, 77–81. [CrossRef]
88. Kohlscheen, J.; Shibata, T. Phase and residual stress evaluation of dual-phase $Al_{70}Cr_{30}N$ and $Al_{80}Cr_{20}N$ PVD films. *Crystals* **2019**, *9*, 362. [CrossRef]
89. Tritremmel, C.; Daniel, R.; Lechthaler, M.; Rudigier, H.; Polcik, P.; Mitterer, C. Microstructure and mechanical properties of nanocrystalline Al-Cr-B-N thin films. *Surf. Coat. Technol.* **2012**, *213*, 1–7. [CrossRef]
90. Gilewicz, A.; Jedrzejewski, R.; Myslinski, P.; Warcholinski, B. Structure, morphology and mechanical properties of AlCrN coatings deposited by cathodic arc evaporation. *J. Mater. Eng. Perform.* **2019**, *28*, 1522–1531. [CrossRef]
91. Warcholinski, B.; Gilewicz, A.; Myslinski, P.; Dobruchowska, E.; Murzynski, D. Structure and Properties of AlCrN Coatings Deposited Using Cathodic Arc Evaporation. *Coatings* **2020**, *10*, 793. [CrossRef]
92. Wang, L.; Zhang, S.; Chen, Z.; Li, J.; Li, M. Influence of deposition parameters on hard Cr-Al-N coatings deposited by multi-arc ion plating. *Appl. Surf. Sci.* **2012**, *258*, 3629–3636. [CrossRef]
93. Vetter, J.; Ishikawa, T. *Developments of Plasma Enhanced Evaporator for High Al Coating ($Al_{70}Cr_{30}$)NO*; Unpublished Report; Metaplas Ionon GmbH: Bergisch Gladbach, Germany, 2003.
94. Gilewicz, A.; Jedrzejewski, R.; Myslinski, P.; Warcholinski, B. Influence of Substrate Bias Voltage on Structure, Morphology and Mechanical Properties of AlCrN Coatings Synthesized Using Cathodic Arc Evaporation. *Tribol. Ind.* **2019**, *41*, 484–497. [CrossRef]
95. Bobzin, K.; Brögelmann, T.; Brugnara, R.H. Aluminum-rich HPPMS $(Cr_{1-x}Al_x)N$ coatings deposited with different target compositions and at various pulse lengths. *Vacuum* **2015**, *122*, 201–207. [CrossRef]
96. Jeager, N.; Klima, S.; Hruby, H.; Julin, J.; Burghammer, M.; Keckes, J.F.; Mitterer, C.; Daniel, R. Evolution of structure and residual stress of a fcc/hex-AlCrN multi-layered system upon thermal loading revealed by cross-sectional Xray nano-diffraction. *Acta Mater.* **2019**, *162*, 55–66. [CrossRef]
97. Forsén, R.; Johansson, M.P.; Odén, M.; Ghafoor, N. Effects of Ti alloying of AlCrN coatings on thermal stability and oxidation resistance. *Thin Solid Film.* **2013**, *534*, 394–402. [CrossRef]
98. Hasegawa, H.; Yamamoto, T.; Suzuki, T.; Yamamoto, K. The effects of deposition temperature and post-annealing on the crystal structure and mechanical property of TiCrAlN films with high Al contents. *Surf. Coat. Technol.* **2006**, *200*, 2864–2869. [CrossRef]
99. Yamamoto, K.; Sato, T.; Takahara, K.; Hanaguri, K. Properties of (Ti,Cr,Al)N coatings with high Al content deposited by new plasma enhanced arc-cathode. *Surf. Coat. Technol.* **2003**, *174*, 620–626. [CrossRef]
100. Xu, Y.X.; Hu, C.; Chen, L.; Pei, F.; Du, Y. Effect of V-addition on the thermal stability and oxidation resistance of CrAlN coatings. *Ceram. Int.* **2018**, *44*, 7013–7019. [CrossRef]
101. Franz, R.; Neidhardt, J.; Sartory, B.; Tessadri, R.; Mitterer, C. Micro- and bonding structure of arc-evaporated AlCrVN hard coatings. *Thin Solid Film.* **2008**, *516*, 6151–6157. [CrossRef]
102. Bobzin, K.; Bagcivan, N.; Ewering, M.; Brugnara, R.H.; Theiß, S. DC-MSIP/HPPMS (Cr,Al,V)N and (Cr,Al,W)N thin films for high-temperature friction reduction. *Surf. Coat. Technol.* **2011**, *205*, 2887–2892. [CrossRef]

103. Iram, S.; Cai, F.; Wang, J.; Zhang, J.; Liang, J.; Ahmad, F.; Zhang, S. Effect of addition of Mo or V on the structure and cutting performance of AlCrN-based coatings. *Coatings* **2020**, *10*, 298. [CrossRef]
104. Sánchez-López, J.C.; Contreras, A.; Domínguez-Meister, S.; García-Luis, A.; Brizuela, M. Tribological behaviour at high temperature of hard CrAlN coatings doped with Y or Zr. *Thin Solid Film.* **2014**, *550*, 413–420. [CrossRef]
105. Klimashin, F.F.; Mayrhofer, P.H. Ab initio-guided development of super-hard Mo-Al-Cr-N coatings. *Scr. Mater.* **2017**, *140*, 27–30 [CrossRef]
106. Bobzin, K.; Brögelmann, T.; Kalscheuer, C. Arc PVD (Cr,Al,Mo)N and (Cr,Al,Cu)N coatings for mobility applications. *Surf. Coat Technol.* **2020**, *384*, 125046. [CrossRef]
107. Tillmann, W.; Dias, N.F.L.; Stangier, D. Effect of Hf on the microstructure, mechanical properties, and oxidation behavior of sputtered CrAlN films. *Vacuum* **2018**, *154*, 208–213. [CrossRef]
108. Hu, C.; Xu, Y.X.; Chen, L.; Pei, F.; Zhang, L.J.; Du, Y. Structural, mechanical and thermal properties of CrAlNbN coatings. *Surf Coat. Technol.* **2018**, *349*, 894–900. [CrossRef]
109. Rovere, F.; Mayrhofer, P.H.; Reinholdt, A.; Mayer, J.; Schneider, J.M. The effect of yttrium incorporation on the oxidation resistance of Cr-Al-N coatings. *Surf. Coat. Technol.* **2008**, *202*, 5870–5875. [CrossRef]
110. Rovere, F.; Music, D.; Schneider, J.M.; Mayrhofer, P.H. Experimental and computational study on the effect of yttrium on the phase stability of sputtered Cr-Al-Y-N hard coatings. *Acta Mater.* **2010**, *58*, 2708–2715. [CrossRef]
111. Qi, Z.B.; Wu, Z.T.; Wang, Z.C. Improved hardness and oxidation resistance for CrAlN hard coatings with Y addition by magnetron co-sputtering. *Surf. Coat. Technol.* **2014**, *259*, 146–151. [CrossRef]
112. Rojas, T.C.; Domínguez-Meister, S.; Brizuela, M.; Sanchez-Lopez, J.C. Influence of Al and Y content on the oxidation resistance of CrAlYN protective coatings for high temperature applications: New insights about the Y role. *J. Alloys Compd.* **2019**, *773* 1172–1181. [CrossRef]
113. Du, H.; Wang, L.; Young, M.; Zhao, H.; Xiong, J.; Wan, W. Structure and properties of lanthanum doped AlCrN coatings. *Surf Coat. Technol.* **2018**, *337*, 439–446. [CrossRef]
114. Shtansky, D.V.; Kiryukhantsev-Korneev, P.V.; Sheveyko, A.N.; Mavrin, B.N.; Rojas, C.; Fernandez, A.; Levashov, E.A. Comparative investigation of TiAlC(N), TiCrAlC(N), and CrAlC(N) coatings deposited by sputtering of MAX-phase $Ti_{2-x}Cr_xAlC$ targets. *Surf Coat. Technol.* **2009**, *203*, 3595–3609. [CrossRef]
115. Tillmann, W.; Stangier, D.; Roese, P.; Shamout, K.; Berges, U.; Westphal, C.; Debus, J. Structural and mechanical properties of carbon incorporation in DC/HiPIMS CrAlN coatings. *Surf. Coat. Technol.* **2019**, *374*, 774–783. [CrossRef]
116. Stueber, M.; Albers, U.; Leiste, H.; Ulrich, S.; Holleck, H.; Barna, P.B.; Kovacs, A.; Hovsepian, P.; Gee, I. Multifunctional nanolaminated PVD coatings in the system Ti-Al-N-C by combination of metastable fcc phases and nanocomposite microstructures. *Surf Coat. Technol.* **2006**, *200*, 6162–6171. [CrossRef]
117. Lee, D.B.; Nguyen, T.D.; Kim, S.K. Air-oxidation of nano-multilayered CrAlSiN thin films between 800 and 1000 °C. *Surf. Coat Technol.* **2009**, *203*, 1199–1204. [CrossRef]
118. Chen, H.W.; Chan, Y.C.; Lee, J.W.; Duh, J.G. Oxidation behavior of Si-doped nanocomposite CrAlSiN coatings. *Surf. Coat. Technol* **2010**, *205*, 1189–1194. [CrossRef]
119. Tritremmel, C.; Daniel, R.; Mayerhofer, P.H.; Lechthaler, M.; Polcik, P.; Mitterer, C. Oxidation behavior of arc evaporated Al-Cr-Si-N thin films. *J. Vac. Sci. Technol. A* **2012**, *30*, 061501. [CrossRef]
120. Wu, W.; Chen, W.; Yang, S.; Lin, Y.; Zhang, S.; Cho, T.-Y.; Lee, G.H.; Kwon, S.-C. Design of AlCrSiN multilayers and nanocomposite coating for HSS cutting tools. *Appl. Surf. Sci.* **2015**, *351*, 803–810. [CrossRef]
121. Cai, F.; Gao, Y.; Zhang, S.; Zhang, L.; Wang, Q. Gradient architecture of Si containing layer and improved cutting performance of AlCrSiN coated tools. *Wear* **2019**, *424*, 193–202. [CrossRef]
122. Wang, Y.X.; Zhang, S.; Lee, J.W.; Lew, W.S.; Sun, D.; Li, B. Toward hard yet tough CrAlSiN coatings via compositional grading *Surf. Coat. Technol.* **2013**, *231*, 346–352. [CrossRef]
123. Sun, S.Q.; Ye, Y.W.; Wang, Y.X.; Liu, M.Q.; Liu, X.; Li, J.L.; Wang, L.P. Structure and tribological performances of CrAlSiN coatings with different Si percentages in seawater. *Tribol. Int.* **2017**, *115*, 591–599. [CrossRef]
124. Haršáni, M.; Ghafoor, N.; Calamba, K.; Zacková, P.; Sahul, M.; Vopát, T.; Satrapinskyy, L.; Čaplovičová, M. Čaplovič Adhesive deformation relationships and mechanical properties of nc-AlCrN/a-SiN_x hard coatings deposited at different bias voltages. *Thin Solid Film.* **2018**, *650*, 11–19. [CrossRef]
125. Gao, Y.; Cai, F.; Zhang, L.; Zhang, S. Structure optimization and cutting performance of gradient multilayer AlCrSiN films with ion source etching pretreatment. *J. Mater. Eng. Perform.* **2020**, *29*, 997–1006. [CrossRef]
126. Jäger, N.; Meindlhumer, M.; Spor, S.; Hruby, H.; Julin, J.; Stark, A.; Nahif, F.; Keckes, J.; Mitterer, C.; Daniel, R. Microstructural evolution and thermal stability of AlCr(Si)N hard coatings revealed by in-situ high-temperature high-energy grazing incidence transmission X-ray diffraction. *Acta Mater.* **2020**, *186*, 545–554. [CrossRef]
127. Kolaklieva, L.; Kakanakov, R.; Stefanov, P.; Kovacheva, D.; Atanasova, G.; Russev, S.; Chitanov, V.; Cholakova, T.; Bahchedjiev C. Mechanical and structural properties of nanocomposite CrAlSiN-AlSiN coating with periodically modulated composition *Coatings* **2020**, *10*, 41. [CrossRef]
128. Wang, Q.M.; Kim, K.H. Microstructural control of Cr-Si-N films by a hybrid arc ion plating and magnetron sputtering process *Acta Mater.* **2009**, *57*, 4974–4987. [CrossRef]

29. Pélisson, A.; Hug, H.J.; Patscheider, J. Morphology, microstructure evolution and optical properties of Al-Si-N nanocomposite coating. *Surf. Coat. Technol.* **2014**, *257*, 114–120. [CrossRef]
30. Schmitt, T.; Steyer, P.; Fontaine, J.; Mary, N.; Esnouf, C.; O'Sullivan, M.; Sanchette, F. Cathodic arc deposited (Cr,Si_x)N coatings: From solid solution to nanocomposite structure. *Surf. Coat. Technol.* **2012**, *213*, 117–125. [CrossRef]
31. Rafaja, D.; Dopita, M.; Růžička, M.; Klemm, V.; Heger, D.; Schreiber, G.; Šíma, M. Microstructure development in Cr-Al-Si-N nanocomposites deposited by cathodic arc evaporation. *Surf. Coat. Technol.* **2006**, *201*, 2835–2843. [CrossRef]
32. Geng, D.; Li, H.; Zhang, Q.; Zhang, X.; Wang, C.; Wu, Z.; Wang, Q. Effect of incorporating oxygen on microstructure and mechanical properties of AlCrSiON coatings deposited by arc ion plating. *Surf. Coat. Technol.* **2017**, *310*, 223–230. [CrossRef]
33. Chen, Y.; Du, H.; Chen, M.; Yang, J.; Xiong, J.; Zhao, H. Structure and wear behavior of AlCrSiN-based coatings. *Appl. Surf. Sci.* **2016**, *370*, 176–183. [CrossRef]
34. Kiryukhantsev-Korneev, P.V.; Pierson, J.F.; Kuptsov, K.A.; Shtansky, D.V. Hard Cr-Al-Si-B-(N) coatings deposited by reactive and non-reactive magnetron sputtering of CrAlSiB target. *Appl. Surf. Sci.* **2014**, *314*, 104–111. [CrossRef]
35. Zhang, L.; Chen, Y.; Feng, Y.-P.; Chen, S.; Wan, Q.-L.; Zhu, J.-F. Electrochemical characterization of AlTiN, AlCrN and AlCrSiWN coatings. *Int. J. Refract. Met. Hard Met.* **2015**, *53*, 68–73. [CrossRef]
36. Tillmann, W.; Fehr, A.; Stangier, D. Powder metallurgic fabricated plug targets for the synthesis of AlCrSiWN multicomponent coating systems. *Int. J. Refract. Met. Hard Mater.* **2019**, *85*, 105081. [CrossRef]
37. Wang, Y.X.; Zhang, S.; Leeb, J.-W.; Lew, W.S.; Li, B. Toughening effect of Ni on nc-CrAlN/a-SiN_x hard nanocomposite. *Appl. Surf. Sci.* **2013**, *265*, 418–423. [CrossRef]
38. Najafi, H.; Karimi, A.; Alexander, D.; Dessarzin, P.; Morstein, M. Effects of Si and Y in structural development of (Al,Cr,Si/Y)O_xN_{1-x} thin films N deposited by magnetron sputtering. *Thin Solid Film.* **2013**, *549*, 224–231. [CrossRef]
39. Chen, M.; Chen, W.; Cai, F.; Zhang, S.; Wang, Q. Structural evolution and electrochemical behaviors of multilayer Al-Cr-Si-N coatings. *Surf. Coat. Technol.* **2016**, *296*, 33–39. [CrossRef]
40. Bakalova, T.; Petkov, N.; Bahchedzhiev, H.; Kejzlar, P.; Louda, P.; Ďurák, M. improving the tribological and mechanical properties of an aluminium alloy by deposition of AlSiN and AlCrSiN coatings. *Manuf. Technol.* **2017**, *17*, 824–830. [CrossRef]
41. Tang, J.-F.; Huang, C.-H.; Lin, C.-Y.; Yang, F.-C.; Chang, C.-L. Effects of substrate rotation speed on structure and adhesion properties of CrN/CrAlSiN multilayer coatings prepared using high-power impulse magnetron sputtering. *Coatings* **2020**, *10*, 742. [CrossRef]
42. Tao, H.; Tsai, M.T.; Chen, H.-W.; Huang, J.C.; Duh, J.-G. Improving high-temperature tribological characteristics on nanocomposite CrAlSiN coating by Mo doping. *Surf. Coat. Technol.* **2018**, *349*, 752–756. [CrossRef]
43. Liu, Z.R.; Peng, B.; Xu, Y.X.; Zhang, Q.; Wang, Q.; Chen, L. Influence of Ni-addition on mechanical, tribological properties and oxidation resistance of AlCrSiN coatings. *Ceram. Int.* **2019**, *45*, 3735–3742. [CrossRef]
44. Liu, Z.R.; Xu, Y.X.; Peng, B.; Wei, W.; Chen, L.; Wang, Q. Structure and property optimization of Ni-containing AlCrSiN coatings by nano-multilayer construction. *J. Alloys Compd.* **2019**, *808*, 151630. [CrossRef]
45. Yamamoto, K.; Kujime, S.; Takahara, K. Structural and mechanical property of Si incorporated (Ti,Cr,Al)N coatings deposited by arc ion plating process. *Surf. Coat. Technol.* **2005**, *200*, 1383–1390. [CrossRef]
46. Chen, W.; Hu, T.; Wang, C.; Xiao, H.; Meng, X. The effect of microstructure on corrosion behavior of a novel AlCrTiSiN ceramic coating. *Ceram. Int.* **2020**, *46*, 12584–12592. [CrossRef]
47. Dai, W.; Kwon, S.-H.; Wang, Q.; Liu, J. Influence of frequency and C_2H_2 flow on growth properties of diamond-like carbon coatings with AlCrSi co-doping deposited using a reactive high power impulse magnetron sputtering. *Thin Solid Film.* **2018**, *647*, 26–32. [CrossRef]
48. Vetter. J. Layer System for the Formation of a Surface Layer on a Surface of a Substrate and Also Arc Vaporization Source for the Manufacture of a Layer System. U.S. Patent 8,119,261 B2, 21 February 2012.
49. Yamamoto, K.; Kujime, S.; Fox-Rabinovich, G. Effect of alloying element (Si,Y) on properties of AIP deposited (Ti,Cr,Al)N coating. *Surf. Coat. Technol.* **2008**, *203*, 579–583. [CrossRef]
50. Shen, W.-J.; Tsai, M.-H.; Yeh, J.-W. Machining Performance of Sputter-Deposited $(Al_{0.34}Cr_{0.22}Nb_{0.11}Si_{0.11}Ti_{0.22})_{50}N_{50}$ High-Entropy Nitride Coatings. *Coatings* **2015**, *5*, 312–325. [CrossRef]
51. Chen, W.; Yan, A.; Meng, X.; Wu, D.; Yao, D.; Zhang, D. Microstructural change and phase transformation in each individual layer of a nano-multilayered AlCrTiSiN high-entropy alloy nitride coating upon annealing. *Appl. Surf. Sci.* **2018**, *462*, 1017–1028. [CrossRef]
52. Qiu, Y.; Zhang, S.; Lee, J.-W.; Li, B.; Wang, Y.; Zhao, D. Self-lubricating CrAlN/VN multilayer coatings at room temperature. *Appl. Surf. Sci.* **2013**, *279*, 189–196. [CrossRef]
53. Wang, Y.; Lee, J.-W.; Duh, J.-G. Mechanical strengthening in self-lubricating CrAlN/VN multilayer coatings for improved high-temperature tribological characteristics. *Surf. Coat. Technol.* **2016**, *303*, 12–17. [CrossRef]
54. Hosokawa, A.; Saito, R.; Ueda, T. Milling characteristics of VN/AlCrN-multilayer PVD coated tools with lubricity and heat resistance. *CIRP Ann. Manuf. Technol.* **2020**, *69*, 49–52. [CrossRef]
55. Liew, W.Y.H.; Jie, J.L.L.; Yan, L.Y.; Dayou, J.; Sipaut, C.S.; Madlan, M.F.B. Frictional and wear behaviour of AlCrN, TiN, TiAlN single-layer coatings, and TiAlN/AlCrN, AlN/TiN nano-multilayer coatings in dry sliding. *Procedia Eng.* **2013**, *68*, 512–517. [CrossRef]

156. Kumar, T.S.; Prabu, S.B.; Manivasagam, G. Metallurgical characteristics of TiAlN/AlCrN coating synthesized by the PVD Proces on a cutting insert. *J. Mater. Eng. Perform.* **2014**, *23*, 2877–2884. [CrossRef]
157. Chang, Y.-Y.; Weng, S.-Y.; Chen, C.-H.; Fu, F.-X. High temperature oxidation and cutting performance of AlCrN, TiVN an multilayered AlCrN/TiVN hard coatings. *Surf. Coat. Technol.* **2017**, *332*, 494–503. [CrossRef]
158. Xiao, B.J.; Li, H.X.; Mei, H.J. A study of oxidation behavior of AlTiN-and AlCrN-based multilayer coatings. *Surf. Coat. Techno* **2018**, *333*, 229–237. [CrossRef]
159. Chen, W.; Lin, Y.; Zheng, J.; Zhang, S.; Liu, S.; Kwon, S. Preparation and characterization of CrAlN/TiAlSiN nano-multilayers b cathodic vacuum arc. *Surf. Coat. Technol.* **2015**, *265*, 205–211. [CrossRef]
160. He, L.; Chen, L.; Xu, Y. Interfacial structure, mechanical properties and thermal stability of CrAlSiN/CrAlN multilayer coating *Mater. Charact.* **2017**, *125*, 1–6. [CrossRef]
161. Seidl, W.M.; Bartosik, M.; Kolozsváric, S.; Bolvardid, H.; Mayrhofer, P.H. Mechanical properties and oxidation resistance o Al-Cr-N/Ti-Al-Ta-N multilayer coatings. *Surf. Coat. Technol.* **2018**, *347*, 427–433. [CrossRef]
162. Ahmad, F.; Zhang, L.; Zheng, J.; Sidra, I.; Zhang, S. Characterization of AlCrN and AlCrON coatings deposited on plasm nitrided AISI H13 steels using ion-source-enhanced arc ion plating. *Coatings* **2020**, *10*, 306. [CrossRef]
163. Raab, R.; Koller, C.M.; Kolozsvári, S.; Ramm, J.; Mayrhofer, P.H. Interfaces in arc evaporated Al-Cr-N/Al-Cr-O multi layers an their impact on hardness. *Surf. Coat. Technol.* **2017**, *324*, 236–242. [CrossRef]
164. Wang, X.; Luo, T.; Li, Q.; Cheng, X.; Li, K. High performance aperiodic metal-dielectric multilayer stacks for solar energy therma conversion. *Sol. Energy Mater. Sol. Cells* **2019**, *191*, 372–380. [CrossRef]
165. Vetter, J. Oxynitrides and Oxides Deposited by Cathodic Vacuum Arc. In *Metallic Oxynitride Thin Films by Reactive Sputtering an Related Deposition Methods: Process, Properties and Applications*; Vaz, F., Martin, N., Fenker, M., Eds.; Bentham Science Publishe Bussum, The Netherlands, 2013; pp. 265–284.
166. Khatibi, A.; Sjölen, J.; Greczynski, G.; Jensen, J.; Eklund, P.; Hultman, L. Structural and mechanical properties of Cr-Al-O-N thi films grown by cathodic arc deposition. *Acta Mater.* **2012**, *60*, 6494–6507. [CrossRef]
167. Khatibi, A.; Genvad, A.; Goethelid, E.; Jensen, J.; Eklund, P.; Hultman, L. Structural and mechanical properties of corundum an cubic $(Al_xCr_{1-x})_{2+y}O_{3-y}$ coatings grown by reactive cathodic arc evaporation in as-deposited and annealed states. *Acta Mate* **2013**, *61*, 4811–4822. [CrossRef]
168. Pilkington, A.; Dowey, S.J.; Toton, J.T.; Doyle, E.D. Machining with AlCr-oxinitride PVD coated cutting tools. *Tribol. Int.* **2013**, *6* 303–313. [CrossRef]
169. Bobzin, K.; Brögelmann, T.; Kalscheuer, C.; Naderi, M. Hybrid dcMS/HPPMS PVD nitride and oxynitride hard coatings fo adhesion and abrasion reduction in plastics processing. *Surf. Coat. Technol.* **2016**, *308*, 349–359. [CrossRef]
170. Bobzin, K.; Brögelmann, T.; Kalscheuer, C.; Liang, T. High-rate deposition of thick (Cr,Al)ON coatings by high speed physica vapor deposition. *Surf. Coat. Technol.* **2017**, *322*, 152–162. [CrossRef]
171. Wang, X.; Zhang, X.; Li, Q.; Min, J.; Cheng, X. Spectral properties of AlCrNO-based multi-layer solar selective absorbing coatin during the initial stage of thermal aging upon exposure to air. *Sol. Energy Mater. Sol. Cells* **2018**, *188*, 81–92. [CrossRef]
172. Najafi, H.; Karimi, A.; Dessarzin, P.; Morstein, M. Correlation between anionic substitution and structural properties i $AlCr(O_xN_{1-x})$ coatings deposited by lateral rotating cathode arc PVD. *Thin Solid Film.* **2011**, *520*, 1597–1602. [CrossRef]
173. Stueber, M.; Diechle, D.; Leiste, H.; Ulrich, S. Synthesis of Al-Cr-O-N thin films in corundum and f.c.c. structure by reactive r magnetron sputtering. *Thin Solid Film.* **2011**, *519*, 4025–4031. [CrossRef]
174. Koller, C.M.; Stueber, M.; Mayrhofer, P.-H. Progress in the synthesis of Al- and Cr-based sesquioxide coatings for protectiv applications. *J. Vac. Sci. Technol. A* **2019**, *37*, 060802. [CrossRef]
175. Brögelmann, T.; Bobzin, K.; Kruppe, N.C.; Carlet, M. (Cr,Al)ON Deposited by Hybrid dcMS/HPPMS. *A study on incorporation Oxygen. Surf. Coat. Technol.* **2019**, *369*, 238–243. [CrossRef]
176. Zhou, H.; Zheng, J.; Gui, B.; Geng, D.; Wang, Q. AlTiCrN coatings deposited by hybrid HIPIMS/D C magnetronco-sputterin *Vacuum* **2017**, *136*, 129–136. [CrossRef]
177. Veprek, S.; Veprek-Heijman, M.J.G. Industrial applications of superhard nanocomposite coatings. *Surf. Coat.Technol.* **2008**, *20* 5063–5073. [CrossRef]
178. Cheng, Y.T.; Qiu, W.Q.; Zhou, K.S.; Jiao, D.L.; Liu, Z.W.; Zhong, X.C.; Zhang, H. The influence of Cr content on the phase structu of the Al-rich Al-Cr-O films deposited by magnetron sputtering at low temperature. *Ceram. Int.* **2019**, *45*, 8175–8180. [CrossRe
179. Najafi, H.; Karimi, A.; Dessarzin, P.; Morstein, M. Formation of cubic structured $(Al_{1-x}Cr_x)_{2+\delta}O_3$ and its dynamic transition t corundum phase during cathodic arc evaporation. *Surf. Coat. Technol.* **2013**, *214*, 46–52. [CrossRef]
180. Almandoz, E.; Fuentes, G.G.; Fernández, J.; de Bujanda, J.M.; Rodríguez, R.J.; Pérez-Trujillo, F.J.; Alcalá, G.; Lousa, A.; Qin, Chemical and mechanical stability of air annealed cathodic arc evaporated CrAlON coatings. *Surf. Coat. Technol.* **2018**, *35* 153–161. [CrossRef]
181. Gao, Y.; Cai, F.; Fang, W.; Chen, Y.; Zhang, S.; Wang, Q. Effect of oxygen content on wear and cutting performance of AlCrO coatings. *J. Mater. Eng. Perform.* **2019**, *28*, 828–837. [CrossRef]
182. Stüber, M.; Albers, U.; Leiste, H.; Seemann, K.; Ziebert, C.; Ulrich, S. Magnetron sputtering of hard Cr-Al-N-O thin films. *Su Coat. Technol.* **2008**, *203*, 661–665. [CrossRef]
183. Diechle, D.; Stueber, M.; Leiste, H.; Ulrich, S. Combinatorial approach to the growth of α-$(Al_{1-x},Cr_x)_{2+\delta}(O_{1-y},N_y)_3$ solid solutio strengthened thin films by reactive r.f. magnetron sputtering. *Surf. Coat. Technol.* **2011**, *206*, 1545–1551. [CrossRef]

184. Gerth, J.; Larsson, M.; Wiklund, U.; Riddar, F.; Hogmark, S. On the wear of PVD-coated HSS hobs in dry gear cutting. *Wear* **2009**, *266*, 444–452. [CrossRef]
185. Cselle, T.; Morstein, M.; Luemkemann, A. QuadCoatings4® new generation of PVD coatings for cutting tools. *Werkzeug Tech.* **2012**, *130*, 2–3.
186. Wang, B. An Investigation of the Adhesion Behavior of Aluminium on Various PVD Coatings Applied to H13 Tool Steel to Minimize Lubrication during High Pressure Die Casting. Ph.D. Thesis, Colorado School of Mines, Golden, CO, USA, 2016.
187. Uestuenyagiz, E.; Nielsen, C.V.; Tiedje, N.S.; Bay, N. Combined numerical and experimental determination of the convective heat transfer coefficient between an AlCrN-coated Vanadis 4E tool and Rhenus oil. *Measurement* **2018**, *127*, 565–570. [CrossRef]
188. Kim, D.; Swan, S.R.; He, B.; Khominich, V.; Bell, E.; Lee, S.-W.; Kim, T.-G. A study on the machinability of advanced arc PVD AlCrN-coated tungsten carbide tools in drilling of CFRP/titanium alloy stacks. *Carbon Lett.* **2020**. [CrossRef]

Article

Application of Adaptive Materials and Coatings to Increase Cutting Tool Performance: Efficiency in the Case of Composite Powder High Speed Steel

Sergey N. Grigoriev, Mars S. Migranov, Yury A. Melnik, Anna A. Okunkova, Sergey V. Fedorov, Vladimir D. Gurin and Marina A. Volosova *

Department of High-Efficiency Processing Technologies, Moscow State University of Technology STANKIN, Vadkovsky per. 1, 127055 Moscow, Russia; prof.s.n.grigoriev@gmail.com (S.N.G.); m.migranov@stankin.ru (M.S.M.); yu.melnik@stankin.ru (Y.A.M.); a.okunkova@stankin.ru (A.A.O.); sv.fedorov@icloud.com (S.V.F.); v.gurin@stankin.ru (V.D.G.)

* Correspondence: m.volosova@stankin.ru; Tel.: +7-916-308-49-00

Abstract: The paper proposes a classification of adaptive materials and coatings for tool purposes, showing the ability to adapt to external heat and power influences, thereby improving tool life. Creating a cutting tool made of composite powder high speed steels containing refractory TiC, TiCN, and Al_2O_3 compounds for milling 41CrS4 steel demonstrated the effectiveness of the adaptive materials. The tool material characteristics under the external loads' influence and the surface layer adaptation to the heat–power exposure conditions were shown by the temperature field study using a semiartificial microthermocouple method (the level of fields is reduced by 20%–25% for 80% HSS + 20% TiCN), frictional interaction high-temperature tribometry (the coefficient of friction did not exceed 0.45 for 80% HSS + 20% TiCN at +20 and 600 °C), laboratory performance tests, and spectrometry of the surface layer secondary structures. Spectral analysis shows the highest spectrum intensity of TiC_2 after 5 min of running in. After 20 min of milling (V = 82 m/min, f = 0.15 mm/tooth), dicarbide decomposes and transits to thermally stable secondary phase films of good lubricity such as TiO (maximum) and TiN (partially). There was an increase in tool life of up to 2 times (>35 min for 80% HSS + 20% TiCN), and a decrease in the roughness of up to 2.9 times (R_a less than 4.5 μm after 25 min of milling).

Keywords: adaptive coating; adaptive material; composite powder HSS; cutting tool; secondary structures; surface layer; thermal-force loads; wear resistance

Citation: Grigoriev, S.N.; Migranov, M.S.; Melnik, Y.A.; Okunkova, A.A.; Fedorov, S.V.; Gurin, V.D.; Volosova, M.A. Application of Adaptive Materials and Coatings to Increase Cutting Tool Performance: Efficiency in the Case of Composite Powder High Speed Steel. *Coatings* **2021**, *11*, 855. https://doi.org/10.3390/coatings11070855

Received: 30 June 2021
Accepted: 13 July 2021
Published: 16 July 2021

1. Introduction

Currently, it is necessary to design, create, and use innovative materials with an improved complex of physical, mechanical, and operational properties based on structural and tool materials, ceramics, and topocomposites (materials with functional coatings) to ensure improved performance of critical engineering products (elements of aviation equipment, power plants, turbocompressors, metalworking tools, and others) operating in difficult and variable conditions associated with the impact of increased thermal and power loads that cause high wear rates of contact surfaces. One of the most promising and dynamically developing scientific approaches aimed at solving the above problem, within the framework of which the most authoritative research groups around the world are working, is the development and use of adaptive materials and coatings, characterized by a particular structural–phase state in bulk and close to the surface layers [1–4].

Scientists and specialists use various terms such as adaptive, self-organizing, self-healing, intelligent, "chameleons" to designate such materials and coatings [5–10]. They have a "common denominator", which means the ability to adapt to operational loads (wear, damage, thermal power conditions, stress–strain state, etc.) with all the variety of existing terms, thereby allowing the functionality of a machine-building product and ensuring the operability of the entire system as a whole [11–14].

Muratore et al., in their research, understand "chameleon" coatings to be adaptive nanocomposite coating materials (solid lubricant with size scales of 10^{-10}–10^{-4} m, MoS coatings) that adjust their surface composition and morphology by multiple mechanism for reducing friction and wear intensity in aerospace applications [5]. They were the first to introduce this term. However, earlier works can be found that did not use these term concerning friction reduction.

Pogrebnyak et al. discussed the structural, phase, and chemical composition of adaptive multicomponent nanocomposite coatings. They conducted review work concerning research of hardness, friction, and wear intensity at elevated temperatures of adaptive coatings and analyzed their adhesive strength [4]. The adaptive coating materials with low friction coefficients and wear rate in a wide temperature range were divided into five groups:

- Diamond-like carbon coatings;
- Metal nitrides (TiN, CrN);
- Transition metal and dichalgogenide materials (MoS_2, WS_2);
- Polymers;
- Soft metals of the copper subgroup (with a filled pre-external $(n-1)d$-sublevel–$3d^{10}4s^1$, $4d^{10}5s^1$, $4f^{14}5d^{10}6s^1$ such as silver, copper, gold).

The main disadvantage of these materials (ceramics and metal ceramics, composites was indicated as poor lubricating properties at room temperature and an unstable friction coefficient in a broad temperature range up to 1000 °C. The authors conclude that th problem can be solved only by application of nanocopmpoite coatings with different internal structures based on the obtained material group, such as Al/Au, Mo_2N/Ag TiN/Ag, NbN/Ag, TaN/Ag, CrN/Ag, ZrN/Ag, VN/Ag, and CrAlN/Ag composites.

Zekonyte et al. [15] researched transition metal dichalcogenides (self-adaptive W–S–C coatings), which belong to a small family of solid lubricants with potential to produce a ultralow friction state using friction force microscopy. These materials form well-oriented films on the sliding surface with a thickness of 10 nm with a decrease in friction coefficient when the load increases. The well-ordered tungsten disulfide (WS_2) layers (tribolayers that peel off as ultrathin flakes) were formed on the coating surface. The observed nanoscal tribological behavior of WSC coatings replicates a deviation of Amonton's law.

Yuan et al. [16] researched a nanomultilayer TiAlCrSiYN/TiAlCrN that was deposited by physical vapor deposition (PVD) on cemented carbide turning inserts for milling DA71 Inconel. It was shown that an initial short-term cutting speed increase during the running in stage noticeably improves tool life by formation of protective/lubricious tribo-cerami films on the friction surface.

Sergevnin [12] studied wear resistance and failure of PVD Ti-Al-Mo-N coatings c 14 μm under different conditions. The coefficient of friction was established as 0.42 and 0.51 at +20 °C and 500 °C and 0.63, 0.71 for TiAlN coating, respectively. The wear of th cutting inserts in 18 min of operation was reduced by 14% comparing TiAlN coating, whe uncoated tool had failure (critical wear of 0.47 mm) after 16 min of turning.

The scientific and technological principles for creating adaptive materials and coating (AMCs) developed by research groups are highly diverse and mainly specialized fc specific engineering products (machine parts and cutting tools) in practice. There ar no universal methods and technological approaches. First, it is necessary to assess th specifics of the operational loads experienced by the product in AMC's developmer and practical application. If the cutting tool is considered an object of research, it shoul be considered a separate class of machine-building products, the operational loads fc which have pronounced distinctive features. The processes of friction and wear in cuttin materials proceed under specific conditions in comparison with friction and wear c machine parts:

- extremely high level of contact pressures on the tool working surfaces, often exceeding 100 MPa;
- the temperature on the friction surfaces during cutting reaches 600 °C and above;
- rubbing surfaces slide at high speed relative to each other, reaching 60 m/min and more [16–19].

Today, a large body of knowledge and results of theoretical and experimental research on AMC development and improvement has accumulated to improve the performance of various metal-cutting tool types in a wide range of operating loads. This work aimed to develop an easy-to-understand and capacious classification of AMC options that can improve the machining performance and cutting tool service life. In addition, the authors of the work aimed to assess the effectiveness of one of the technological approaches developed at Moscow State University of Technology "STANKIN" for creating adaptive materials using composite powder high speed steels (CPHSSs) containing refractory compounds such as TiC, TiCN, and Al_2O_3 for milling 41CrS4 steel. The change in the material characteristics under the influence of external loads and the formation of secondary stable structures (oxides and nitrides), i.e., operating conditions, was shown by conducting metallophysical and tribological studies and operational tests in laboratory conditions.

The novelty of the work is research of adaptive coatings of HSS + TiC, TiCN, and Al_2O_3 additives systems under conditions of 41CrS4 steel milling at normal temperature and in the conditions of elevated temperatures (600 °C), classification of adaptive materials and coatings, and creating a cutting tool made of composite powder high speed steels containing refractory compounds.

The following four types of tool materials:

- 100% HSS (HS6-5-2C steel powder);
- CPHSS samples of 80% HSS, 20% TiC (option 1);
- CPHSS samples of 80% HSS, 15% TiC, 5% Al_2O_3 (option 2);
- CPHSS samples of 80% HSS, 20% TiCN (option 3).

were researched by temperature field studies using a semiartificial microthermocouple method, frictional interaction high-temperature tribometry, laboratory performance tests, and spectrometry of the surface layer secondary structures. The main findings are:

- The coefficient of friction did not exceed 0.45 for 80% HSS + 20%TiCN at +20 °C and 600 °C);
- A graphical visualization of temperature fields in the tool cutting wedge section showed development of the maximum temperature from the tip to the flank face over time for HSS material. The zone of maximum temperatures remains at the top of the cutting wedge for CPHSS material. The temperature rise at the very beginning of the working stroke is less intense, and the level of fields is reduced by 20%–25%.
- Spectral analysis shows the highest spectrum intensity of TiC_2 after 5 min of running in. After 20 min of milling (V = 82 m/min, f = 0.15 mm/tooth, B = 5 mm, and t = 0.5 mm), dicarbide decomposes and transits to thermally stable secondary phase films of good lubricity such as TiO (maximum spectrum intensity) and TiN (partially);
- There was an increase in tool life of up to 2 times (>35 min for 80% HSS + 20% TiCN);
- There was a decrease in the roughness of up to 2.9 times (R_a less than 4.5 μm after 25 min of milling).

1.1. Cutting Tool Performance and AMC Challenges

New materials and coatings for tool purposes with improved physical, mechanical, and operational properties are of particular interest for modern machine-building enterprises. A cutting tool with higher efficiency can significantly improve the quality of machined parts and minimize energy and resource consumption for cutting when using expensive multi-axis computer numerical control (CNC) machine tools for high speed machining [20–24]. The operation of such machines is characterized by a high machine hour cost, increased cutting tool heat and power loads due to the need to intensify cutting conditions, and an increase in tool consumption per output unit. The acting loads level increases many times when difficult-to-machine structural materials are cut. The processes of adhesion–friction interaction on the tool's working surfaces are enhanced, and accelerated wear occurs with a simultaneous deterioration in the dimensional accuracy of the workpiece being processed and an increase in its roughness. In such conditions, the cutting tool requires increased efficiency and operational stability, mainly determining the efficiency of metalworking [25–28].

Although the types of cutting tools, tool materials, operating conditions, and the natures of the acting heat and power loads are highly diverse, they are united by common characteristic features. A wide range of acting loads is primarily determined by the type of material to be machined (carbon and low-alloy structural steels, cast and ductile irons, hardened steels and hard cast irons, heat-resistant Fe-, Ni-, or Co-based and titanium alloys, aluminum alloys, Zn-, Mg-, or Cu-based alloys and non-ferrous metals). Additionally, the most critical operating conditions are the cutting shear thickness or cutting feed, which determines the force impact level (increased during preliminary roughing and low during finishing and semifinishing) and the value of the cutting speed (increased and moderate), which determines the thermal effect level on the cutting tool contact pads. The stability of the loads acting on the cutting tool is also important: constant (with continuous contact of the cutting edges of the tool with the workpiece, for example, during turning) and intermittent (during milling, planing, chiseling, intermittent turning, etc.).

An analysis of the wear options and the reasons for the tool operability loss in practice can assist in formulating a problem set to be addressed by developing and applying AMCs. Thus, despite the variety of operating conditions, all AMC practical tasks can be reduced to the following:

- reducing the possibility of the development of creep and dynamic recrystallization processes, which weaken the cutting edge at the flank face and lead to catastrophic wear (typical for high speed steels);
- a decrease in the intensity of crater formation on the rake face, which reduces the strength of the cutting edge and leads to its destruction and rapid catastrophic wear of the cutting part of the tool;
- a decrease in the intensity of abrasive and adhesive wear on the tool flank, leading to an increase in temperature and catastrophic wear;
- reducing the possibility of plastic deformation of the cutting part of the tool (lowering or indentation), which worsens the removal of chips from the cutting zone, and reduces the workpiece dimensional accuracy and surface quality;
- prevention of the formation of or obstacle to the development of thermal microcrack networks on the tool cutting edge, which increases the risk of the cutting part chipping and reduces the tool resource;
- a decrease in the intensity of build-up on the rake face, which reduces the dimensional accuracy of the part and the quality of the surface (periodic detachment of the build-up leads to the formation of cracks near the cutting edge, reducing the tool life);
- minimization of the possibility of tool material chipping of microvolumes, which causes intensive wear of the flank face and a decrease in the accuracy of the workpiece dimensional accuracy and surface quality.

A decrease in the intensity of adhesion of the machined material to the tool's rake and flank faces significantly increases the machined part roughness and reduces the tool resource.

1.2. AMC Type Classification for Cutting Tools

The options for wear and the reasons for cutting tool efficiency loss when varying within a wide range of their operating conditions were considered above. All AMCs for cutting tool purposes must withstand high contact pressures and temperatures to function directly in the cutting zone or near it. Authoritative research groups develop and apply in practice various technological approaches considering these specific conditions, thanks to which the working sections of the cutting tool can adapt to the external environment effects. The authors of this work propose an array of available scientific and technological approaches in the field of AMCs for cutting tools to classify them into three main groups (Figure 1 shows a variant of the graphical visualization of the proposed classification):

1. Materials and coatings with active control function;
2. Self-organizing materials and coatings;
3. Self-repairing materials and coatings.

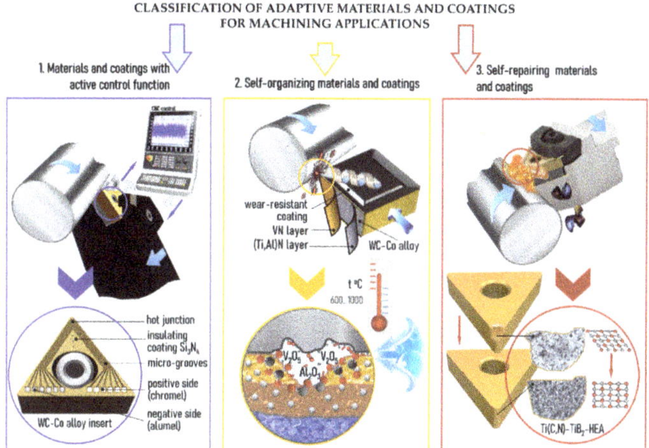

Figure 1. Classification of AMC types to improve the performance of the cutting tool.

1.3. Materials and Coatings with Active Control Function

The group of materials and coatings with active control function should include AMCs, which in their essence play the role of sensors that, when exposed to external influences, recognize (detect) critical changes during the cutting tool operation and transmit the corresponding signals wirelessly to the control system of the metal-cutting machine. The system carries out a control (adaptive) effect on cutting according to the built-in algorithms, depending on the information received and the technological problems to be solved. For example, it adjusts the cutting speed, feed, or depth to reduce the likelihood of brittle fracture of the cutting part or reduce tool excessive wear. In emergency cases, it stops machining, thus realizing the active control principle.

When implementing this approach, coatings can be applied to the cutting tool (for example, thin-film thermocouples), or heat-sensitive and elastic-sensitive inserts can be used in the tool design, located near the cutting zone, which demonstrate deformations or other physical parameters characterizing the machining and cutting tool state [29–31]. It is advisable to use these AMCs in the manufacture of the most critical products in the aerospace, nuclear, and other industries when there is a technological task of manufacturing either an extended (large) part or a complex-shaped part for which a cutting tool failure during machining will inevitably lead to a failure of expensive products. Another application area is special alloys, difficult-to-machine composites, and machining of other materials when there is a high probability of unexpected cutting tool failure.

There has been research on online monitoring of cutting temperature during continuous turning of titanium and nickel alloys using interchangeable triangular and tetrahedral cutting inserts made of WC–Co hard alloy with built-in chromel–alumel thermocouple sensors. Microgrooves with a depth of about 100 μm are etched on the flank face using laser action to accommodate thin-film thermocouples on carbide inserts near the cutting edge, and the surface of the cutting tool is covered with a thick insulating layer [29]. The sensors can be staggered concerning the direction of the chip flow to increase the measurement accuracy. Detailed technological principles and options for thin-film thermocouples formin on the tool surface are described in [29–32]. Well-known studies have demonstrated th possibility of high-precision measurement in cutting temperature values in the range c up to 1000 °C. In this case, the natural wear of the cutting edge of the carbide inserts doe not affect the performance of the sensors and the reliability of the received diagnostic dat There are experimental data on the efficiency of using the above-described cutting plat sensors during operation under intermittent cutting with difficult-to-machine materia conditions [30]. Under such conditions, sensors based on chromel–alumel thermocouple demonstrate high sensitivity and low inertia. There has been some research in whic thin-film thermocouples were embedded in polycrystalline cubic boron nitride cuttin inserts by diffusion bonding to control turning of very hard materials [31].

Additionally, it should be noted that it is promising to use sensors based on flexibl piezoelectric polymer films to design structures and for cutting tool production. Thi gives an excellent response under mechanical stress action and allows for assessing th components of the cutting force according to the degree of their deformation, which is a informative diagnostic (troubleshooting) sign of wear and tear of tool surfaces and th state of the entire metalworking system as a whole [33,34]. For these purposes, it is mos convenient to use PVDF (polyvinylidene fluoride) sensors (film) with a multilayer structur consisting of a PVDF film sandwiched between two electrodes and protective coatings [35 The locations for installing the sensors are variable, but they should be as close as possibl to the processing area (in this case, it is necessary to consider that the working temperatur of the PVDF film is 700 °C). They are installed in seats under each cutting plate in th tool body as a rule (in the case of using a multiblade assembly tool) or in special recesse made in the tool holder, into which the cutting plate abuts with the working surfaces an where elastic deformations arise under the action of cutting forces. The timely response c the control system to the signals transmitted by the sensors allows the cutting tool to b adapted to changed conditions, thereby preserving its performance for a longer time.

1.4. Self-Organizing Materials and Coatings

This group should include AMCs capable of providing the structural self-organizatio of the surface layer of the cutting tool during contact interaction with the environmer and workpiece being machined under conditions of heat–power loads typical for cuttin processes due to secondary structure (phase) formation. It is the most extensive AMC grou in terms of implementation options and research depth, with which many authoritativ world-class teams work [15,36–45]. A classic example of secondary structures is thos formed at elevated cutting temperatures in conditions of interaction with the natura environment (air) or artificially introduced external media of oxide and sometimes nitrid compounds of metals of IV–VI groups of the periodic table. Secondary structures hav thermal stability and improved lubricity in the tribocontact zone, significantly reduce th intensity of frictional interaction, and increase the wear resistance of the tool contact pad

A schematic diagram demonstrating the stages of secondary structures forming durin physicochemical processes in the contact zone in cutting materials is proposed by th authors in Figure 2. At the initial stage (stage 1, Figure 2), when the cutting wedg separates the chips from the workpiece, chemically clean surfaces (often called juvenil are formed on the tool and the workpiece, on which there are no oxides and adsorbe films. "Freshly" formed surfaces are of a very active energetic and electronic state, an they are ready to interact with the components of the external environment (stage Figure 2), which can be characterized by various sorption processes, primarily physic and chemical adsorption [46–48]. The energy of the broken molecular bonds of the juvenil surface is such that the external environment molecules can undergo destruction an decay into atoms, ions, and radicals. These formed particles are also chemically activ and can enter into chemical interaction with freshly formed metal surfaces. At the ne moment, the formation of new chemical compounds (secondary structures) is observe in the "tool-processed material" contact zone due to reactions between chemically pur surfaces and components of the external environment, that is, the processes of ion and chemisorption (stage 3, Figure 2). The physicochemical processes in the contact zone with protective (lubricating) films and secondary structures forming at the interface with thermal stability and improved lubricity passivate the adhesion and frictional interaction between chemically pure surfaces. The secondary structures represent a stable zone with an increased internal energy level from the thermodynamics point [49–51]. The transformation of the conditions for contacting the cutting wedge working surfaces of the tool with the machined material is the consequence. The formed lubricating films and secondary structure function are integral parts of the tool material and participate in cutting. The formed protective films cannot be exclusively continuous. In addition, their inevitable local abrasion and destruction occur together with the tool material with the formation of chemically pure freshly formed surfaces, which again interact with the external environment in the contact zone at increased heat and power loads (stage 4, Figure 2) and the formation of secondary structures occurs. These stages are cyclically repeated several times throughout machining the part until the cutting tool reaches critical wear.

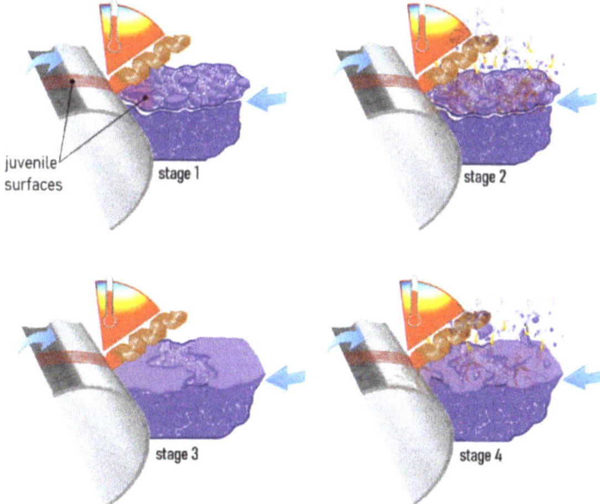

Figure 2. Schematic diagram of the secondary structures' physicochemical formation in the contact zone of the cutting wedge and the workpiece to be machined.

The understanding of physicochemical processes described above for contacting a tool and a workpiece provides researchers with an excellent toolbox for creating self-organizing materials and coatings. It is possible to purposefully design and implement the composition and structure of the tool material and/or the surface coating applied to the tool and predict the formation of certain secondary structures necessary for specific operating conditions, taking into account the thermal effect. It is important to emphasize that obtaining reliable diagnostic information about the changes that have occurred in the ultrathin surface layer of the tool material or coating is possible only by high-precision techniques based on spectroscopy principles. Only by systematically investigating and understanding the nature of the changes and the formation of secondary structures can a scientifically grounded choice of architecture and chemical and phase composition of the tool material and coating be made for high speed and carbide tools and also for ceramic tools and for specific operating conditions to ensure the maximum possible wear resistance.

The basic concept in self-organizing AMCs is based on the elements in their composition, which form oxide phases with weak bonds between atomic planes during heating [52,53]. These phases have good thermal stability and resistance to oxidation at the tribocontact spot with the workpiece being machined (for example, W, Mo, V, and Ti form a series of oxides with a layered structure). One of the most studied compounds that forms stable oxide phases upon oxidation is vanadium nitride. Since VN is a compound potentially capable of imparting antifriction properties to existing coatings during oxidation, there are solutions for creating superlattice coatings [54], in which layers of nanosized thickness TiAlN/VN alternate. The coefficient of friction of coatings of this type is relatively low at temperatures of 700 °C and is no more than 0.5. TiAlN/VN coatings show a lower friction coefficient than hard coatings with superlattice TiAlN/CrN or CrN/NbN [55,56].

The mechanism of formation and the type of secondary structures formed are separate for each coating in their temperature range, and a transition from one joint to another is possible during cutting with an increase in temperature on the contact pads of the tool [57]. For example, the beginning of the formation of oxide phases in the tribocontact zone is noted for a TiAlN/VN coating at 500 °C, and a complex structural analysis reveals the presence of vanadium pentoxide V_2O_5 in the near-surface layer. The specified compound melts with its removal from the surface layer at 685 °C, forming high-temperature modifications of other vanadium oxides.

The well-known studies of the mechanism of self-organization in milling with a tool made of WC-Co substrate with multilayer nanosized coatings based on complex nitrides of TiAlCrSiYN/TiAlCrN [16] are very indicative. This coating provides mass transfer of many elements, primarily Al and Si, to the friction surface, accompanied by the formation of dissipative structures due to a high nonequilibrium state and a complex nanocrystalline/layered structure. Ultimately, this leads to a substantial decrease in entropy production and the corresponding wear rate because of self-organization already occurring at the initial stage of wear. The formation of synergistic complex tribooxides endows the surface layer with increased thermal barrier properties. It creates the effect when the intense heat flux accompanying cutting is minimally accumulated in the surface layer of the cutting tool and is maximally transferred to the environment, ensuring the working condition of the tool for a longer time.

The authors of this project comprehensively studied the processes of wear and changes in the structure of the tribocontact zone of the surface of tools from high speed steels and WC-Co substrate and oxide–carbide and oxide–nitride ceramics, including those with TiNbAlN/TiZrN coatings, etc. which were deposited by magnetron sputtering bombarded with fast argon atoms [58–61]. Forming nonstoichiometric amorphous oxide films on the surface of ceramic samples was revealed during the research. Nitride and carbide phases from the surface layer were partially transformed into simple or complex oxides in the contact zone due to tribooxidation. The formation of these secondary structures made it possible to increase the wear resistance of ceramic tools up to two times during turning and milling.

As a typical example of the practical use of the adaptation mechanism in the conditions of frictional and adhesive contact interaction, it will be interesting to consider the results of research of the scientific group of MSTU "STANKIN" devoted to the multilateral study of the structural state and performance properties of improved tool high speed steels.

1.5. Self-Repairing Materials and Coatings

This group should include AMCs which can provide a spontaneous response to external mechanical, thermal and physical impacts by changing (restoring) the structure and microgeometry. Examples of the implementation of this approach are self-sharpening cutting edges of metalworking tools with blades, for example, equipped with hard laminar or laminated coatings [62]. Nevertheless, the greatest research interest for scientists today is tool materials and coatings that have a shape memory effect (high damping alloys). This effect consists of restoring the original shape of the plastically deformed material, which occurs after its heating to a specific temperature [63,64].

The self-healing phenomenon is associated with martensitic transformations in the crystal lattice of the material, during which an ordered movement of atoms occurs. Martensite in shape memory materials is thermoelastic and consists of crystals in the form of thin plates that stretch in the outer layers and shrink in the inner layers [65]. The sources of deformation are interphase, twinning, and intercrystalline boundaries. After heating the deformed alloy, internal stresses appear, tending to return the metal to its original shape. The nature of spontaneous recovery depends on the mechanism of the previous exposure and temperature conditions.

Among some known shape memory alloys used in mechanical engineering, an intermetallic compound based on titanium nickelide (Ni-Ti system) has the maximum thermal cyclic strength, but its hardness and heat resistance are insufficient for use as a material for the cutting part of the tool [66]. At the same time, there are examples of using Ni-Ti or analogs as inserts in the construction of cutting tools, particularly for the machining of rocks and building materials. For example, there are known variants of using such materials as compact drives for bringing the worn cutting part of the cutters of drill bits to the working position due to the form restoration of the alloy under a particular thermal effect. There are technical solutions for the manufacture of tool bases from shape memory alloys for fixing diamond elements. In the event of overheating and an unacceptable increase in temperature, the contact area with the workpiece material decreases and the position of the cutting diamond inserts is corrected to minimize their abrasion. Recently, new high-entropy alloys of the Fe-Ni-Co-Al-Nb-Ti type have come to replace traditional alloys of the Ni-Ti system, which have certain advantages and significant prospects for application in tool production. There are examples of assembled tool constructions with CBN inserts including damping elements. This ensures improved vibration resistance and durability of the tool [67].

An analysis of the latest progressive research results shows that the use of high-entropy alloys (HEAs) in their manufacture has attracted the most significant interest in creating tool materials with a self-healing effect today [68–70]. In particular, HEAs based on Fe-Co-Cr-Ni-Al and Co-Cr-Fe-Ni-Ti-Al are already effectively used as binders (mass fraction up to 20%) in the production of TiCN-TiB_2 composite cermets by vacuum hot pressing. After receiving the samples, the binder structure based on the HEA is a solid solution, which tightly binds TiCN and TiB_2. The use of the HEA makes it possible to obtain a fundamentally new tool material with a unique set of physical and mechanical properties compared to using a traditional binder (Ni-Co) in the manufacture of cermets. There is research on the creation of tungsten-containing hard alloys of the WC group using the Al-Co-Cr-Cu-Fe-Ni binder, which makes it possible to operate the material at higher thermal loads compared to samples in which Co is used as a binder [71]. The use of an HEA makes it possible to obtain a new class of tool material for a broader range of technological applications than traditional materials.

In addition, development is currently underway to use Fe-Co-Cr-Ni-Mo-based HEAs to create diamond and ceramic composites through spark plasma sintering and other advanced technologies [72–75]. The use of an HEA as a tool matrix in which diamond particles are embedded allows, by adjusting the technological conditions of sintering, various microstructures at the HEA/diamond interface and sufficient interfacial bond strength to be obtained. Researchers have discovered an interstitial strengthening effect, which has a very beneficial effect on the mechanical properties of the new tool material.

2. Materials and Methods

2.1. CPHSS-Based Tool Materials

The object of the study was composite powder high speed steels (CPHSSs), the components of which have a powder composition according to the chemical composition corresponding to the widespread industrial high speed HS6-5-2C steel (AISI M2 according to ISO 4957 [76]) with the addition of 20% refractory compounds TiC, TiCN, and Al_2O_3. CPHSS occupies an intermediate position between the classic tool materials such as high speed steels (HSSs) and carbide alloys in their properties. The introduction of refractory carbides, carbonitrides, and oxides into the CPHSS composition provides them with high hardness and heat resistance and the high speed powder base with high viscosity anc ductility. CPHSS can be classified as a generic class of adaptive tool materials [77–80].

The industrial technology of hot isostatic pressing of powder compositions followe by pressure treatment by heated blank extrusion was used to obtain experimental sample (Figure 3). Sixteen samples of each tool material were tested. All experiments were repeate five times to obtain data. The following variants of samples of tool materials based on HS with and without additives of refractory compounds were studied:

- 100% HSS (basic version made of HS6-5-2C steel powder);
- CPHSS samples from 80% HSS, 20% TiC (option 1), 80% HSS, 15% TiC, 5% Al_2O (option 2), 80% HSS, 20% TiCN (option 3).

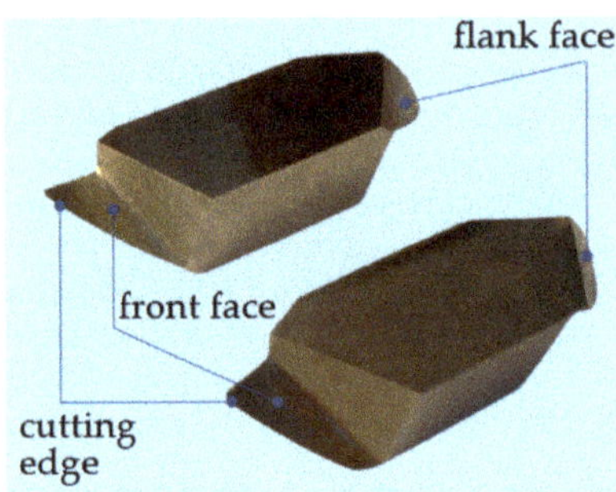

Figure 3. Produced samples of HSS and CPHSS compounds.

Preliminary studies carried out at MSTU "STANKIN" showed that an increase in th content of refractory compounds over 20% significantly reduces the strength characteristic of the material [47,49]. The research was limited to the specified content of alloyin additives since it was focused on the process of interrupting cutting (milling) when the ris of brittle fracture of the tool material increases.

Different values of the fundamental physical and mechanical properties characterize the samples prepared for research, precontrolled in the laboratories of the universit (Table 1).

Table 1. Physical and mechanical properties of research samples.

Material	Rockwell Hardness HRC	Flexural Strength, MPa	Average Coefficient of Friction on Steel
100% HSS	63–65	3300	0.75
CPHSS	70–72	2400–2500	0.5–0.6

2.2. Study of the Structure, Elemental Composition, and Tribological Properties

The study of the samples' structure was carried out on a scanning electron microscop VEGA 3 LMH manufactured by Tescan (Brno, Czech Republic). This instrument wa equipped with an Oxford Instruments INCA Energy energy-dispersive X-ray spectroscop (EDX) system.

Two spectroscopic techniques were used for an analysis and study of the elementa composition of the near-surface layer of the samples of the cutting tool and the workpiec being processed:

- Secondary ion mass spectrometry (SIMS) through analysis on an XT 300M mas spectrometer manufactured by Extorr Inc. (New Kensington, PA, USA);
- Auger electron spectroscopy (AES) by analysis on a JAMP-10S spectrometer manufa tured by JEOL, Ltd. (Tokyo, Japan).

The evaluation of the coefficient of friction (COF) of the surface layer of the sample was carried out on a TNT tribometer manufactured by Anton Paar TriTec SA (Corcelle

Cormondrèche, Switzerland) according to a pin-on-disc method (ASTM G99 17 standard [81]) under the following test conditions (Table 2).

Table 2. Test conditions for evaluation of the friction coefficient (a pin-on-disc method).

Parameter	Value and Description
Motion pattern	Rotation of the sample (disc) relative to a stationary counter body
Counter body	A pin made of 100Cr6 steel with a diameter of 6 mm
Applied load, N	1
Radius of movement, mm	2
Sliding speed, cm/s	10
Test temperature, °C	+20.0, 600.0

Samples were made from various types of tool materials such as discs with a diameter of 19 mm and a height of 8 mm for tribological tests. The same samples were used to evaluate the resistance of the surface layer of the samples to abrasion by testing on a CALOTEST device manufactured by CSM Instruments (Needham Heights, MA, USA) under the conditions of force acting on the sample of a rotating sphere when a water-based abrasive slurry was fed into the contact zone. Optical analysis of the geometric dimensions of the wear hole and its measurement on a Dektak XT stylus profilometer manufactured by Bruker AXS GmbH (Karlsruhe, Germany) to quantify the intensity of samples' wear was carried out.

2.3. Investigation of Temperature Fields in the Cutting Wedge

The semiartificial clamped microthermocouple method was used [82] to study the temperature fields in the cutting wedge of the tool. Composite plates of various tool materials were used, divided by the main intersection plane (cutting edge normal plane) into two parts, tightly (without a gap) compressed by pins before sharpening, excluding the possibility of their relative displacement to lay semiartificial microthermocouples. The articulating surfaces of each part of the plate were carefully lapped until a roughness R_a of no more than 0.04 µm was achieved. The position of the interface plane corresponds to the cutting zone at the tool–chip interface section position, in which the temperatures caused by the contact interaction of the tool and the workpiece are maximum. A constantan wire with a diameter of 0.1 mm, insulated with spacers, was placed at a given point of the cutting zone at the tool–chip interface section along a groove in one of the parts of the composite plate. An uninsulated flattened end with a wire thickness of less than 10 µm was compressed between two ground surfaces of the composite plate portions. The wire contact area with the plate (thermocouple hot junction) was adjusted to a size of 0.15 mm × 0.15 mm. The studies were carried out on a ZMM-Sliven CU500/1000 screw-cutting lathe (ZMM, Nova Zagora, Bulgaria) when simulating the interrupted machining in longitudinal turning with a constant cut section of a workpiece made of 41CrS4 steel (HB 240), on which the grooves were previously milled. It provided the conditions for intermittent operation of the tool close to the milling process. The tests were carried out at a cutting speed V = 68 m/min, feed f = 0.15 mm/rev, depth t = 2.0 mm with a working stroke length of 90 mm, and an idle stroke of 45 mm.

2.4. Durability Testing of Cutting Tools

Durability (fatigue) tests were carried out with the milling of 41CrS4 structural steel (DIN 17210 standard [83]) common in mechanical engineering. It was decided to use an end mill model since wear testing involves a tremendous amount of experimentation and increased consumption of tool material. The designed and manufactured model of the tool was a holder, in which the original insert was fixed (Figure 4), and the operating conditions of which were as close as possible to the operation of a real end mill. The tests were carried out at a cutting speed V = 82 m/min, feed f = 0.15 mm/tooth, milling width B = 5 mm, and depth t = 0.5 mm. In all cases, the criterion for the cutters' failure was the achievement of wear on the flank surface of a limiting value of 350 μm. Service tool life was defined as the time it took to cut before the cutter reached the wear limit. A Stereo Discovery V12 metallographic optical microscope (ZEISS GmbH, Jena, Germany) was used to quantify wear. The curves of the wear with the cutting time were plotted based on the data obtained (the work shows averaged curves obtained based on tests of five samples of each type of tool material). A portable Accretech Handysurf profilometer (ZEISS GmbH, Jena, Germany) was used to obtain reliable data on the roughness of the processed workpiece throughout the entire test cycle. Laboratory resistance tests were carried out on a vertical milling machine with an infinitely variable feed drive VM 127M of the Russian production of JSC Votkinskiy Zavod (Votkinsk, Republic of Udmurtia, Russia)

(a)

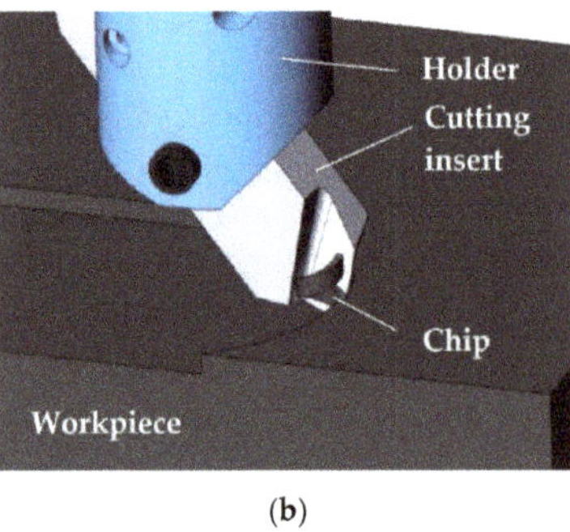

(b)

Figure 4. General view of the working area of the machine during performance testing (**a**) and the design of the model of the tool for milling (**b**).

3. Results and Discussion

3.1. Material Structure Analysis

The cross-sections of the material's structure based on 100% HSS and experimental CPHSS samples were studied for comparative analysis. The fine structure of 100% HSS consists of martensite in which carbides of standard alloying elements are distributed (Figure 5a). There is practically no carbide inhomogeneity in the CPHSS with an 80% HSS, 15% TiC, 5% Al_2O_3 structure (Figure 5b). The pronounced inclusions of refractory compounds TiC and Al_2O_3 have an irregular shape and size from 0.5 to 4.0 μm. SEM images of structures of other studied CPHSSs are not shown since they turned out to be similar to Figure 5b.

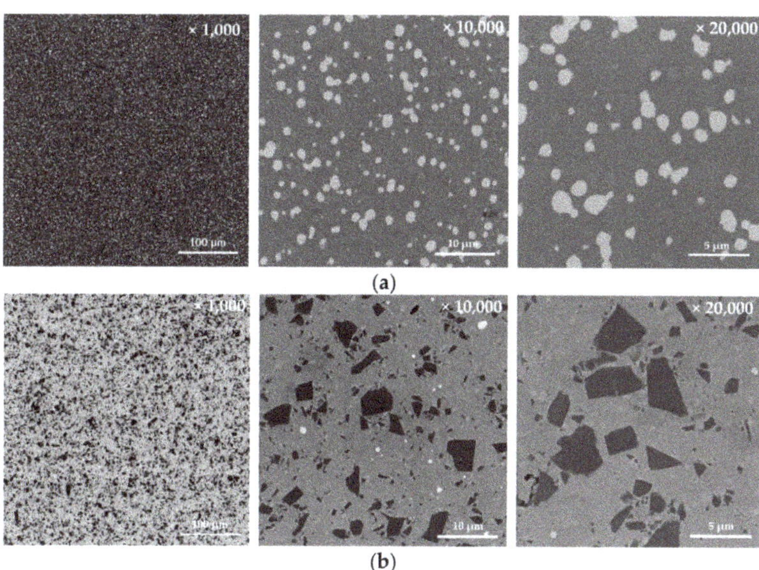

Figure 5. SEM images of cross-sections of HS6-5-2C industrial steel (**a**) and experimental CPHSS sample of 80% HSS, 15% TiC, 5% Al_2O_3 (**b**) (× 1000).

3.2. Testing of Tool Materials under Stationary Loads

Subsequently, four variants of tool materials were tested under stationary loads (under conditions of abrasion during calostesting). Figure 6 presents experimentally obtained data illustrating the dependence of the volume of worn material on testing time for resistance to abrasive wear; the volume of worn-out material was calculated using the formulas given in [84]. The presented dependences give us a specific idea of the kinetics of the development of abrasive wear of the contact areas of various tool materials over time. The CPHSS samples differ in many respects from the standard HSS, which was expected since the hardness of the latter is lower than that of CPHSS, and the average friction coefficient over steel is higher than that of CPHSS.

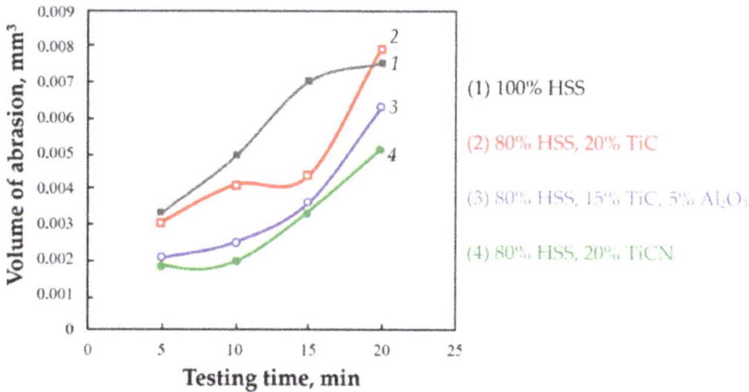

Figure 6. Dependences of the abrasion of the contact areas of various samples of tool materials on the time of testing for abrasive wear resistance: (1) 100% HSS; (2) 80% HSS, 20% TiC; (3) 80% HSS, 15% TiC, 5% Al_2O_3; (4) 80% HSS, 20% TiCN.

Figure 7 shows the characteristic 3D profiles of well-shaped wear spots for various samples, measured after 20 min of testing. It can be seen that the lowest abrasion rate is characteristic of the sample containing 80% HSS, 20% TiCN. The sample containing 80% HSS, 15% TiC, 5% Al_2O_3 had a slightly higher intensity. The minimum wear resistance o all CPHSSs is shown by a sample of 80% HSS, 20% TiC.

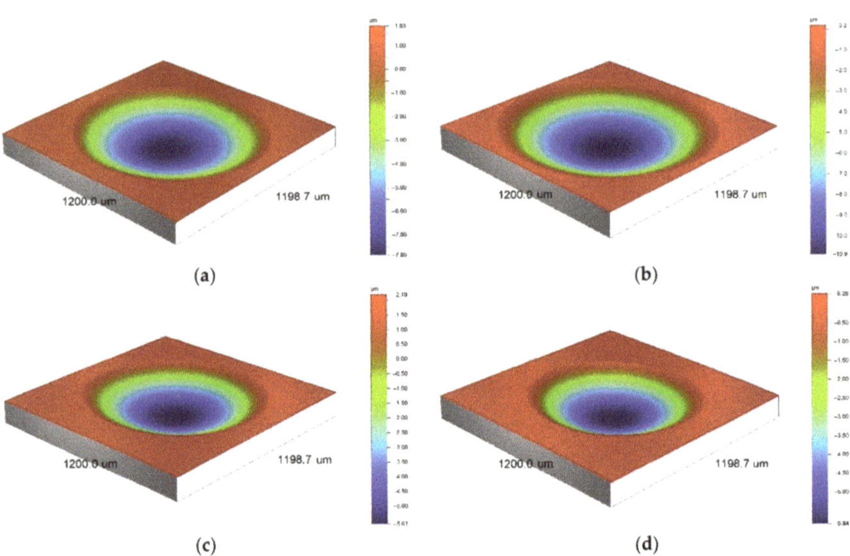

Figure 7. The 3D profiles of wear centers of various samples of tool materials after 20 min of testing under conditions of force action of a rotating sphere in an abrasive environment: (**a**) 100% HSS; (**b**) 80% HSS, 20% TiC; (**c**) 80% HSS, 15% TiC, 5% Al_2O_3; (**d**) 80% HSS, 20% TiCN.

The above results are demonstrative but not informative enough since they do no give an idea of the possible adaptive ability of experimental materials and the effectivenes of their use in the manufacture of cutting tools. They require the impact of heat–powe loads at a level as close as possible to the operational loads (in machining).

3.3. *Quantitative Assessment of the Cutting Part Wear and Roughness*

Figure 8 shows the laboratory tests' results of the tool in milling 41CrS4 steel and give a quantitative assessment of the dimensional wear of the cutting part and the roughness o the processed surface of the workpiece. It can be seen that the curves of the developmen of wear for CPHSS samples (Figure 8a) over time are of a classical nature, and they mainl show three stages: running-in (I); steady-state wear (II); critical wear (III). At the sam time, zone II is noticeably narrowed for a tool made from 100% HSS, and there is n pronounced transition to zone III. Curves of changes in the roughness of the workpiec (Figure 8b) correlate well with the curves of wear development: at the initial stage of testin during running-in (natural blunting) of the tool, no change in roughness is observed; upo transition to the normal wear zone, there is a gradual increase in the contact area of th rear surface of the tool with the surface of the workpiece, accompanied by an increas in the adhesion component of the friction force and a gradual increase in the roughnes of both the tool and the workpiece; contact interaction processes are rapidly intensifie at the last (critical) stage of wear, and cutting forces increase, which causes a noticeabl deterioration in the quality of the surface layer of the workpiece [85]. The tool life i significantly improved for CPHSS compared to a standard tool material and is 17 min fo HS6-5-2C and 29–35 min for CPHSS samples. It can be seen that the highest durabilit (35 min) among the investigated experimental materials is shown by a sample containin 80% HSS and 20% TiCN, and the least (29 min) is shown by the sample containing 80%

HSS and 20% TiC. At the same time, it is evident that the increase in resistance achieved for all CPHSS samples can be regarded as a technologically significant result. The results of testing a sample of 80% HSS and 20% TiC (Figure 6, curve 2) showed the maximum increase in the wear of the contact zone after 15 min of milling and by the end of the tests. The volume of abrasion even slightly exceeds the wear of a sample made from traditional HSS. The intensity of material destruction under an abrasive action is largely determined by the ratio of the counterbody hardness and the test metal surface hardness in the contact zone. A sharp increase in the wear rate of an 80% HSS and 20% TiC sample can be associated with a decrease in the hardness of its surface layer at a certain moment. The specific structure of CPHSS (Figure 5), containing a significant number of refractory inclusions of polyhedral and fragmentary shapes, undissolved in the steel bond, suggests that the level of hardness (and the intensity of abrasive wear directly related to it) of such a material is determined by the degree of preservation of the high-hardness carbide phase. It is most likely that over time and with an increase in the volume of abrasion, hard and, at the same time, very brittle TiC particles (their ductility increases at temperatures above 600 °C) are not retained in the steel bond and crumble, significantly changing the conditions of contact interaction between the tool material and counterbody. For the other two studied CPHSS samples (Figure 6, curves 3 and 4), a high degree of safety in the steel bond of high-hardness components is ensured throughout the entire test cycle. However, it is too early to conclude the prospects of one or another CPHSS variant based on evaluating the resistance to abrasion without thermal action.

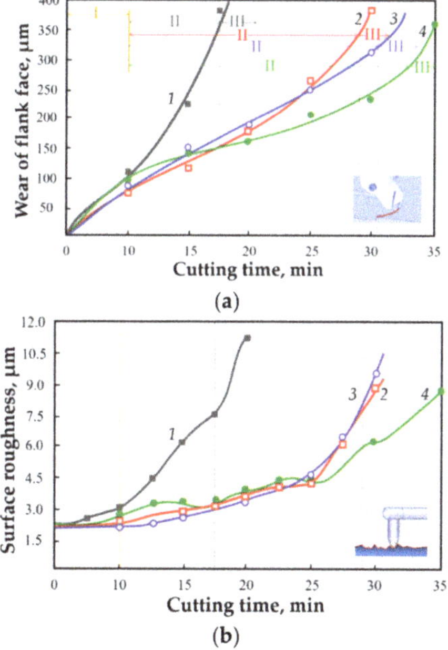

Figure 8. Dependences of cutting wear of flank face (**a**) and 41CrS4 steel workpiece's surface roughness (**b**) on the time for various tool materials in milling at V = 82 m/min, f = 0.15 mm/tooth, B = 5 mm, and t = 0.5 mm: (1) 100% HSS; (2) 80% HSS, 20% TiC; (3) 80% HSS, 15% TiC, 5% Al_2O_3; (4) 80% HSS, 20% TiCN.

It is well known that the temperature on its contact surfaces has a decisive influence on the tool life during machining [86,87]. The contact areas of the rubbing surfaces of the tool, chips, and workpiece being processed are small, and the pressure and friction rate exerted on them are extremely high. The high temperature in the cutting zone is the cause of structural changes in the material of the cutting tool and the reason for the rapid loss of its performance [88–90]. Therefore, to explain the physical nature of the increase in the tool material durability, it is crucial to study the changes in temperature fields in the cutting wedge of the tool during operation.

The best values for samples containing 20% TiCN can be explained by the specific physical and mechanical properties of this two-phase compound, primarily determined by the nature of interatomic bonds depending on the ratio of carbon and nitrogen atoms [91] Some nitrogen atoms are replaced by carbon atoms in titanium carbonitride, forming unlimited TiN-TiC solid solutions. Titanium nitride has the same crystal lattice as titanium carbide, and it is capable of replacing it isomorphically [92]. TiC_xN_y compounds [93,94] formed in the structure of CPHSS samples combine the advantages of carbide and nitride phases. The hardness of titanium carbonitride even slightly exceeds the value for titanium carbide when the plasticity is not inferior to that for titanium nitride (which is extremely important for the tool's operation during milling when the risk of chipping of the cutting edges increases). Titanium carbonitride inclusions have high thermodynamic stability and are closer to HSS in terms of thermal expansion coefficient. In addition, the improved performance for specimens with 20% TiCN compared to 20% TiC results from a lower affinity for iron-containing steels and alloys and a lower adhesive activity when heated [92–94]. The fact that the introduction of Al_2O_3 particles into a powder composition based on TiC somewhat improves the characteristics of CPHSS, but at the same time it is inferior to samples with TiCN additives, indicates the need for separate studies related to the optimization of the compositions of powder mixtures during sintering of experimental instrumental materials.

3.4. Temperature Fields of the Cutting Wedge and Tribological Tests

Figure 9 shows a graphical visualization of temperature fields in the cutting zone at the tool–chip interface section of the cutting wedge of the tool during interrupted machining 41CrS4 steel. It can be seen that the maximum temperatures develop at the tip of the cutting wedge for a tool made of HS6-5-2C material at the beginning of the working stroke (Figure 9a), but over time the region of maximum temperatures moves from the tip along the flank face. It was revealed that the presence of refractory compounds of the TiCN type (Figure 9b) in the composition of the experimental CPHSS material has a significant effect on the development of temperature fields during the working stroke of the tool. Unlike tools made of standard material, the zone of maximum temperatures remains at the top of the cutting wedge for CPHSS, the temperature rise at the very beginning of the working stroke is less intense, and their level is reduced by 20%–25%. This pattern remains for different periods of the tool's working stroke due to a significant reduction in the contact area of the material to be machined and the flank face, which has a higher hardness and a lower friction coefficient in CPHSS specimens. As can be seen from the data presented the temperature field in the contact zone of the cutting wedge stabilizes by the end of the working stroke (at $T = 0.1$ s), and the temperature values for the two types of tool material become close. However, with a decrease in the working stroke, as shown in Figure 9, the use of CPHSS material dramatically affects the contact processes on the flank face and significantly reduces the temperature at the tip of the cutting wedge. The influence of the presence of refractory joints in the material structure on the temperature level on the main flank face is less significant and not as noticeable. However, their total effect on the development of temperature fields and the temperature level in the cutting wedge during the working stroke significantly contribute to the tool's performance, increasing its durability, as shown by performance tests under laboratory conditions (Figure 8).

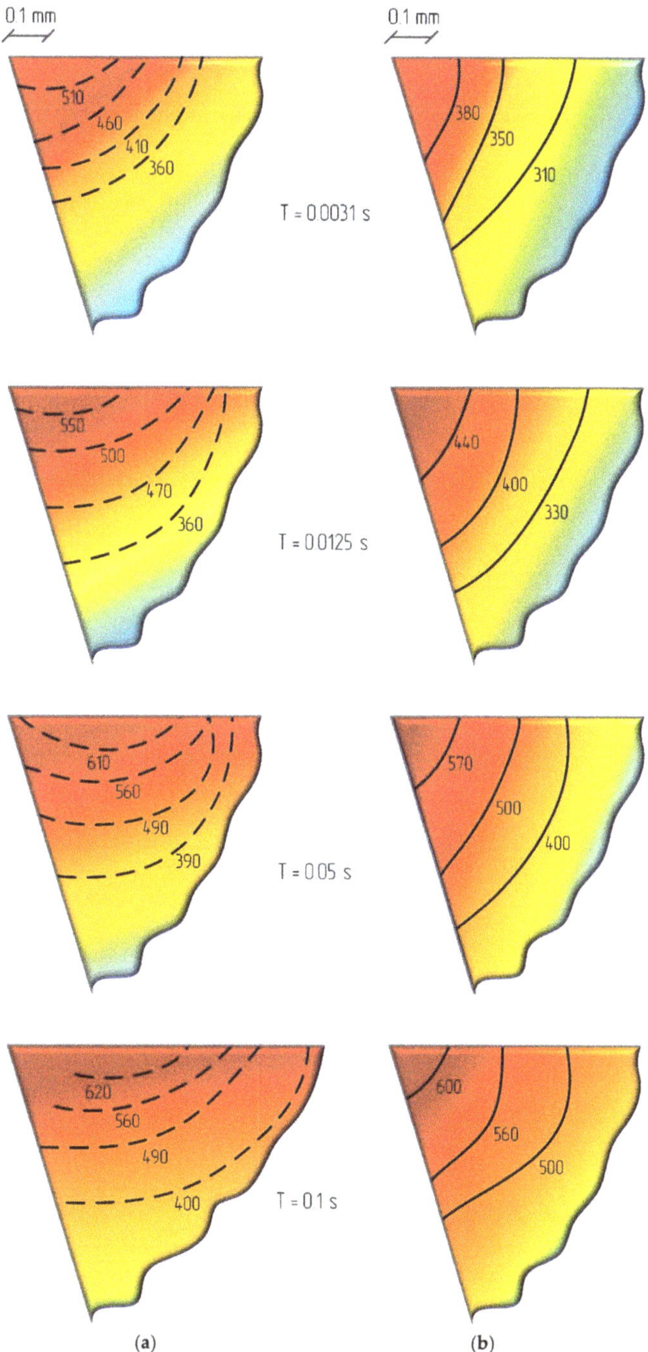

Figure 9. Temperature fields in the cutting zone at the tool–chip interface section of the cutting wedge of a tool made of HS6-5-2C steel (**a**) and experimental CPHSS (80% HSS, 20% TiCN) (**b**), obtained at different periods T of the working stroke during interrupted cutting at V = 68 m/min, f = 0.15 mm/rev, t = 2 mm.

3.5. Tribological Tests

Figure 10 demonstrates the processes occurring in the contact zone of the tool wit the workpiece. At room temperature (Figure 10a), the curves of change in the coefficient o friction (COF) of various tool materials in contact with a steel pin have a classical trenc and the coefficient of friction throughout the entire path demonstrates a stable characte with small peaks at the initial stage of the tests for all investigated material samples. A the same time, the coefficient of friction is different for different materials: the COF ha a maximum value of 0.75 for a sample made of HS6-5-2C (100% HSS), while the CO remains at the level of 0.45–0.6 for samples of CPHSS after running-in and stabilization o the contact interaction. The smallest COF (0.45) among the experimental CPHSS material under study is shown by a sample containing 80% HSS and 20% TiCN, and the larges (0.6) by a sample containing 80% HSS and 20% TiC. The results of tribological tests unde high-temperature heating conditions (Figure 10b) have a completely different form, i.e a level similar to the thermal effect that tool contact pads undergo in the real operatin conditions. The curves of the friction coefficient change for a number of the samples unde study have sharp peaks and a pronounced unstable character. It is typical of a sample mad of HS6-5-2C material (the nature of the curves was repeatedly checked to eliminate errors for which COF "jumps" from 0.6 to 1.1 are observed over 70 m of the friction path, and the relative stabilization is observed with a gradual increase (COF is 0.9 at 200 m). A sampl containing 80% HSS and 20% TiC shows a somewhat stable character and for the highes COF value of all CPHSS samples, the COF develops abruptly, and its values range from 0. to 0.75 at a distance of 40–140 m; a pronounced stabilization of the friction conditions i observed, and the average COF value is 0.65 at a distance of 150 m. The most favorabl friction conditions are demonstrated by a sample containing 80% HSS and 20% TiCN:

- The COF curve does not have sharp bursts throughout the entire path, and its valu remains at a low level from 0.3 to 0.5 (COF value upon heating has even lower value than those of a similar sample when tested without heating);
- The COF stabilizes even more and amounts to a little less than 0.45 after passing a pat of 150 m. The sample containing 80% HSS, 20% TiC, and 5% Al_2O_3 demonstrates a intermediate value among the CPHSS samples in terms of tribological characteristic

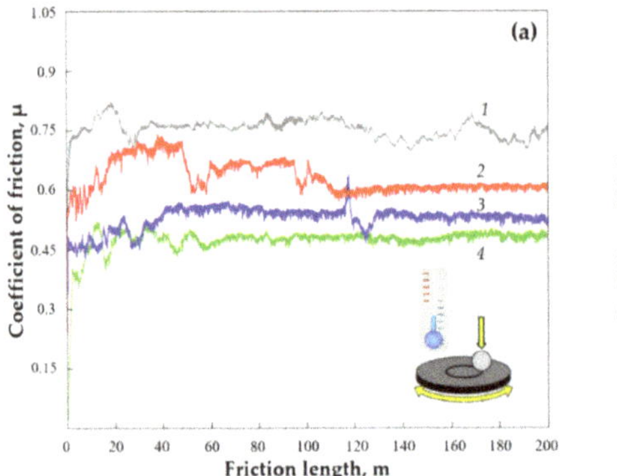

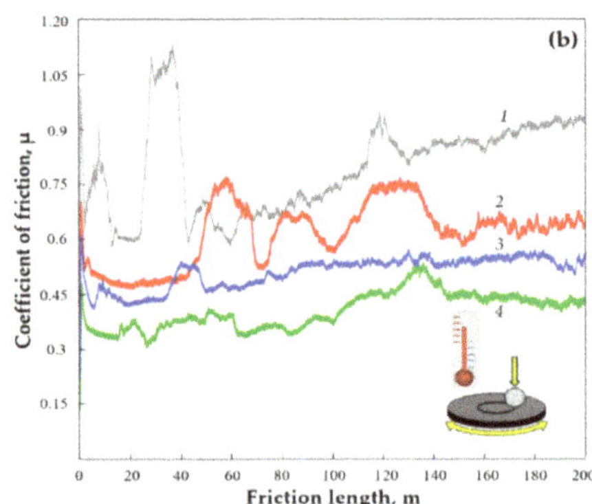

Figure 10. Dependences of the coefficient of friction (COF) of various tool materials on the friction path in contact with a steel pin without heating (**a**) and when heated at +600 °C (**b**): 100% HSS (1); 80% HSS, 20% TiC (2); 80% HSS, 15% TiC, 5% Al_2O_3 (3); 80% HSS, 20% TiCN (4).

COF behavior under high-temperature exposure (Figure 10) is directly related to th structural features of the tool materials. The primary structural component is tungsten an molybdenum carbide $Fe_3(W, Mo)_3C$ for 100% HSS. The available results of the behavio of various grades of HSS when studying at high-temperature frictional contact with low carbon steels [46,47,59] show that under thermal action with an increase in the numbe of test cycles, there is gradual coagulation and shape change of carbides from spherical to drop-shaped and multifaceted, the crystal lattice of the material changes, and a layer consisting of martensite and austenite forms on the surface. At the same time, a decrease in the hardness and strength of the HSS surface layer is observed. The stability of the crystal lattice significantly affects the coefficient of friction, and the decrease in the tool steel hardness significantly increases the friction force adhesive component.

A different mechanism takes place at high-temperature frictional contact of CPHSS samples. The refractory compounds TiC, TiCN, and Al_2O_3 present in the samples' structures upon heating contribute to an increase in the density of defects, and a weakening of interatomic bonds occurs, which improves the oxygen access to the surface layer, and the rate of formation of very hard oxide films with metals increases (in particular, titanium begins to interact with oxygen at temperatures of 600 °C and above). Three stages of contact interaction (Figure 3) explain the described mechanism well. In addition, with an increase in temperature, an increase in the plasticity of refractory compounds is observed. Their lubricating effect is manifested. The specific surface energy in the surface layer decreases, and the work expended on surface deformation decreases. This explains the decrease in the COF for the CPHSS samples. Similar phenomena were observed by the authors of other works when studying the frictional interaction of various tool materials with carbides and nitrides of refractory metals-based coatings [12–14,16,22,37].

A comparison of the operational and tribological characteristics based on the test results of various tool materials described above show strong correlations, where the maximum wear rate during cutting and the worst tribological characteristics during high-temperature testing were shown by a sample of HS6-5-2C powder, and the best performance indicators and tribological properties by the sample containing 80% HSS and 20% TiCN. There is no doubt that the presence of refractory compounds of the TiCN type in the structure of the tool material can provide structural adaptation (self-organization) of the surface layer under external heat–power action due to the formation of secondary structures with thermodynamic stability and improved lubricity.

The most informative is spectroscopic methods to identify and prove the formation of secondary phases. In this case, it is important to analyze the thin surface layer of the tool's contact pads and the workpiece being processed. Figure 11 shows the results of secondary ion mass spectrometry of worn-out contact pads of the tool made of CPHSS material containing 80% HSS and 20% TiCN (Figure 11a) and auger electron spectroscopy of the workpiece's processed surface (Figure 11b) after specific time intervals of the intermittent cutting process. A CPHSS sample that showed maximum wear resistance in service was deliberately analyzed.

The data obtained show that in machining, the tool material containing refractory inclusions of TiCN interacts with environmental components and elements that make up the workpiece to be processed, forming new (secondary) phases. In turn, the chemical composition of the surface layer of the workpiece also undergoes a noticeable transformation (Figure 11).

Spectral analysis shows very different compositions of the surface of the tool contact pads in the initial period of cutting (during running in) and in the time interval of steady-state wear after 20 min of operation (Figure 11a). Spectral analysis revealed the highest intensity of the spectrum of the metastable compound based on TiC_2 (titanium dicarbide) and less intense spectra of TiN and TiO (in decreasing order of intensity) after 5 min of running in. The wear rate of the contact pads stabilizes with the further operation of the tool under the influence of heat and power loads of the cutting process and the external environment's effect. Spectrometry shows a sharp decrease in the intensity of the TiC_2 spectrum in comparison with the running-in zone, and one can observe the decomposition of titanium dicarbide and its transition to the oxygen-containing TiO phase (maximum spectrum intensity) and partially to TiN, since the intensity of its spectrum increases markedly in comparison with the running-in zone.

It is precisely thin surface films based on secondary phases such as thermally stable compounds of titanium with oxygen and nitrogen, which have good lubricity, that significantly reduce the intensity of the adhesive and frictional interaction of the tool and the workpiece during milling and provide the previously established (Figure 8) twofold increase in resistance and improve the workpiece surface quality.

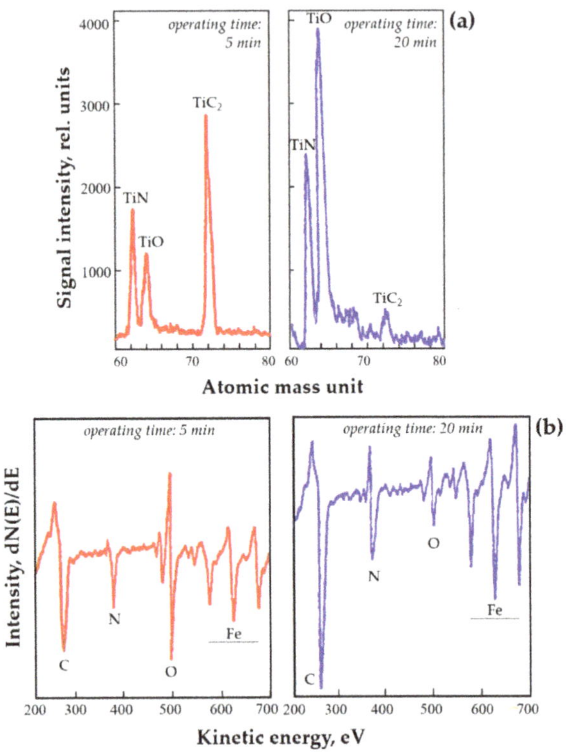

Figure 11. Results of secondary ion mass spectrometry of worn tool pads made of CPHSS containing 80% HSS and 20% TiCN (**a**) and auger electron spectroscopy of the surface of a machined workpiece made of 41CrS4 steel (**b**) after 5 and 20 min of operation in interrupted machining at V = 68 m/min f = 0.15 mm/rev, t = 2 mm.

4. Conclusions

All available technological approaches can be divided into three groups following the classification proposed by the authors of the types of adaptive materials and coatings (AMCs) for tool purposes, showing the ability to adapt to external negative conditions:

- materials and coatings with active control function;
- self-organizing materials and coatings;
- self-repairing materials and coatings.

The complex analytical studies carried out in this work allow us to judge the current directions of work of the leading research teams in the development and application of AMCs, aimed at maintaining the working capacity of the cutting tool for a longer time. The developed technological approaches are compared with the options of tool wear in practice under various operating conditions, and the areas of their practical application are determined.

It can be concluded that the most extensive group is AMCs from the point of view of implementation options, scale, and depth of research. They can provide structural self-organization of the surface layer of the cutting tool due to the formation of secondary structures in contact with the environment and the workpiece being machined under the conditions of heat and power loads acting during cutting. The data presented in the paper illustrate how a properly designed surface layer can effectively adapt to external negative conditions.

Based on the experimental results of the authors of the work, composite powder high speed steels (CPHSSs) were proposed and studied as a typical example of a material with adaptive abilities under operational loads, the components of which are refractory compounds such as TiC, TiCN, and Al_2O_3. It was instrumentally confirmed that the thin surface films formed when cutting based on secondary phases, such as thermally stable compounds of titanium with oxygen and nitrogen, significantly reduce the frictional interaction intensity and redistribute the thermal fields in the cutting zone at the tool–chip interface section of the tool cutting wedge. The established changes in the surface layer provide a twofold increase in tool life during milling, which was confirmed in laboratory conditions and improve the workpiece's surface quality.

Author Contributions: Conceptualization, S.N.G.; methodology, M.S.M. and M.A.V.; validation, S.V.F. and A.A.O.; formal analysis, Y.A.M., A.A.O., and V.D.G.; investigation, S.V.F., M.S.M., and M.A.V.; resources, Y.A.M. and V.D.G.; data curation, S.V.F. and A.A.O.; writing—original draft preparation, M.S.M.; writing—review and editing, M.A.V.; visualization, Y.A.M. and M.A.V.; supervision, V.D.G. and S.N.G.; project administration, S.N.G. All authors have read and agreed to the published version of the manuscript.

Funding: The study was supported by a grant of the Russian Science Foundation (project No. 21-79-30058). The work was carried out on the equipment of the Center of Collective Use of MSUT "STANKIN".

Institutional Review Board Statement: Not applicable.

Informed Consent Statement: Not applicable.

Data Availability Statement: Data are available in a publicly accessible repository.

Acknowledgments: The authors are grateful to Pavel Podrabinnik for his help in timely obtaining the most relevant data regarding the study of samples.

Conflicts of Interest: The authors declare no conflict of interest.

References

Zhu, S.; Cheng, J.; Qiao, Z.; Yang, J. High temperature solid-lubricating materials. *Tribol. Int.* **2019**, *133*, 206–223. [CrossRef]

Fox-Rabinovich, G.S.; Gershman, I.S.; Veldhuis, S. Thin-film PVD coating metamaterials exhibiting similarities to natural processes under extreme tribological conditions. *Nanomaterials* **2020**, *10*, 1720. [CrossRef]

Grigoriev, S.N.; Sobol, O.V.; Beresnev, V.M.; Serdyuk, I.V.; Pogrebnyak, A.D.; Kolesnikov, D.A.; Nemchenko, U.S. Tribological characteristics of (TiZrHfVNbTa)N coatings applied using the vacuum arc deposition method. *J. Frict. Wear* **2014**, *35*, 359–364. [CrossRef]

Pogrebnjak, A.D.; Bagdasaryan, A.A.; Pshyk, A.; Dyadyura, K. Adaptive multicomponent nanocomposite coatings in surface engineering. *Physics-Uspekhi* **2017**, *60*, 586–607. [CrossRef]

Muratore, C.; Voevodin, A.A. Chameleon coatings: Adaptive surfaces to reduce friction and wear in extreme environments. Annu. *Rev. Mater. Res.* **2009**, *39*, 297–324. [CrossRef]

Aouadi, S.M.; Gu, J.; Berman, D. Self-healing ceramic coatings that operate in extreme environments: A review. *J. Vac. Sci. Technol. A* **2020**, *38*, 050802. [CrossRef]

Bouzakis, K.-D.; Michailidis, N.; Skordaris, G.; Bouzakis, E.; Biermann, D.; M'Saoubi, R. Cutting with coated tools: Coating technologies, characterization methods and performance optimization. *CIRP Ann.* **2012**, *61*, 703–723. [CrossRef]

Gershenson, C.; Fernández, N. Complexity and information: Measuring emergence, self-organization, and homeostasis at multiple scales. *Complexity* **2012**, *18*, 29–44. [CrossRef]

Yeh, J.; Lin, S. Breakthrough applications of high-entropy materials. *J. Mater. Res.* **2018**, *33*, 3129–3137. [CrossRef]

10. Feistel, R.; Ebeling, W. *Introduction to the Field of Self-Organization, in Physics of Self-Organization and Evolution*; Wiley-VCH Verlag GmbH & Co. KGaA: Weinheim, Germany, 2011; pp. 1–33.

11. Grigoriev, S.N.; Vereschaka, A.A.; Vereschaka, A.S.; Kutin, A.A. Cutting tools made of layered composite ceramics with nano-scale multilayered coatings. *Proc. CIRP* **2012**, *1*, 301–306. [CrossRef]

12. Sergevnin, V.S.; Blinkov, I.V.; Volkhonskii, A.O.; Belov, D.S.; Chernogor, A.V. Structure formation of adaptive arc-PVD Ti-Al-Mo-N and Ti-Al-Mo-Ni-N coatings and their wear-resistance under various friction conditions. *Surf. Coat. Technol.* **2019**, *376*, 38–4 [CrossRef]
13. Shen, W.-J.; Tsai, M.-H.; Yeh, J.-W. Machining performance of sputter-deposited $(Al_{0.34}Cr_{0.22}Nb_{0.11}Si_{0.11}Ti_{0.22})_{50}N_{50}$ high-entrop nitride coatings. *Coatings* **2015**, *5*, 312–325. [CrossRef]
14. Kovalev, A.; Wainstein, D.; Rashkovskiy, A. Investigation of anomalous physical properties of multilayer nanolaminat (TiAl)N/Cu coatings by electron spectroscopy techniques. *Surf. Interface Anal.* **2010**, *42*, 1361–1363. [CrossRef]
15. Zekonyte, J.; Polcar, T. Friction force microscopy analysis of self-adaptive W-S-C coatings: Nanoscale friction and wear. *ACS App Mater. Interfaces* **2015**, *7*, 21056–21064. [CrossRef] [PubMed]
16. Yuan, J.; Yamamoto, K.; Covelli, D.; Tauhiduzzaman, M.; Arif, T.; Gershman, I.S.; Veldhuis, S.C.; Fox-Rabinovich, G.S. Tribo-film control in adaptive TiAlCrSiYN/TiAlCrN multilayer PVD coating by accelerating the initial machining conditions. *Surf. Coa Technol.* **2016**, *294*, 54–61. [CrossRef]
17. Vereschaka, A.S.; Grigoriev, S.N.; Tabakov, V.P.; Sotova, E.S.; Vereschaka, A.A.; Kulikov, M.Y. Improving the efficiency of th cutting tool made of ceramic when machining hardened steel by applying nano-dispersed multi-layered coatings. *Key Eng. Mate* **2014**, *581*, 68–73. [CrossRef]
18. Durmaz, Y.M.; Yildiz, F. The wear performance of carbide tools coated with TiAlSiN, AlCrN and TiAlN ceramic films in intelliger machining process. *Ceram. Int.* **2018**, *45*, 3839–3848. [CrossRef]
19. Vereschaka, A.; Tabakov, V.; Grigoriev, S.; Aksenenko, A.; Sitnikov, N.; Oganyan, G.; Seleznev, A.; Shevchenko, S. Effect c adhesion and the wear-resistant layer thickness ratio on mechanical and performance properties of ZrN-(Zr,Al,Si)N coating *Surf. Coat. Technol.* **2019**, *357*, 218–234. [CrossRef]
20. Li, Y.; Zheng, G.; Cheng, X.; Yang, X.; Xu, R.; Zhang, H. Cutting performance evaluation of the coated tools in high-speed millin of AISI 4340 steel. *Materials* **2019**, *12*, 3266. [CrossRef]
21. Santecchia, E.; Hamouda, A.M.S.; Musharavati, F.; Zalnezhad, E.; Cabibbo, M.; Spigarelli, S. Wear resistance investigation c titanium nitride-based coatings. *Ceram. Int.* **2015**, *41*, 10349–10379. [CrossRef]
22. Wang, B.; Liu, Z.; Cai, Y.; Luo, X.; Ma, H.; Song, Q.; Xiong, Z. Advancements in material removal mechanism and surface integrit of high speed metal cutting: A review. *Int. J. Mach. Tool. Manuf.* **2021**, *166*, 103744. [CrossRef]
23. Chaus, A.S.; Sahul, M.; Moravcik, R.; Sobota, R. Role of microstructural factor in wear resistance and cutting performance c high-speed steel end mills. *Wear* **2021**, *474*, 203865.
24. Malekan, M.; Bloch-Jensen, C.D.; Zolbin, M.A.; Orskov, K.B.; Jensen, H.M.; Aghababaei, R. Cutting edge wear in high-spee stainless steel end milling. *Int. J. Adv. Manuf. Technol.* **2021**, *114*, 2911–2928. [CrossRef]
25. Kuzin, V.V.; Grigoriev, S.N.; Volosova, M.A. The role of the thermal factor in the wear mechanism of ceramic tools: Part Macrolevel. *J. Frict. Wear* **2014**, *35*, 505–510. [CrossRef]
26. Meng, X.; Lin, Y.; Mi, S. The research of tool wear mechanism for high-speed milling ADC12 aluminum alloy considering th cutting force effect. *Materials* **2021**, *14*, 1054. [CrossRef] [PubMed]
27. Bremer, F.; Matthiesen, S. High-speed cutting with involute blades: Experimental research on cutting forces. *J. Food Eng.* **202** *293*, 110380.
28. Kuzin, V.V.; Grigoriev, S.N.; Fedorov, M.Y. Role of the thermal factor in the wear mechanism of ceramic tools. Part 2: Microlevel. *Frict. Wear* **2015**, *36*, 40–44. [CrossRef]
29. Li, J.; Tao, B.; Huang, S.; Yin, Z. Cutting tools embedded with thin film thermocouples vertically to the rake face for temperatur measurement. *Sens. Actuators A Phys.* **2019**, *296*, 392–399. [CrossRef]
30. Sugita, N.; Ishii, K.; Furusho, T.; Harada, K.; Mitsuishi, M. Cutting temperature measurement by a micro-sensor array integrate on the rake face of a cutting tool. *CIRP Ann.* **2015**, *64*, 77–80. [CrossRef]
31. Li, J.; Tao, B.; Huang, S.; Yin, Z. Built-in thin film thermocouples in surface textures of cemented carbide tools for cuttin temperature measurement. *Sens. Actuators A Phys.* **2018**, *279*, 663–670. [CrossRef]
32. Plogmeyer, M.; González, G.; Schulze, V.; Bräuer, G. Development of thin-film based sensors for temperature and tool wea monitoring during machining. *tm-Technisch. Messen* **2020**, *87*, 768–776. [CrossRef]
33. Cheng, K.; Niu, Z.C.; Wang, R.C.; Rakowski, R.; Bateman, R. Smart cutting tools and smart machining: Development approache and their implementation and application perspectives. *Chin. J. Mech. Eng.* **2017**, *30*, 1162–1176. [CrossRef]
34. Ting, Y.; Suprapto; Nugraha, A.; Chiu, C.W.; Gunawan, H. Design and characterization of one-layer PVDF thin film for a 3D forc sensor. *Sens. Actuators A Phys.* **2016**, *250*, 129–137. [CrossRef]
35. Luo, M.; Luo, H.; Axinte, D.; Liu, D.; Mei, J.; Liao, Z. A wireless instrumented milling cutter system with embedded PVD sensors. *Mech. Syst. Signal. Process.* **2018**, *110*, 556–568. [CrossRef]
36. Sobol, O.V.; Andreev, A.A.; Grigoriev, S.N.; Gorban, V.F.; Volosova, M.A.; Aleshin, S.V.; Stolbovoi, V.A. Effect of high-voltag pulses on the structure and properties of titanium nitride vacuum-arc coatings. *Met. Sci. Heat Treat.* **2012**, *54*, 195–203. [CrossRe
37. Sergevnin, V.S.; Blinkov, I.V.; Volkhonskii, A.O.; Belov, D.S.; Kuznetsov, D.V.; Gorshenkov, M.V.; Skryleva, E.A. Wear behaviour c wear-resistant adaptive nano-multilayered Ti-Al-Mo-N coatings. *Appl. Surf. Sci.* **2016**, *388*, 13–23. [CrossRef]
38. Wainstein, D.; Kovalev, A. Tribooxidation as a way to improve the wear resistance of cutting tools. *Coatings* **2018**, *8*, 223. [CrossRe
39. Voitov, V.; Stadnychenko, V.; Varvarov, V.; Stadnychenko, N. Mechanisms of self-organization in tribosystems operating unde conditions of abnormally low friction and wear. *Adv. Mech. Eng.* **2020**, *12*, 1687814020963843. [CrossRef]

40. Podgursky, V.; Bogatov, A.; Yashin, M.; Sobolev, S.; Gershman, I.S. Relation between Self-Organization and wear mechanisms of diamond films. *Entropy* **2018**, *20*, 279. [CrossRef] [PubMed]
41. Erdemir, A.; Eryilmaz, O. Achieving superlubricity in DLC films by controlling bulk, surface, and tribochemistry. *Friction* **2014**, *2*, 140–155. [CrossRef]
42. Fominski, V.Yu.; Grigoriev, S.N.; Gnedovets, A.G.; Romanov, R.I. Pulsed laser deposition of composite Mo–Se–Ni–C coatings using standard and shadow mask configuration. *Surf. Coat. Technol.* **2012**, *206*, 5046–5054. [CrossRef]
43. Barszcz, M.; Pashechko, M.; Dziedzic, K.; Jozwik, J. Study on the self-organization of an Fe-Mn-C-B Coating during friction with surface-active lubricant. *Materials* **2020**, *13*, 3025. [CrossRef]
44. Liu, Z.L.; Messer-Hannemann, P.; Laube, S.; Greiner, C. Tribological performance and microstructural evolution of alpha-brass alloys as a function of zinc concentration. *Friction* **2020**, *8*, 1117–1136. [CrossRef]
45. Sobol, O.V.; Andreev, A.A.; Grigoriev, S.N.; Gorban, V.F.; Volosova, M.A.; Aleshin, S.V.; Stolbovoy, V.A. Physical characteristics, structure and stress state of vacuum-arc TiN coating, deposition on the substrate when applying high-voltage pulse during the deposition. *Probl. Atom. Sci. Tech.* **2011**, *4*, 174–177.
46. Gnyusov, S.F.; Ignatov, A.A.; Durakov, V.G.; Tarasov, S. Yu. The effect of thermal cycling by electron-beam surfacing on structure and wear resistance of deposited M2 steel. *Appl. Surf. Sci.* **2012**, *263*, 215–222. [CrossRef]
47. Vereschaka, A.A.; Volosova, M.A.; Grigoriev, S.N.; Vereschaka, A.S. Development of wear-resistant complex for high-speed steel tool when using process of combined cathodic vacuum arc deposition. *Procedia CIRP* **2013**, *9*, 8–12. [CrossRef]
48. Naumov, A.G.; Latyshev, V.N.; Novikov, V.V.; Afanasyeva, O.V.; Komelkov, V.A. On the kinetics of forming an oxide film for lubrication when cutting steel in a controlled atmosphere. *J. Frict. Wear* **2017**, *38*, 364–368. [CrossRef]
49. Gershman, I.; Mironov, A.; Podrabinnik, P.; Kuznetsova, E.; Gershman, E.; Peretyagin, P. Relationship of secondary structures and wear resistance of antifriction aluminum alloys for journal bearings from the point of view of self-organization during friction. *Entropy* **2019**, *21*, 1048. [CrossRef]
50. Brizmer, V.; Stadler, K.; van Drogen, M.; Han, B.; Matta, C.; Piras, E. The tribological performance of black oxide coating in rolling/sliding contacts. *Tribol. Trans.* **2017**, *60*, 557–574. [CrossRef]
51. Sobol, O.V.; Andreev, A.A.; Grigoriev, S.N.; Volosova, M.A.; Gorban, V.F. Vacuum-arc multilayer nanostructured TiN/Ti coatings: Structure, stress state, properties. *Met. Sci. Heat. Treat.* **2012**, *54*, 28–33. [CrossRef]
52. Fergus, J.W.; Mallipedi, C.V.S.; Edwards, D.L. Silver/silver-oxide composite coating for intrinsically adaptive thermal regulation. *Compos. Part B-Eng.* **1998**, *29*, 51–56. [CrossRef]
53. Volosova, M.A.; Grigor'ev, S.N.; Kuzin, V.V. Effect of titanium nitride coating on stress structural inhomogeneity in oxide-carbide ceramic. Part 4. Action of heat flow. *Refract. Ind. Ceram.* **2015**, *56*, 91–96. [CrossRef]
54. Zhou, Z.; Rainforth, W.M.; Rother, B.; Ehiasarian, A.P.; Hovsepian, P.E.; Munz, W.D. Elemental distributions and substrate rotation in industrial TiAlN/VN superlattice hard PVD coatings. *Surf. Coat. Technol.* **2004**, *183*, 275–282. [CrossRef]
55. Hahn, R.; Bartosik, M.; Soler, R.; Kirchlechner, C.; Dehm, G.; Mayrhofer, P.H. Superlattice effect for enhanced fracture toughness of hard coatings. *Scr. Mater.* **2016**, *124*, 67–70. [CrossRef]
56. Comakli, O. Influence of CrN, TiAlN monolayers and TiAlN/CrN multilayer ceramic films on structural, mechanical and tribological behavior of beta-type Ti45Nb alloys. *Ceram. Int.* **2020**, *46*, 8185–8191. [CrossRef]
57. Li, M.L.; Wang, E.Q.; Yue, J.L.; Huang, X.Z. Microstructure, mechanical and tribological property of TiAlN/VN nano-multilayer films. *J. Inorg. Mater.* **2017**, *32*, 1280–1284.
58. Metel, A.; Bolbukov, V.; Volosova, M.; Grigoriev, S.; Melnik, Y. Equipment for deposition of thin metallic films bombarded by fast argon atoms. *Instrum. Exp. Tech.* **2014**, *57*, 345–351. [CrossRef]
59. Grigoriev, S.N.; Metel, A.S.; Fedorov, S.V. Modification of the structure and properties of high-speed steel by combined vacuum-plasma treatment. *Met. Sci. Heat. Treat.* **2012**, *54*, 8–12. [CrossRef]
60. Metel, A.S.; Grigoriev, S.N.; Melnik, Y.A.; Bolbukov, V.P. Characteristics of a fast neutral atom source with electrons injected into the source through its emissive grid from the vacuum chamber. *Instrum. Exp. Tech.* **2012**, *55*, 288–293. [CrossRef]
61. Grigoriev, S.N.; Melnik, Y.A.; Metel, A.S.; Panin, V.V.; Prudnikov, V.V. A compact vapor source of conductive target material sputtered by 3-keV ions at 0.05-Pa pressure. *Instrum. Exp. Tech.* **2009**, *52*, 731. [CrossRef]
62. Lakhotkin, Y.V.; Aleksandrov, S.; Zhuk, Y. Self-Sharpening Cutting Tool with Hard Coating. US Patent 7166371-B2, 23 January 2007.
63. Robertson, S.W.; Mehta, A.; Pelton, A.R.; Ritchie, R.O. Evolution of crack-tip transformation zones in superelastic Nitinol subjected to in situ fatigue: A fracture mechanics and synchrotron X-ray microdiffraction analysis. *Acta Mater.* **2007**, *55*, 6198–6207. [CrossRef]
64. Silva, N.J.; De Araujo, C.J.; Gonzalez, C.H.; Grassi, E.N.D.; Oliveira, C.A.N. Comparative study of dynamic properties a NiTi alloy with shape memory and classical structural materials. *Materials* **2011**, *16*, 830–835.
65. Roytburd, A.L. Principal concepts of martensitic theory. *J. Phys. IV* **1995**, *05*, 21–30. [CrossRef]
66. Razov, A.I. Application of titanium nickelide–based alloys in engineering. *Phys. Met. Metallogr.* **2004**, *97*, 97–126.
67. Devin, L.N.; Osadchii, A.A. Improving performance of cBN cutting tools by increasing their damping properties. *J. Superhard. Mater.* **2012**, *34*, 328–335. [CrossRef]
68. Fu, Z.; Koc, R. Processing and characterization of TiB_2-TiNiFeCrCoAl high-entropy alloy composite. *J. Am. Ceram. Soc.* **2017**, *100*, 2803–2813. [CrossRef]

69. Li, Z.; Zhu, C.; Cai, B.; Chang, F.; Liu, X.; Naeem, M.Z.; Dai, P. Effect of TiB_2 content on the microstructure and mechanical properties of Ti(C,N) TiB2-FeCoCrNiAl high-entropy alloys composite cermets. *J. Ceram. Soc. Jpn.* **2020**, *128*, 66–74. [CrossRef]
70. Du, R.; Gao, Q.; Wu, S.; Lu, S.; Zhou, X. Influence of TiB2 particles on aging behavior of in-situ TiB2/Al-4.5Cu composites. *Mater. Sci. Eng. A Struct. Mater.* **2018**, *721*, 244–250. [CrossRef]
71. Chen, C.S.; Yang, C.C.; Chai, H.Y.; Yeh, J.W.; Chau, J.L.H. Novel cermet material of WC/multi-element alloy. *Int. J. Refract. Hard Mater.* **2014**, *43*, 200–204. [CrossRef]
72. Zhang, W.; Zhang, M.Y.; Peng, Y.B.; Wang, L.; Liu, Y.; Hu, S.H.; Hu, Y. Interfacial structures and mechanical properties of a high entropy alloy-diamond composite. *Int. J. Refract. Met. Hard Mater.* **2020**, *86*, 105109. [CrossRef]
73. Grigoriev, S.; Peretyagin, P.; Smirnov, A.; Solis, W.; Diaz, L.A.; Fernandez, A.; Torrecillas, R. Effect of graphene addition on the mechanical and electrical properties of Al_2O_3-SiCw ceramics. *J. Eur. Ceram. Soc.* **2017**, *37*, 2473–2479. [CrossRef]
74. Fominski, V.Yu.; Grigoriev, S.N.; Celis, J.P.; Romanov, R.I.; Oshurko, V.B. Structure and mechanical properties of W-Se-C/diamond-like carbon and W-Se/diamond-like carbon bi-layer coatings prepared by pulsed laser deposition. *Thin Solid Films* **2012**, *520*, 6476–6483. [CrossRef]
75. Metel, A.S.; Grigoriev, S.N.; Melnik, Y.A.; Panin, V.V. Filling the vacuum chamber of a technological system with homogeneous plasma using a stationary glow discharge. *Plasma Phys. Rep.* **2009**, *35*, 1058–1067. [CrossRef]
76. BS EN ISO 4957; *Tool Steels*; BSI: London, UK, 2018.
77. Pellizzari, M.; Fedrizzi, A.; Zadra, M. Spark plasma co-sintering of hot work and high speed steel powders for fabrication of a novel tool steel with composite microstructure. *Powder Technol.* **2011**, *214*, 292–299. [CrossRef]
78. Krioni, N.K.; Migranov, M.S.; Fox-Rabinovich, G.S.; Shuster, L.S. Study of the tribotechnical properties of a cutting tool made of sintered powder tool materials. *J. Frict. Wear* **2018**, *39*, 12–18. [CrossRef]
79. Hadian, A.; Zamani, C.; Clemens, F.J. Effect of sintering temperature on microstructural evolution of M48 high speed tool steel bonded NbC matrix cemented carbides sintered in inert atmosphere. *Int. J. Refract. Met. Hard Mater.* **2018**, *74*, 20–27. [CrossRef]
80. Zhang, F.; Luo, P.; Ouyang, Q.; He, Q.; Hu, M.; Li, S. Microstructure and mechanical properties of b4c-blended M3:2 high-speed steel powders consolidated by sintering and heat treatment. *J. Mater. Eng. Perform.* **2019**, *28*, 6145–6156. [CrossRef]
81. ASTM G99–17; *Standard Test Method for Wear Testing with a Pin-on-Disk Apparatus*; ASTM International: West Conshohocken, PA, USA, 2017; Volume 03.02.
82. Xin, H.; Shi, Y.; Ning, L. Tool wear in disk milling grooving of titanium alloy. *Adv. Mech. Eng.* **2016**, *8*, 1687814016671620 [CrossRef]
83. DIN 17210; *Case Hardening Steels Technical Delivery Conditions*; Deutsches Institut fur Normung E.V. (DIN): Berlin, Germany, 1986
84. Kropotkina, E.; Zykova, M.; Shein, A.; Kapustina, N. Application of roller burnishing technologies to improve the wear resistance of submerged pump parts made of powder alloys. *Mech. Ind.* **2018**, *19*, 705. [CrossRef]
85. Fan, X.; Loftus, M. The influence of cutting force on surface machining quality. *Int. J. Prod. Res.* **2007**, *45*, 899–911. [CrossRef]
86. Zhao, J.; Liu, Z. Influences of coating thickness on cutting temperature for dry hard turning Inconel 718 with PVD TiAlN coated carbide tools in initial tool wear stage. *J. Manuf. Process.* **2020**, *56*, 1155–1165. [CrossRef]
87. Kuzin, V.V.; Grigor'ev, S.N.; Volosova, M.A. Effect of a TiC coating on the stress-strain state of a plate of a high-density nitride ceramic under nonsteady thermoelastic conditions. *Refract. Ind. Ceram.* **2014**, *54*, 376–380. [CrossRef]
88. Akbar, F.; Arsalan, M. Thermal modelling of cutting tool temperatures and heat partition in orthogonal machining of high-strength alloy steel. *Proc. Inst. Mech. Eng. B J. Eng.* **2021**, *235*, 1309–1326. [CrossRef]
89. Metel, A.; Grigoriev, S.; Melnik, Y.; Panin, V.; Prudnikov, V. Cutting tools nitriding in plasma produced by a fast neutral molecule beam. *Jpn. J. Appl. Phys.* **2011**, *50*, 08JG04. [CrossRef]
90. Zhao, J.; Liu, Z.; Wang, B.; Hu, J.; Wan, Y. Tool coating effects on cutting temperature during metal cutting processes: Comprehensive review and future research directions. *Mech. Syst. Signal. Process.* **2021**, *150*, 107302. [CrossRef]
91. Ivashchenko, V.I.; Turchi, P.E.A.; Gonis, A.; Ivashchenko, L.A.; Skrynskii, P.L. Electronic origin of elastic properties of titanium carbonitride alloys. *Metall. Mater. Trans. A* **2006**, *37*, 3391–3396. [CrossRef]
92. Soković, M.; Bahor, M. On the inter-relationships of some machinability parameters in finish machining with cermet TiN (PVD) coated tools. *J. Mater. Process. Technol.* **1998**, *78*, 163–170. [CrossRef]
93. Chen, R.; Tu, J.P.; Liu, D.G.; Mai, Y.J.; Gu, C.D. Microstructure, mechanical and tribological properties of TiCN nanocomposite films deposited by DC magnetron sputtering. *Surf. Coat. Technol.* **2011**, *205*, 5228–5234. [CrossRef]
94. Lengauer, W.; Scagnetto, F. Ti(C,N)-based cermets: Critical review of achievements and recent developments. *Solid State Phenom* **2018**, *274*, 53–100. [CrossRef]

rticle

·ombined Processing of Micro Cutters Using a Beam of Fast ·rgon Atoms in Plasma

·lexander Metel *, Yury Melnik, Enver Mustafaev, Ilya Minin and Petr Pivkin

Department of High-Efficiency Processing Technology, Moscow State University of Technology STANKIN, Vadkovskiy per. 3A, 127055 Moscow, Russia; yu.melnik@stankin.ru (Y.M.); e.mustafaev@stankin.ru (E.M.); i.minin@stankin.ru (I.M.); p.pivkin@stankin.ru (P.P.)

* Correspondence: a.metel@stankin.ru; Tel.: +7-903-246-43-22

Abstract: We present a new method for coating deposition on micro cutters without an increase in their cutting edges radii caused by the deposition. For this purpose, the cutting edges are sharpened before the coating deposition with a concentrated beam of fast argon atoms. The sharpening decreases the initial radius and, hence, limits its value after the coating deposition. The concentrated beam of fast argon atoms is generated using an immersed in the gas discharge plasma concave grid under a negative high voltage. Ions accelerated from the plasma by the grid pass through the grid holes and are concentrated in the focal point of the grid. As a result of the charge exchange in the space charge sheaths of the grid, they are transformed into fast atoms. A uniform sputtering by the fast atoms of the micro-cutter surface reduces the radius of its cutting edge.

Keywords: micro cutters; cutting edges; wear-resistance; coating deposition; adhesion; plasma; ions; charge exchange collisions; fast gas atoms; etching; sharpening

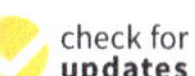

·tation: Metel, A.; Melnik, Y.; ·ustafaev, E.; Minin, I.; Pivkin, P. ·ombined Processing of Micro ·utters Using a Beam of Fast Argon ·oms in Plasma. *Coatings* **2021**, *11*, ·5. https://doi.org/10.3390/ ·atings11040465

·cademic Editor: Cecilia Bartuli

·eceived: 16 March 2021
·ccepted: 13 April 2021
·ublished: 16 April 2021

1. Introduction

The useful life of cutting tools substantially increases after the deposition of thin wear-resistant films [1,2]. This is due to the film hardness of about 25–30 GPa [3] exceeding by many times the hardness of the bulk material. Magnetrons [4] and vacuum arc [5] are mainly used to produce the metal vapor needed for coating synthesis. The density of magnetron discharge plasma near the product surface and the sputtering rate are substantially enhanced when pulsed DC magnetrons [6–13] are used. Metal vapor can be also produced due to sputtering a target at the bottom of a hollow cathode [14].

To prevent the formation of metal droplets in the coatings, the vacuum arc plasma can be filtered using magnetic field. An overview of filtered vacuum arc deposition systems based on magnetic ducts is presented in [15]. The droplets can be also removed from the plasma by transformation of radial plasma streams emitted by the arc cathode spots on the side surface of a cylindrical cathode into an axial stream by means of "single bottle neck" magnetic field [16]. Filtered vacuum arc with ion-species-selective bias has been applied to the synthesis of metal-doped diamond-like carbon films [17]. A magnetic island filter comprising three external coils, which generate a uniform magnetic field, and an internal coaxial coil with a cylindrical permanent magnet in its core, placed within the magnetic island, that generate a field in the opposite direction to the external field is described in [18]. Instead of coils, an internal curvilinear spiral was used in [19], which was transparent and allowed observation of the plasma movement from the arc cathode to the substrate. Analysis of the workpiece surface in [20] revealed that its roughness and the number of cathode spots show no direct relation because the current density per cathode spot does not change according to the number of cathode spots. In a pulsed vacuum arc discharge, the microparticles can be charged, and a roughly quadratic dependence of particle charge on the particle diameter was observed [21] with a 1-μm-diameter particle having a positive charge of ≈1000 electronic charges and a 5-μm-diameter particle having a charge of ≈25,000 electronic charges. Highly adherent CrN films were magnetron sputte deposited in [22] after the filtered cathodic arc etching pretreatment.

Generally, the tools are preliminarily heated and activated. Such pretreatment allow an appreciable improvement of the coating adhesion. More improvements of the adhesio and other coating properties are available when instead of DC biasing, high-voltage pulse are used [23–26].

The thicker the coating, the longer is its useful life. Nevertheless, on roughing tool with the cutting edge radii amounting to ≈15–20 μm, the coating thickness usually doe not exceed 5–7 μm. At higher values of the coating thickness, the cutting edge radii grow to inadmissible values exceeding 30 μm, which are characteristic of blunted tools [27 The 30-μm radius of the cutting edge exceeds the uncut chip thickness of micro-milling Therefore, the material deformation mechanisms are often dominated by the plowing effect [28], and the quality of the surface is dissatisfactory [29].

The problem of cutting tools blunting caused by the coating deposition is especiall acute for micro cutters with a cutting edge radius of less than 10 μm. For instance, cylin drical end-milling micro cutters work in cutting conditions, at which the depth of cut i commensurate with the cutting edge radius. It is impossible to increase the cutting stabi ity during micro-milling through an increase in the depth of cut because of a significan degree of the tool bending. Therefore, keeping the minimum cutting edge radius is ver important [30].

The minimum radius of the cutting edge will improve the cutting conditions an increase the likelihood of the cutting into the material of the whole cutting wedge, not jus its tip, which can reduce cutting forces [31] and vibration [32], improve the chip formatio use cutting patterns with a minimum depth of cut at a relatively high feed per tooth, an reduce the likelihood of the workpiece abrasion.

To date, a coating with a thickness exceeding 2–3 μm cannot be deposited on m cro cutters because after grinding, the radius of the cutting edge is in the range of chi thickness: about 10 μm. The wear-resistant coatings can highly improve the cutting tool efficiency [33–36]. However, in the case of micro tools, their cutting edge radii will increas by 2–2.5 times, which will not allow a stable cutting process. In this case, an increase in th thickness of the cut chips is impossible, because micro cutters are small and have a lowe ability to resist bending, which promises their destruction when bending.

Thus, there is a limitation on the maximum feed per tooth value, and there is minimum feed per tooth limit by the minimum chip thickness. The latter relates to th minimum allowable cutting edge radius that must be met for a stable cutting proces Chips simply will not be formed if their thickness is less than the minimum. Therefor the only way to increase the useful life of micro cutters without losing the cutting proces stability in conditions of parametric tool failure is to reduce the cutting edge radius.

Experimental investigations regarding the formation of the cutting edge geometr during an immersed tumbling process are presented in [27]. It was found that the process ing time is mainly responsible for the size of the cutting edge radius. The lapping mediun defines the chipping of the edge. During the experiments, micro cutters were manufacture with a cutting edge radius of ≈4.0 μm, which is lower than the values available with trad tional grinding methods. Another method for diminishing the cutting edge radius base on the ELID (electrolytic in-process dressing) technique allows manufacturing cutters wit the radius of approximately 3 μm [37]. However, it cannot be applied for small-sized tool

We propose in the present study to reduce the cutting edge radius by etching the too surface with a beam of fast argon atoms [38–41]. The etching will allow reduction of th coated tool radius lower than the radius of the initial tool.

2. Experimental

We processed end mills by Cerin (IT) under catalog number 105.030031240 with th nominal diameter of the working part amounting to 3 mm (h10) and length of 12 mm as well as the nominal angle of inclination of the chip groove equal to 30 degrees fo machining light alloys, aluminum, and titanium (Figure 1). The material of the end mills is solid carbide type HM CK10-20-MG Micrograin: monocarbide (WC-94%; Co-6%) with the carbide grain size of 0.5–0.8 μm, hardness Rockwell 91.8 HRA, and transverse rupture strength 3000 N/mm^2. The work surfaces of the end mills are polished and uncoated. According to the catalog, their application areas are light alloys, plastics, reinforced plastics, and titanium alloys.

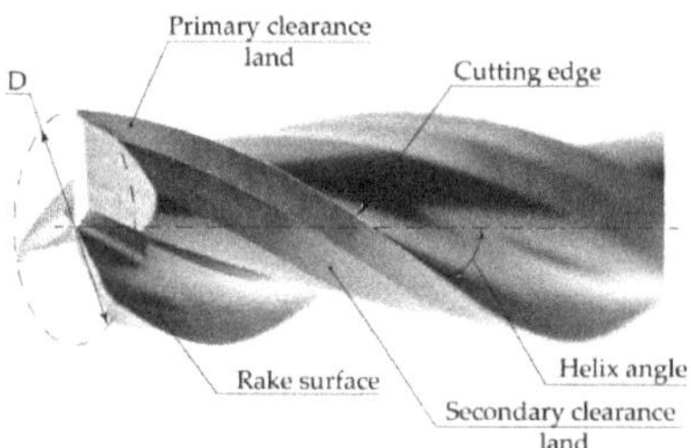

Figure 1. Photograph of a 3-mm-diameter end mill with a 30-degree angle of the chip groove inclination.

While studying the bombardment by a beam of fast argon atoms, it is necessary to compare the initial geometric parameters of the cutting tools with the parameters obtained after processing. Analysis of the changes in the controlled parameters will make it possible to conclude the effectiveness of the fast atom beam. The controlled parameters were the profiles of the front surface located at different distances from the end of the cutter, as well as the angular parameters of the cutting edge, the front, and rear corners, and the cutting edge radius.

During the study, three samples of this type of end mill were measured. To control the initial geometric parameters, the algorithms of the Walter Helicheck Plus (Walter, Germany) measuring system was used, which are based on the use of machine vision methods. The measuring was carried out with cameras of transmitted and reflected light. Measurement inaccuracy by positioning stability for measuring diametrical and linear parameters does not exceed 0.32 μm. The measurement was carried out in a non-contact manner using a CCD (Charge-Coupled Device) matrix camera of transmitted light and a CCD camera of reflected light with ×400 magnification.

A hydraulic chuck with an SK 50 tool cone was used for fixing the measured samples of cutters. The obtained results of primary measurements for the working geometry of three samples are presented in Table 1.

Table 1. Results of primary measurements for the working geometry of three samples.

Parameters	End Mill 1	End Mill 2	End Mill 3
Cutting diameter, mm	2.9865	2.9550	2.9728
Rake angle (radial section at a depth of 0.2 mm), deg.	12.201	12.243	12.184
Helix angle, deg.	29.879	29.724	29.829
Clearance angle, deg.	29.174	29.060	29.365
Condition of Work Surfaces			
Depth of chip groove 1st tooth, mm	0.0105	0.0248	0.0160
Depth of chip groove 2nd tooth, mm	0.0228	0.0138	0.0135
Depth of chip groove 3rd tooth, mm	0.0250	0.0110	0.0160

The main geometrical parameters of a solid carbide end micro mill and radius of cutting on the radial section are shown in Figure 2. The main geometric parameters of the cutting edge were measured in three areas: A, B, and C. In the range of each area, the cutting edge radius was controlled in sets of sections: r_{eA1}, r_{eA2}, r_{eA3}, and r_{eAn} for A area, r_{eB1}, r_{eB2}, r_{eB3}, and r_{eBn} for B, and r_{eC1}, r_{eC2}, r_{eC3}, and r_{eCn} for C.

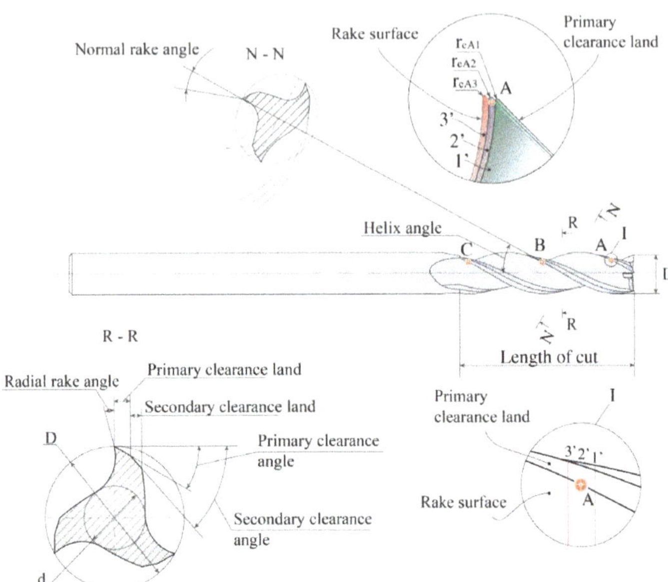

Figure 2. Main geometrical parameters of a solid carbide end micro mill.

Cutting edge radius was measured with an optical 3D measuring system MicroCAD Premium plus. Its measuring algorithm is to measure a strip of light using a micromirror projector. Figure 3a shows a photograph of the measuring system and Figure 3b elucidates the measurement of the cutting edge radius.

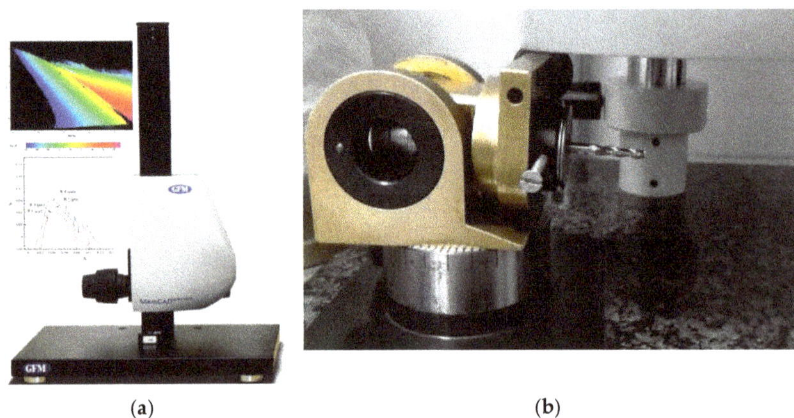

(**a**) (**b**)

Figure 3. Optical measuring system MikroCAD premium+ manufactured by GFMesstechnik GmbH (**a**) and a 3-mm-diameter end mill installed under the system (**b**).

During the measurements, the radius of the cutting edge was estimated in four radial sections distanced from each other at 0.05 mm and distanced from the mill end at 2 mm (Figure 4), 7 mm, or 12 mm.

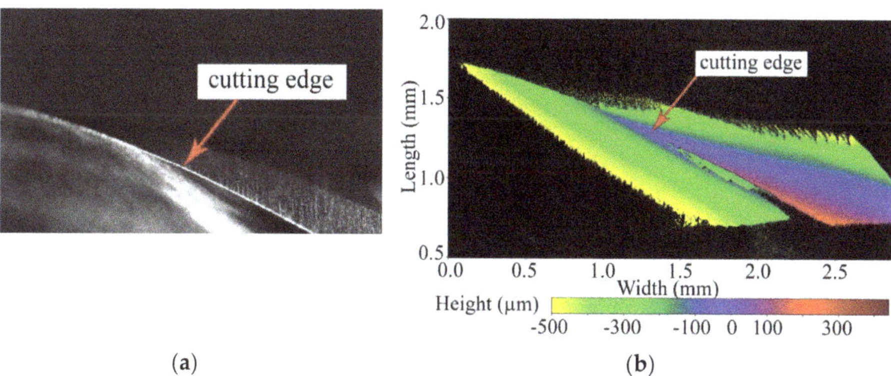

(a) (b)

Figure 4. The end mill tooth at a distance of 7 mm from the mill end. (**a**) Image from optical 3D scanning camera; (**b**) topography of the cutting edge.

The radius of the cutting edge was measured in three areas: A, B, and C, which were located from each other with an equal step of 5 mm at a distance of 2, 7, and 12 mm. In each of the areas, the value of the radius r_{eA}, r_{eB}, and r_{eC} is found as the arithmetic mean of the measured values for four sections located at 0.05 mm distance from each other (Equations (1)–(3)).

$$r_{eA} = \frac{r_{eA1} + r_{eA2} + r_{eA3} + \ldots + r_{eAn}}{n} \quad (1)$$

$$r_{eB} = \frac{r_{eB1} + r_{eB2} + r_{eB3} + \ldots + r_{eBn}}{n} \quad (2)$$

$$r_{eC} = \frac{r_{eC1} + r_{eC2} + r_{eC3} + \ldots + r_{eCn}}{n} \quad (3)$$

The radius of the cutting edge was measured for the basic type of end mill with grinding clearance and rake surfaces. The average value of the cutting edge radius per 2-mm-distanced area (A) was equal to r_{eA} 10.8 µm (Figure 5); the average value of the cutting edge radius per 7-mm-distanced area (B) was equal to r_{eB} 10.5 µm; and the average value of the cutting edge radius per 12-mm-distanced area (C) was equal to r_{eC} 10.25 µm.

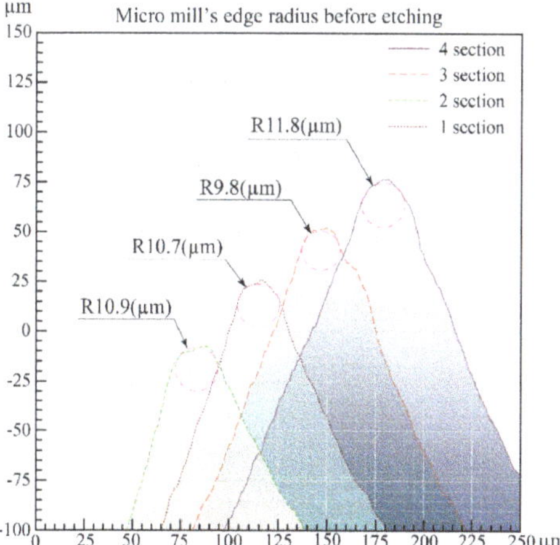

Figure 5. Cutting edge radii r_{eA1}, r_{eA2}, r_{eA3}, and r_{eA4} in four sections distanced from the mill end at 2 mm and distanced from each other at 0.05 mm.

The complete experimental process of the present work comprises:

1. Measurement of cutting edge radii, geometry, and roughness of micro cutters.
2. Etching the micro cutters to reduce their cutting edge radii.
3. Measurement of the cutting edge radii after the etching.
4. Measurement of the geometry and roughness of micro cutters after etching.
5. Deposition of a wear-resistant coating.
6. Final measurement of the cutting edge radii of coated micro cutters.

3. Etching the End Mills with Fast Atoms

The experimental system for processing the end mills is presented in Figure 6. A 20-cm-diameter concave grid with a surface curvature radius of 20 cm is fixed to th high-voltage feedthrough in the center of the vacuum chamber.

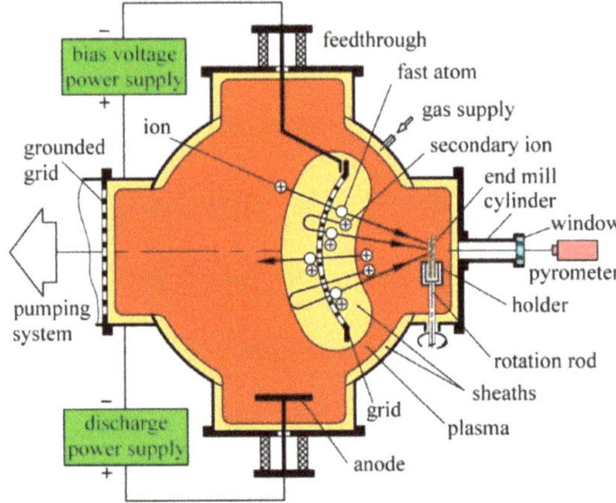

Figure 6. Scheme of a beam generation by a grid immersed in plasma.

On the grid surface, holes with a diameter of 7 mm are evenly distributed at a distanc of 8 mm between their centers. At a distance of 6 cm from the chamber wall is installed rotating holder for the tools being processed by the beam. The axis of the holder rotatio passes through the focal point of the concave grid surface.

At an argon pressure in the chamber $p \approx 0.5$ Pa, an increase in the voltage between th anode and the chamber to several hundred volts leads to establishing a gas discharge wit a current in the anode circuit I_a up to 4 A and a discharge voltage of U_d = 400–500 V [42 The chamber is filled with a brightly glowing homogeneous plasma, which is separate from its walls by a cathode sheath and from the grid by a grid sheath of the ion spac charge. The sheath width d can be calculated using the derived from the Child–Langmui law [43] following expression

$$d = (2/3)\varepsilon_o{}^{1/2}(2e/M)^{1/4}(U + U_d)^{3/4}/j^{1/2}, \quad (4)$$

where ε_o is the electric constant equal to 0.885×10^{-11} F/m, e is the electron charge, M is the ion mass, U is the grid bias voltage, and j is the ion current density. When the io mass $M = AM_A$ is measured in atomic mass units $A = 1.66 \times 10^{-27}$ kg, the sheath width i equal to

$$d = 2.34 \times 10^{-4}(M_A)^{-1/4}(U + U_c)^{3/4}/j^{1/2}. \quad (5)$$

When $U = 0$, the chamber and the grid are equipotential, and both sheaths are of th same width. At p ranging from 0.2 to 1 Pa, the voltage U_d between the chamber and th anode is virtually independent of the argon pressure p at a constant current I_a in the anod circuit. However, at $p < 0.2$ Pa, a decrease in pressure causes an increase in the discharg voltage U_d, and at $p = 0.02$ Pa, it reaches a value of $U_d \approx 1$ kV. With an increase in the voltage U between the chamber and the grid from zero to 5 kV, the current I in the grid circuit at a constant current I_a in the anode circuit approximately doubles, the width of the sheath d between the plasma and the grid grows to 5–10 cm, and the discharge voltage U_d decreases approximately by two times.

The density of the gas atoms at room temperature and pressure $p = 0.02$ Pa amounts to $n = 0.5 \times 10^{19}$ m^{-3} [44], and the average path of argon ions between charge exchange collisions, called the charge exchange length, is equal to $\lambda = 1/n\sigma = 1$ m. We took into account that for argon ions with an energy of 5 keV, the charge exchange cross-section is equal to $\sigma = 2 \times 10^{-19}$ m^2 [45,46].

At a pressure of $p = 0.02$ Pa, the width of the grid sheath $d \approx 0.05$ m is much lower than the charge exchange length λ, and therefore, no formation of fast atoms in the sheath occurs. All ions extracted from the plasma bombard the grid and cause the emission from its surface of the secondary electrons [47]. Hence, the grid emits only two beams of electrons with energy $e(U + U_d)$, propagating in opposite directions.

With a pressure increase to 0.2 Pa, accelerated ions turn in the sheath into fast atoms escaping from the sheath. The energy of each fast atom, $e\varphi$, corresponds to the potential φ of the point in the sheath where it appears. With a pressure increase from 0.2 to 2 Pa, the number of fast atoms increases tenfold, their energy decreases from 5 to 500 eV, and the neutral beam current grows up.

For an estimate of the beam diameter, the distribution of the etching rate by fast atoms of a polished titanium target covered with a mask on its surface was measured. The straight boundary between them is located horizontally (Figure 7). The target was etched with fast atoms, and then, the height of the step between the covered by the mask and open substrate surfaces was measured along the border between them using a mechanical profiler Dektak XT. The etching rate distribution at various distances Z between the target and the grid revealed the beam diameter dependence on Z. As Z increases, the diameter D decreases from 45 mm at $Z = 17$ cm to 6 mm at $Z = 20$ cm, and as the distance increases from $Z = 21$ cm to $Z = 24$ cm, the diameter D increases from 7 mm up to 52 mm.

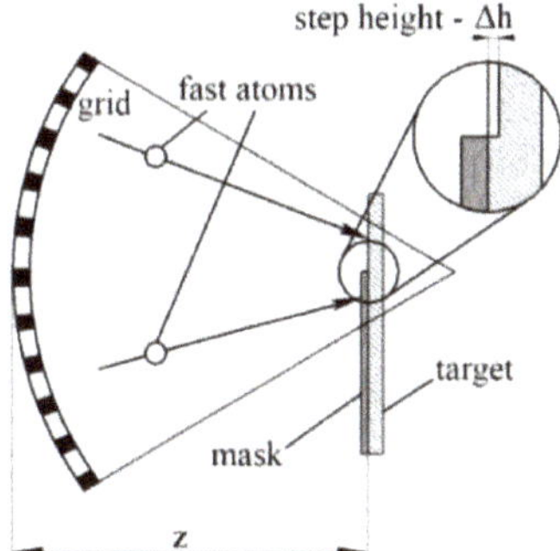

Figure 7. Schematic of the step formation on a target at the distance Z from the grid due to etching with fast argon atoms.

Figure 8a presents a photograph of the accelerating grid fastened to the high-voltage feedthrough and the end mill placed on the rotating holder. The grid was distanced at 22 cm from the axis of the rotating holder where the beam diameter amounted to 2 cm. For this reason, the 12-mm-long cutting part of the end mill was etched quite homogeneously by the fast atoms. The beam of fast atoms was generated by the grid immersed in the argon plasma and negatively biased to 5000 V (Figure 8b).

(**a**)

(**b**)

Figure 8. Photographs of the accelerating grid and 3-mm-diameter end mill on the rotating holder (**a**) as well as the accelerating grid immersed in plasma and generating a fast atom beam (**b**).

The results of the mills etching are presented in Table 2. They show that at the beginning, for the sections distanced from the mill end at 7 mm, the etching rate is maximum and amounts to 3 μm/h. Further on, the etching rate falls down, and after a 3-h-long etching, it is close to zero for the section distant from the mill end at 7 mm and to 0.2 μm/h for the sections distant from the mill end at 2 and 12 mm.

Table 2. Cutting edge radii after etching the end mills.

Etching Time	Cutting Edge Radius (μm) for the Section Distant from the Mill end at 2 mm (A)	7 mm (B)	12 mm (C)
Before the etching	10.8	10.5	10.25
1.5 h	6.8	6.0	6.4
3 h	4.55	3.7	4.4
4.5 h	4.25	3.65	4.1

The average value of the cutting edge radius after etching per 2-mm-distanced area (A) was equal to r_{eA} 4.25 μm; the average value of the cutting edge radius per 7-mm-distanced area (B) was equal to r_{eB} 3.65 μm (Figure 9); the average value of the cutting edge radius per 12-mm-distanced area (C) was equal to r_{eC} 4.1 μm.

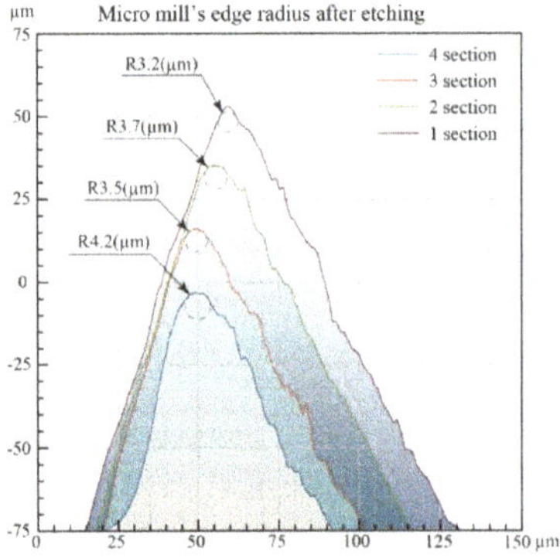

Figure 9. Profiles of the cutting edge radii after etching in four sections distanced from the mill end at 7 mm and distanced from each other at 0.05 mm.

The analysis of the entire set of measured cutting edge radii showed that the best results were obtained for the 3-h-long etching.

A further increase in the etching time had no benefit for the tool sharpening. It should be mentioned that after quite long etching of the end mills, no significant increase in the surface roughness has been observed (Figure 10).

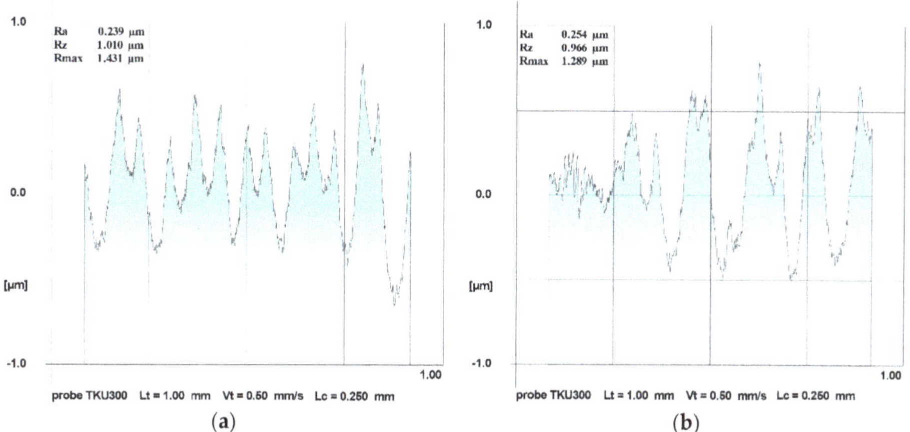

Figure 10. Measurements of the end mill roughness before (**a**) and after the etching (**b**).

After a 3-h-long etching of the end mill, a wear-resistant coating was synthesized on its surface. The synthesis was carried out according to the standard technology on a Platit π 311 system for the deposition of wear-resistant coatings manufactured by Platit (Switzerland). The deposited diamond-like coating (DLC) was a two-layer composition: an adhesive sublayer based on a complex nitride (CrAlSi)N and an outer wear-resistant DLC layer (Figure 10b). The choice of this particular coating (CrAlSi)N/DLC as an object for research is not accidental. This compound can be used to improve the performance of hard-alloy tools, including small-sized ones. To estimate the thickness of the synthesized coating, a cylindrical mask with an inner diameter of 3 mm was preliminary put on the end mill.

Within an hour, a DLC coating was synthesized on the end mill and its mask. After removing them from the chamber, the mask was detached from the end mill. Using a Dektak XT mechanical profilometer, the height of the step between the coating surface and the end mill surface screened with the mask was measured (Figure 11a). The thickness of the wear-resistant coating synthesized on the end mill is equal to the step height of 2.4 µm.

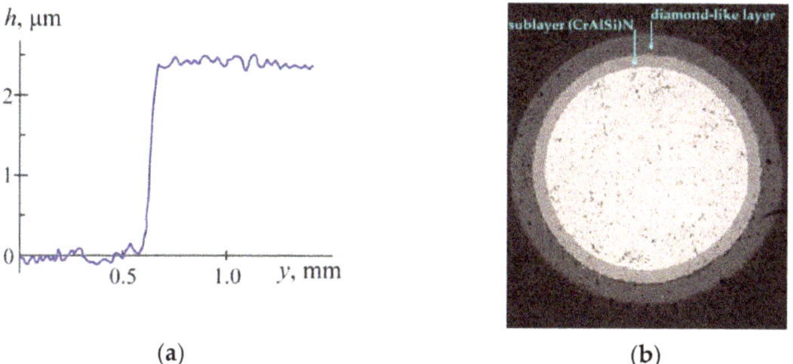

Figure 11. Profilogram of the end mill surface without the detached mask (**a**) and construction of a two-layer DLC coating deposited on a 3-mm-diameter end mill (**b**).

Figure 12 shows the experimentally obtained measurement results, which were performed in four sections of the cutting edge of a 3-mm diameter end mill after a 3-h-long etching and deposition of a two-layer DLC coating. The average value of the cutting edge radii after etching and deposition of a two-layer DLC coating per 2-mm-distanced area (A) was equal to r_{eA} 6.25 µm; the average value of the cutting edge radius per 7-mm-distanced area (B) was equal to r_{eB} 5.5 µm; and the average value of the cutting edge radius per 12-mm-distanced area (C) was equal to r_{eC} 6.2 µm (Figure 12) (Table 3).

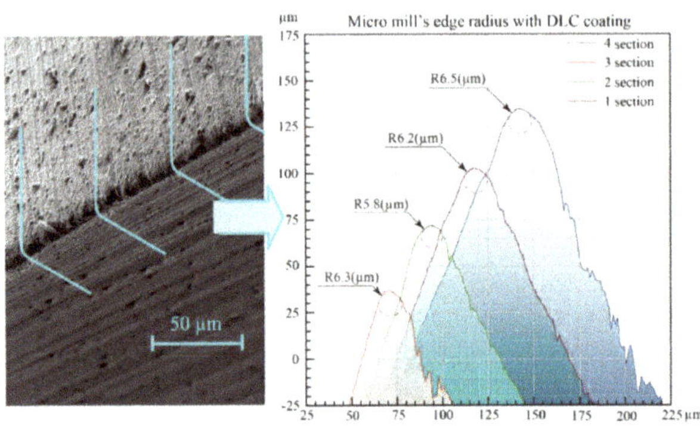

Figure 12. Profiles of cutting edge radii after etching and deposition of a two-layer DLC coating in four sections distanced from the mill end at 12 mm and distanced from each other at 0.05 mm.

Table 3. Cutting edge radii end mills.

Etching Time	Cutting Edge Radius (µm) for the Section Distant from the Mill End at 2 mm (A)	7 mm (B)	12 mm (C)
Before the etching	10.8	10.5	10.25
After the etching	4.25	3.65	4.1
After the etching and deposition of a two-layer DLC coating	6.25	5.5	6.2
Grinding and deposition of a two-layer DLC coating	12.7	12.55	12.25

Figure 13 presents SEM images (magnification 5000 times) of the end mill cutting edges after the diamond-like coatings deposition on the end mills sharpened through grinding (a) and by fast argon atoms (b).

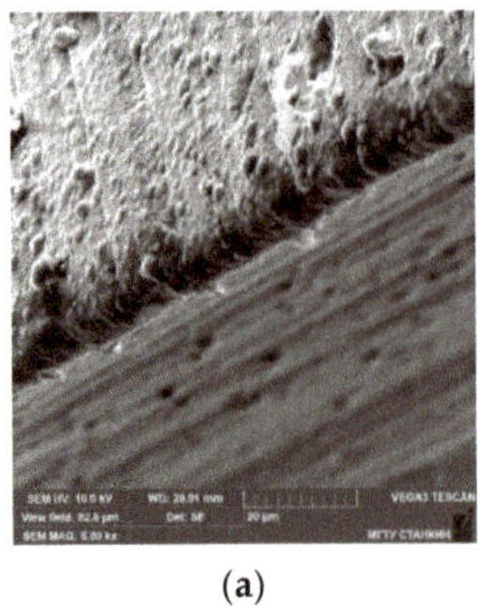

(a)

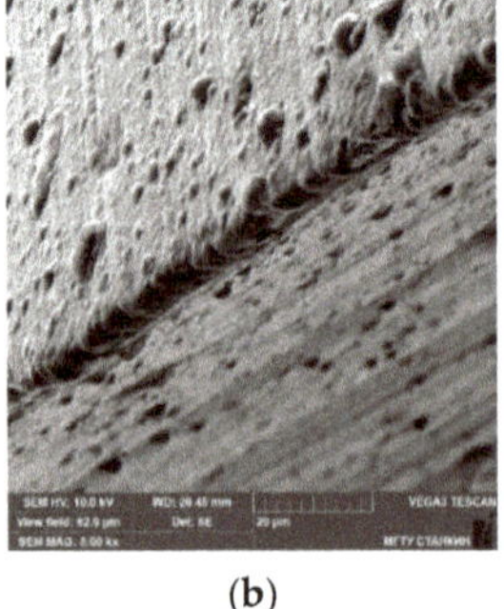

(b)

Figure 13. SEM images of the cutting edges coated with DLC films after sharpening through grinding (**a**) and by fast argon atoms (**b**).

4. Discussion

The above results showed that the etching of end mills with a fast atom beam allows an appreciable sharpening of the tools. Due to the etching, the cutting edge radius diminishes to ≈3−4 µm from the minimum value of ≈10−11 µm available with the tool sharpening through grinding. The sharpening occurs due to a quite homogeneous sputtering of the tool surface near the edge at low gas pressure $p < 1$ Pa. Homogeneous plasma inside the chamber is generated by the glow discharge with a large hollow cathode [48], which is the vacuum chamber itself.

It was discovered in [49] that the lower limit of the glow discharge operating pressure is proportional to the aperture of electron losses from the hollow cathode, and due to a decrease in the aperture, it can be diminished to ≈0.01 Pa. This finding made it possible to use the discharge for the plasma immersion processing of products [50,51], in broad beam sources of gaseous ions [52–55] and electron beam sources [56–58]. The present research is the first attempt to use the discharge for the micro tools processing.

In our case, the aperture of electron losses is equal to the surface area of the anode immersed in the discharge plasma. The volume of the vacuum chamber with a diameter of 50 cm and length of 55 cm amounts to $V = 0.12$ m^3, and its internal surface area is equal to $S = 1.5$ m^2. When the anode area S_a is less than a critical value

$$S^* = (\pi/e)^{1/2}(2m/M)^{1/2}S \approx (2m/M)^{1/2}S \quad (6)$$

where S is the chamber surface area, m and M are the electron mass and the ion mass, and e is the Naperian base, a positive anode fall of potential U_a occurs [42]. With a pressure decrease from 0.1 to 0.01 Pa, the anode fall can grow from $U_a \approx 10$ to $U_a \approx 500$ V. It results in the anode overheating and melting by electrons accelerated in the negative space charge sheath near the anode surface.

For the discharge in argon and $S = 1.5$ m^2, the critical value amounts to $S^* = 0.008$ m^2. Not to have problems with the discharge anode, its surface area was chosen to be equal to $S_a = 0.02$ m^2. This area exceeds the critical anode area of $S^* = 0.008$ m^2 and, hence, it prevents the positive anode fall of potential. The discharge plasma uniformity at the gas pressure of 0.01–1 Pa makes it possible to accelerate ions from the main plasma volume in the center of the chamber using a concave grid (Figure 6) and transform the concentrated ion beam into a fast atom beam. The decrease in the cutting edge radius of the end mills after etching with such a beam can be explained in Figure 14.

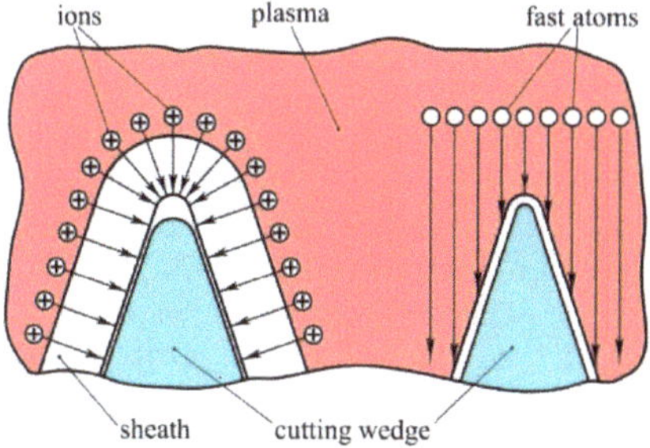

Figure 14. Scheme of etching cutting tools in plasma by accelerated particles.

When a negative voltage is applied to the cutting edge in the plasma, ions are extracted from the plasma and enter the sheath. The width of the sheath at a voltage from 100 to 1000 V exceeds 1 mm. Therefore, the radius of the plasma boundary emitting ions onto the cutting edge surface exceeds about one hundred times the cutting edge radius of 10 µm. In a homogeneous plasma, the ion current density is constant at the entire plasma boundary. However, the area of the plasma emitting ions on the cutting edge is one hundred times larger than the edge area.

Therefore, the ion current density at the edge is a hundred times higher than at the rest of the wedge surface. Etching of the edge leads to a significant increase in its radius, and the tool becomes blunt. To avoid bluntness of the tool, it is necessary to etch not by ions accelerated from the plasma by a voltage applied to the tool but by a broad beam of accelerated ions or fast atoms. They sputter the cutting edge of the tool with the same intensity as the rest of its surface. When removing a surface layer of the same thickness, the radius of the cutting edge decreases. Therefore, the tool is sharpened, when it is etched by a broad beam.

In our experiments, the cutting edge radius diminished to $\approx 3-4$ μm from the minimum value of $\approx 10-11$ μm available at the tool sharpening through grinding. In the beginning, the etching rate amounted to ≈ 2 μm/h, and after a 3-h-long etching, it diminished to zero. The reason for this phenomenon can be related to the structure of the end mill material. In our case, it was a carbide type HM CK10-20-MG Micrograin of carbide group: tungsten-cobalt single-corpus with the carbide grain size of 0.8 μm. One can hardly imagine a tool with a cutting edge radius of ≈ 1 μm made of a material with a grain size of 0.8 μm. In our experiments, the minimum cutting edge radius exceeds the grain size by five times.

The results of the end mills etching presented in Table 2 exhibit another specific feature to be explained. They reveal a difference in the dependencies on the etching time of the cutting edge radius for the end mill sections distant from the mill end at 2, 7, and 12 mm The 12-mm-long cutting part of the end mill rotated in the center of the 20-mm-diameter beam and should be etched quite homogeneously. Nevertheless, the etching rate for the section distant from the mill end at 7 mm was a little higher than for the sections distant from the mill end at 2 and 12 mm. It could be caused by a radial distribution of the fast atom flow density with a maximum at the beam axis.

The use of wear-resistant coatings prevents the intensive wear of the cutting tool, however, the coating affects the microgeometry of the cutting edge by increasing the radius of its rounding. With an increase in the radius of the cutting edge about the thickness of the cut layer, the deformation area of the workpiece increases, and as a result, the force load on the tool tooth increases [31]. Cutting forces have a great influence on the chip formation nature, the process of material cutting, vibration characteristics of cutting, the integrity of the cutting tool, and the surface quality [59–62]. Additional surface treatment by fast atoms will reduce the cutting edge radius and prepare surfaces for the deposition of a wear-resistant coating while providing a relatively smaller radius of the cutting edge than when coating the surface without preparation of this kind.

Unlike the constant undeformed chip thickness (UCT) and direction of cutting speed in orthogonal micro-cutting, both vary in a time-dependent manner in micro-milling The micro-milling process can be considered as the composition of varied orthogonal micro-cutting in the time domain [63–65]. In end micro-milling (Figure 15a), the thickness (i) initially increases from zero, (ii) reaches the maximum value (approximately equal to the feed per tooth), and (iii) reduces to zero, while in a side micro-milling (Figure 15b), the thickness decreases from maximum to zero. The above stages may occur sequentially for a single cutting edge, or, in most common scenarios, simultaneously for multiple cutting edges depending on the employed tool geometries (the cutting edge radius and tooth number) and the machining parameter (feed per tooth) [66]. This would result in very complex material behaviors in comparison with orthogonal micro-cutting.

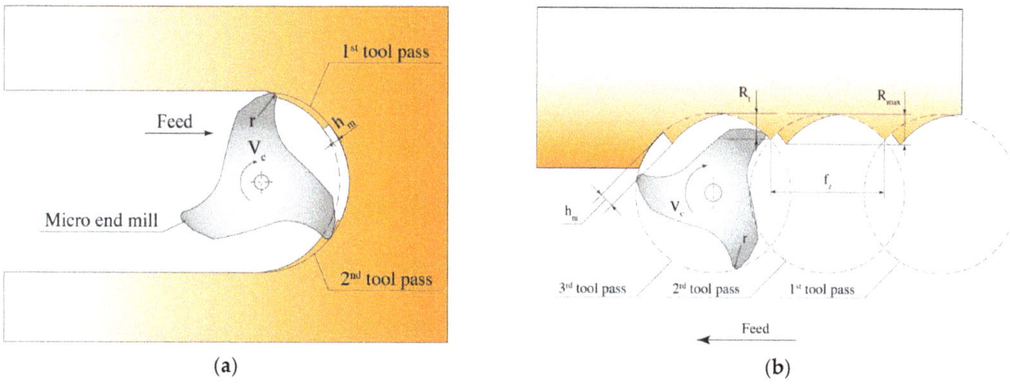

Figure 15. Tool pass in full immersion (**a**) and side micro-milling process (**b**).

Consider the side micro-milling process based on the two-tooth micro milling cutter as an example. If the UCT at the entrance is larger than the minimum undeformed chip thickness (MUCT), the chip begins to form (Figure 15b). However, when the instantaneous UCT becomes smaller than the MUCT, the chip generation would stop. A part of the workpiece material will elastically recover, while other material will undergo plastic deformation after the micro-milling cutter is passed by. The machined surface in a side micro-milling includes both the theoretical residual height (R_t) and the residual height (R_{max}) left by the existing MUCT (Figure 16b) [62]. R_{max} in micro-milling can be expressed as Equation (7) [62], respectively, and they are primarily related to the feed per tooth (f_z) and MUCT (h_m), as shown in Equation (8).

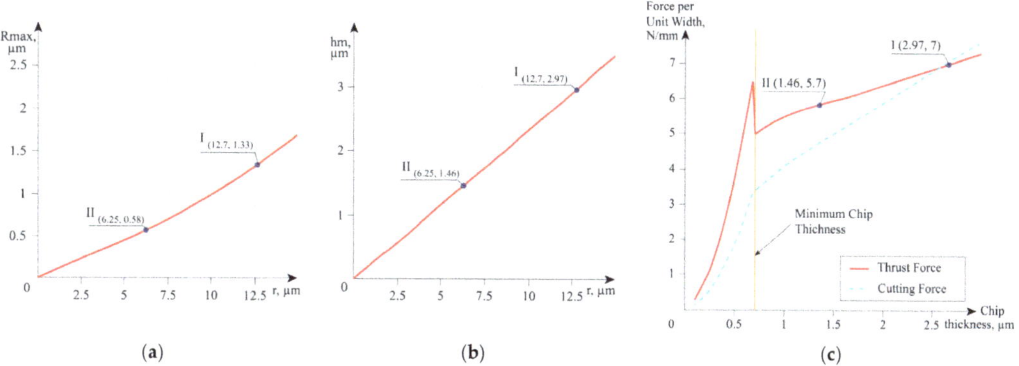

Figure 16. (**a**) Residual height of residual microroughness (R_{max}); (**b**) minimum undeformed chip thickness h_m; (**c**) forces in the ploughing-dominated region: point I—micro end mill coated after etching by fast atoms, point II—micro end mill coated after grinding.

As a result of using the etching operation before coating, the radius of the cutting edge can be reduced to 6.25 microns compared to 12.7 after coating a ground micro mill.

Reducing the cutting edge radius allows decreasing the height of residual microroughness formed during cutting in the area of the plunger zone from 1.33 µm (point I in Figure 16a) to 0.58 µm (point II in Figure 16a), minimum undeformed chip thickness h_m from 2.97 µm (point I in Figure 16b) to 1.46 µm (point II in Figure 16b), forces in the ploughing-dominated region determined on the base of ref. [65,66] from 7 N/mm (point I in Figure 16c) to 5.7 N/mm (point II in Figure 16c) when machining a steel part with the following parameters β 40°, Poisson's ratio k 0.25 (for steel), mill's radius R_f 1.5 mm, feed per tooth 0.005 mm/tooth. After the chip thickness of 3.0 μm, where the effect of ploughin becomes small and that of shearing is more significant, forces for pearlite became highe but increasing these parameters are actual only for the machining process with small dep of cut [67–69].

$$R_{\max} = \frac{f_z^2}{8R_f} + \frac{(1-k)h_m}{2}\left(1 + \frac{[(1-k)h_m]^2}{f_z^2}\right) \quad (7$$

$$h_m = r(1 - \cos(\beta)) \quad (8$$

5. Conclusions

1. The etching of micro tools with a concentrated beam of fast atoms allows an apprecia ble sharpening of the tools. Due to the etching, the cutting edge radius diminishes t ≈3–4 μm from the minimum value of ≈10–11 μm available at the tool sharpenin through grinding.
2. The sharpening of micro tools with a fast atom beam makes it possible to avoid bluntin of their cutting edges caused by an increase in their radii after the coating deposition.
3. The new method of cutting tool processing improves the cutting conditions an property of the cutting wedge.

Author Contributions: Conceptualization, A.M. and P.P.; methodology, A.M. and P.P.; softwar I.M. and E.M.; validation, A.M., P.P. and Y.M.; formal analysis, Y.M.; investigation, Y.M. and I.M resources, Y.M., I.M. and E.M.; data curation, P.P., Y.M. and I.M.; writing—original draft preparatio A.M. and P.P.; writing—review and editing, A.M.; visualization, I.M.; supervision, A.M.; projec administration, P.P.; funding acquisition, A.M. All authors have read and agreed to the publishe version of the manuscript.

Funding: This work is funded by the state assignment of the Ministry of Science and Highe Education of the Russian Federation, project No. 0707-2020-0025.

Institutional Review Board Statement: Not applicable.

Informed Consent Statement: Not applicable.

Data Availability Statement: Not applicable.

Acknowledgments: The work was carried out using the equipment of the Center of collective use c "MSUT "STANKIN".

Conflicts of Interest: The authors declare no conflict of interest.

References

1. Grigoriev, S.; Metel, A. Plasma- and Beam-Assisted Deposition Methods; Nato Science Series, Series Ii: Mathematics, Physic and Chemistry; Nanostructured Thin Films and Nanodispersion Strengthened Coatings. In *NATO Science Series II: Mathematic Physics and Chemistry*; Voevodin, A.A., Shtansky, D.V., Levashov, E.A., Moore, J.J., Eds.; Springer: Dordrecht, The Netherland 2004; Volume 155, pp. 147–154. [CrossRef]
2. Sobol', O.V.; Andreev, A.A.; Grigoriev, S.N.; Gorban', V.F.; Volosova, M.A.; Aleshin, S.V.; Stolbovoy, V.A. Physical char-acteristic structure and stress state of vacuum-arc TiN coating, deposition on the substrate when applying high-voltage pulse during th deposition. *Probl. Atom. Sci. Technol.* **2011**, *4*, 174–177.
3. Zhitomirsky, V.; Grimberg, I.; Rapoport, L.; Boxman, R.; Travitzky, N.; Goldsmith, S.; Weiss, B. Bias voltage and incidenc angle effects on the structure and properties of vacuum arc deposited TiN coatings. *Surf. Coat. Technol.* **2000**, *133–134*, 114–12 [CrossRef]
4. Musil, J. High-rate magnetron sputtering. *J. Vac. Sci. Technol. A* **1996**, *14*, 2187–2191. [CrossRef]
5. Boxman, R.L.; Zhitomirsky, V.N. Vacuum arc deposition devices. *Rev. Sci. Instrum.* **2006**, *77*, 021101. [CrossRef]
6. Musil, J.; Leština, J.; Vlček, J.; Tölg, T. Pulsed dc magnetron discharge for high-rate sputtering of thin films. *J. Vac. Sci. Technol.* **2001**, *19*, 420–424. [CrossRef]
7. Horwat, D.; Anders, A. Spatial distribution of average charge state and deposition rate in high power impulse magnetro sputtering of copper. *J. Phys. D Appl. Phys.* **2008**, *41*, 135210. [CrossRef]
8. Oks, E.; Anders, A. Evolution of the plasma composition of a high power impulse magnetron sputtering system studied with time-of-flight spectrometer. *J. Appl. Phys.* **2009**, *105*, 093304. [CrossRef]
9. Anders, A. Discharge physics of high power impulse magnetron sputtering. *Surf. Coat. Technol.* **2011**, *205*, S1–S9. [CrossRef]

10. Greczynski, G.; Hultman, L. Time and energy resolved ion mass spectroscopy studies of the ion flux during high power pulsed magnetron sputtering of Cr in Ar and Ar/N2 atmospheres. *Vacuum* **2010**, *84*, 1159–1170. [CrossRef]
11. Sarakinos, K.; Alami, J.; Konstantinidis, S. High power pulsed magnetron sputtering: A review on scientific and engineering state of the art. *Surf. Coat. Technol.* **2010**, *204*, 1661–1684. [CrossRef]
12. Hala, M.; Viau, N.; Zabeida, O.; Klemberg-Sapieha, J.E.; Martinu, L. Dynamics of reactive high power impulse magnetron sputtering discharge studied by time- and space-resolved optical emission spectroscopy and fast imaging. *J. Appl. Phys.* **2010**, *107*, 043305. [CrossRef]
13. Musil, J.; Baroch, P. High-rate pulsed reactive magnetron sputtering of oxide nanocomposite coatings. *Vacuum* **2013**, *87*, 96–102. [CrossRef]
14. Metel, A.S.; Bolbukov, V.P.; Volosova, M.A.; Grigoriev, S.N.; Melnik, Y.A. Equipment for deposition of thin metallic films bombarded by fast argon atoms. *Instrum. Exp. Tech.* **2014**, *57*, 345–351. [CrossRef]
15. Boxman, R.; Zhitomirsky, V.; Alterkop, B.; Gidalevich, E.; Beilis, I.; Keidar, M.; Goldsmith, S. Recent progress in filtered vacuum arc deposition. *Surf. Coat. Technol.* **1996**, *86–87*, 243–253. [CrossRef]
16. Aksenov, I.I.; Aksyonov, D.S. Vacuum arc plasma source with rectilinear filter for deposition of functional coatings. *Funct. Mater.* **2008**, *15*, 442–447. Available online: https://scholar.google.com/scholar?q=Vacuum%20arc%20plasma%20source%20with%20rectilinear%20filter%20for%20deposition%20of%20functional%20coatings (accessed on 14 April 2021).
17. Anders, A.; Pasaja, N.; Sansongsiri, S. Filtered cathodic arc deposition with ion-species-selective bias. *Rev. Sci. Instrum.* **2007**, *78*, 063901. [CrossRef]
18. Kleiman, A.; Márquez, A.; Boxman, R.L. Performance of a magnetic island macroparticle filter in a titanium vacuum arc. *Plasma Sources Sci. Technol.* **2007**, *17*, 015008. [CrossRef]
19. Sanginés, R.; Bilek, M.M.M.; McKenzie, D.R. Optimizing filter efficiency in pulsed cathodic vacuum arcs operating at high currents. *Plasma Sources Sci. Technol.* **2009**, *18*, 045007. [CrossRef]
20. Sato, A.; Iwao, T.; Yumoto, M. Relation between surface roughness and number of cathode spots of a low-pressure arc. *Plasma Sources Sci. Technol.* **2008**, *17*, 045007. [CrossRef]
21. Rysanek, F.; Burton, R.L. Charging of Macroparticles in a Pulsed Vacuum Arc Discharge. *IEEE Trans. Plasma Sci.* **2008**, *36*, 2147–2162. [CrossRef]
22. Ehiasarian, A.P.; Anders, A.; Petrov, I. Combined filtered cathodic arc etching pretreatment–magnetron sputter deposition of highly adherent CrN films. *J. Vac. Sci. Technol. A* **2007**, *25*, 543–550. [CrossRef]
23. Metel, A.S.; Bolbukov, V.P.; Volosova, M.A.; Grigoriev, S.N.; Melnik, Y.A. Source of metal atoms and fast gas molecules for coating deposition on complex shaped dielectric products. *Surf. Coat. Technol.* **2013**, *225*, 34–39. [CrossRef]
24. Ruset, C.; Grigore, E. The influence of ion implantation on the properties of titanium nitride layer deposited by magnetron sputtering. *Surf. Coat. Technol.* **2002**, *156*, 159–161. [CrossRef]
25. Ruset, C.; Grigore, E.; Collins, G.; Short, K.; Rossi, F.; Gibson, N.; Dong, H.; Bell, T. Characteristics of the Ti_2N layer produced by an ion assisted deposition method. *Surf. Coat. Technol.* **2003**, *174–175*, 698–703. [CrossRef]
26. Grigore, E.; Ruset, C.; Short, K.T.; Hoeft, D.; Dong, H.; Li, X.Y.; Bell, T. In situ investigation of the internal stress within the nc-Ti_2N/nc-TiN nanocomposite coatings produced by a combined magnetron sputtering and ion implantation method. *Surf. Coat. Technol.* **2005**, *200*, 744–747. [CrossRef]
27. Uhlmann, E.; Oberschmidt, D.; Kuche, Y.; Löwenstein, A. Cutting Edge Preparation of Micro Milling Tools. *Procedia CIRP* **2014**, *14*, 349–354. [CrossRef]
28. Aurich, J.C.; Bohley, M.; Reichenbach, I.G.; Kirsch, B. Surface quality in micro milling: Influences of spindle and cutting parameters. *CIRP Ann.* **2017**, *66*, 101–104. [CrossRef]
29. Dib, M.; Duduch, J.; Jasinevicius, R. Minimum chip thickness determination by means of cutting force signal in micro endmilling. *Precis. Eng.* **2018**, *51*, 244–262. [CrossRef]
30. Biermann, D.; Baschin, A. Influence of cutting edge geometry and cutting edge radius on the stability of micromilling processes. *Prod. Eng.* **2009**, *3*, 375–380. [CrossRef]
31. Wojciechowski, S.; Matuszak, M.; Powałka, B.; Madajewski, M.; Maruda, R.; Królczyk, G. Prediction of cutting forces during micro end milling considering chip thickness accumulation. *Int. J. Mach. Tools Manuf.* **2019**, *147*, 103466. [CrossRef]
32. Mian, A.J.; Driver, N.; Mativenga, P.T. Identification of factors that dominate size effect in micro-machining. *Int. J. Mach. Tools Manuf.* **2011**, *51*, 383–394. [CrossRef]
33. Volosova, M.A.; Grigor'Ev, S.N.; Kuzin, V.V. Effect of Titanium Nitride Coating on Stress Structural Inhomogeneity in Oxide-Carbide Ceramic. Part 4. Action of Heat Flow. *Refract. Ind. Ceram.* **2015**, *56*, 91–96. [CrossRef]
34. Grigoriev, S.N.; Sobol, O.V.; Beresnev, V.M.; Serdyuk, I.V.; Pogrebnyak, A.D.; Kolesnikov, D.A.; Nemchenko, U.S. Tribological characteristics of (TiZrHfVNbTa) N coatings applied using the vacuum arc deposition method. *J. Frict. Wear* **2014**, *35*, 359–364. [CrossRef]
35. Kuzin, V.V.; Grigoriev, S.N.; Fedorov, M.Y. Role of the thermal factor in the wear mechanism of ceramic tools. Part 2: Microlevel. *J. Frict. Wear* **2015**, *36*, 40–44. [CrossRef]
36. Kuzin, V.V.; Grigor'Ev, S.N.; Volosova, M.A. Effect of a TiC Coating on the Stress-Strain State of a Plate of a High-Density Nitride Ceramic Under Nonsteady Thermoelastic Conditions. *Refract. Ind. Ceram.* **2014**, *54*, 376–380. [CrossRef]

37. Lee, S.W.; Choi, H.J.; Lee, H.W.; Choi, J.Y.; Jeong, H.D. A Study of Micro-Tool Machining Using Electrolytic In-Process Dressing and an Evaluation of its Characteristics. *Key Eng. Mater.* **2003**, *238–239*, 35–42. [CrossRef]
38. Grigoriev, S.N.; Melnik, Y.A.; Metel, A.S.; Panin, V.V. Broad beam source of fast atoms produced as a result of charge exchange collisions of ions accelerated between two plasmas. *Instrum. Exp. Tech.* **2009**, *52*, 602–608. [CrossRef]
39. Metel, A.S.; Grigoriev, S.N.; Melnik, Y.A.; Bolbukov, V.P. Broad beam sources of fast molecules with segmented cold cathodes and emissive grids. *Instrum. Exp. Tech.* **2012**, *55*, 122–130. [CrossRef]
40. Metel, A.S.; Grigoriev, S.N.; Melnik, Y.A.; Bolbukov, V.P. Characteristics of a fast neutral atom source with electrons injected into the source through its emissive grid from the vacuum chamber. *Instrum. Exp. Tech.* **2012**, *55*, 288–293. [CrossRef]
41. Metel, A.; Grigoriev, S.; Melnik, Y.; Panin, V.; Prudnikov, V. Cutting Tools Nitriding in Plasma Produced by a Fast Neutral Molecule Beam. *Jpn. J. Appl. Phys.* **2011**, *50*, 08JG04. [CrossRef]
42. Metel, A.S.; Grigoriev, S.N.; Melnik, Y.A.; Prudnikov, V.V. Glow discharge with electrostatic confinement of electrons in a chamber bombarded by fast electrons. *Plasma Phys. Rep.* **2011**, *37*, 628–637. [CrossRef]
43. Langmuir, I. The Interaction of Electron and Positive Ion Space Charges in Cathode Sheaths. *Phys. Rev.* **1929**, *33*, 954–989 [CrossRef]
44. McDaniel, E.W. *Collision Phenomena in Ionized Gases*; Willey: New York, NY, USA, 1964. Available online: https://scholar.google.com/scholar_lookup?title=Collision+Phenomena+in+Ionized+Gases&author=McDaniel,+E.W.&publication_year=1964 (accessed on 14 April 2021).
45. Phelps, A.V. Cross Sections and Swarm Coefficients for Nitrogen Ions and Neutrals in N_2 and Argon Ions and Neutrals in Ar for Energies from 0.1 eV to 10 keV. *J. Phys. Chem. Ref. Data* **1991**, *20*, 557–573. [CrossRef]
46. Phelps, A.V.; Greene, C.H.; Burke, J.P. Collision cross sections for argon atoms with argon atoms for energies from 0.01 eV to 10 keV. *J. Phys. B At. Mol. Opt. Phys.* **2000**, *33*, 2965–2981. [CrossRef]
47. Kaminsky, M. *Atomic and Ionic Impact Phenomena on Metal. Surfaces*; Springer: Berlin, Germany, 1965. Available online: https://scholar.google.com/scholar_lookup?title=Atomic+and+Ionic+Impact+Phenomena+on+Metal+Surfaces&author=Kaminsky,+M.&publication_year=1965 (accessed on 14 April 2021).
48. Korolev, Y.D.; Koval, N.N. Low-pressure discharges with hollow cathode and hollow anode and their applications. *J. Phys. D Appl. Phys.* **2018**, *51*, 323001. [CrossRef]
49. Metel, A. Ionization effect in cathode layers on glow-discharge characteristics with oscillating electrons.1. Discharge with the hollow-cathode. *Zhurnal Tekhnicheskoi Fiz.* **1985**, *55*, 1928–1934.
50. Koval, N.; Ryabchikov, A.; Sivin, D.; Lopatin, I.; Krysina, O.; Akhmadeev, Y.; Ignatov, D. Low-energy high-current plasma immersion implantation of nitrogen ions in plasma of non-self-sustained arc discharge with thermionic and hollow cathodes *Surf. Coat. Technol.* **2018**, *340*, 152–158. [CrossRef]
51. Gavrilov, N.V.; Mamaev, A.S.; Chukin, A.V. Nitriding of Stainless Steel in Electron-Beam Plasma in the Pulsed and DC Generation Modes. *J. Surf. Investig. X ray Synchrotron Neutron Tech.* **2017**, *11*, 1167–1172. [CrossRef]
52. Gavrilov, N.V. New broad beam gas ion source for industrial application. *J. Vac. Sci. Technol. A* **1996**, *14*, 1050–1055. [CrossRef]
53. Gavrilov, N.; Mesyats, G.; Radkovski, G.; Bersenev, V. Development of technological sources of gas ions on the basis of hollow-cathode glow discharges. *Surf. Coat. Technol.* **1997**, *96*, 81–88. [CrossRef]
54. Oks, E.M.; Vizir, A.V.; Yushkov, G.Y. Low-pressure hollow-cathode glow discharge plasma for broad beam gaseous ion source *Rev. Sci. Instrum.* **1998**, *69*, 853–855. [CrossRef]
55. Vizir, A.V.; Yushkov, G.Y.; Oks, E.M. Further development of a gaseous ion source based on low-pressure hollow cathode glow *Rev. Sci. Instrum.* **2000**, *71*, 728–730. [CrossRef]
56. Favre, M.; Chuaqui, H.; Wyndham, E.; Choi, P. Measurements on electron beams in pulsed hollow-cathode discharges. *IEEE Trans. Plasma Sci.* **1992**, *20*, 53–56. [CrossRef]
57. Larigaldie, S.; Caillault, L. Dynamics of a helium plasma sheet created by a hollow-cathode electron beam. *J. Phys. D Appl. Phys.* **2000**, *33*, 3190–3197. [CrossRef]
58. Gleizer, J.Z.; Krokhmal, A.; Krasik, Y.E.; Felsteiner, J. High-current electron beam generation by a pulsed hollow cathode. *J. Appl Phys.* **2002**, *91*, 3431–3443. [CrossRef]
59. Son, S.M.; Han, S.L.; Ahn, J.H. Effects of the friction coefficient on the minimum cutting thickness in micro cutting. *Int. J. Mach Tools Manuf.* **2005**, *45*, 535. [CrossRef]
60. Balázs, B.Z.; Geier, N.; Takács, M.; Davim, J.P. A review on micro-milling: Recent advances and future trends. *Int. J. Adv. Manuf Technol.* **2021**, *112*, 655–684. [CrossRef]
61. O'Toole, L.; Kang, C.-W.; Fang, F.-Z. Precision micro-milling process: State of the art. *Adv. Manuf.* **2020**, 1–33. [CrossRef]
62. Davoudinejad, A.; Tosello, G.; Parenti, P.; Annoni, M. 3D Finite Element Simulation of Micro End-Milling by Considering the Effect of Tool Run-Out. *Micromachines* **2017**, *8*, 187. [CrossRef]
63. Chen, N.; Li, H.N.; Wu, J.; Li, Z.; Li, L.; Liu, G.; He, N. Advances in micro milling: From tool fabrication to process outcomes. *Int J. Mach. Tools Manuf.* **2021**, *160*, 103670. [CrossRef]
64. Weule, H.; Hüntrup, V.; Tritschler, H. Micro-Cutting of Steel to Meet New Requirements in Miniaturization. *CIRP Ann.* **2001**, *50*, 61–64. [CrossRef]
65. Jun, M.B.G.; Devor, R.E.; Kapoor, S.G. Investigation of the Dynamics of Microend Milling—Part II: Model Validation and Interpretation. *J. Manuf. Sci. Eng.* **2006**, *128*, 901–912. [CrossRef]

5. Zhang, X.; Ehmann, K.F.; Yu, T.; Wang, W. Cutting forces in micro end milling processes. *Int. J. Mach. Tool Manuf.* **2016**, *107*, 21–40. [CrossRef]
7. Vereschaka, A.; Grigoriev, S.; Popov, A.; Batako, A. Nano-scale Multilayered Composite Coatings for Cutting Tools Operating under Heavy Cutting Conditions. *Procedia CIRP* **2014**, *14*, 239–244. [CrossRef]
3. Fominski, V.; Grigoriev, S.; Gnedovets, A.; Romanov, R. Pulsed laser deposition of composite Mo Se Ni C coatings using standard and shadow mask configuration. *Surf. Coat. Technol.* **2012**, *206*, 5046–5054. [CrossRef]
9. Vereschaka, A.; Volosova, M.; Grigoriev, S. Development of Wear-resistant Complex for High-speed Steel Tool when Using Process of Combined Cathodic Vacuum Arc Deposition. *Procedia CIRP* **2013**, *9*, 8–12. [CrossRef]

 coatings

MDPI

Article

The Geometric Surface Structure of EN X153CrMoV12 Tool Steel after Finish Turning Using PCBN Cutting Tools

Michał Ociepa [1], Mariusz Jenek [1] and Piotr Kuryło [2,*]

[1] Institute of Machine Construction and Operations Engineering, University of Zielona Góra, 65-516 Zielona Góra, Poland; M.Ociepa@ibem.uz.zgora.pl (M.O.); M.Jenek@iim.uz.zgora.pl (M.J.)
[2] Faculty of Mechanical Engineering, University of Zielona Góra, 65-516 Zielona Góra, Poland
* Correspondence: P.Kurylo@ibem.uz.zgora.pl

Abstract: The article presents the results of studying the effects of coated (TiN, TiAlN) and uncoated polycrystalline cubic boron nitride (PCBN) machining blades on the key geometric structure parameters of the surface of hardened and tempered EN X153CrMoV12 steel after finish turning. A comparative analysis of the use of coated and coated cutting tools in finish turning of hardened steels was made. Tool materials based on polycrystalline cubic boron nitride PCBN (High-CBN; Low-CBN) have been described and characterized. The advantages of using TiN and TiAlN-coated cutting tools compared to uncoated were demonstrated. The lowest influence of the feed on the values of all tested roughness parameters was noted for surfaces treated with TiN- and TiAlN-coated tools (both with 50 vol.% of CBN). For uncoated tools (60 vol.% of CBN) for feeds f = 0.2 and 0.3 mm/rev., the highest values of R_a and R_z roughness parameters were found. Moreover, the lack of protective coating contributed to the occurrence of intense adhesive wear on the flank surface, which was also in the range of the feed values f = 0.2 and 0.3 mm/rev. The analysis of material surface after treatment with the uncoated tools with the feed f = 0.2 mm/rev. showed the occurrence of the phenomenon of lateral material flow and numerous chip deflections.

Keywords: roughness; coatings; finish turning; PCBN; tempered steel; tool wear

Citation: Ociepa, M.; Jenek, M.; Kuryło, P. The Geometric Surface Structure of EN X153CrMoV12 Tool Steel after Finish Turning Using PCBN Cutting Tools. *Coatings* **2021**, *11*, 428. https://doi.org/10.3390/coatings11040428

Received: 4 February 2021
Accepted: 3 April 2021
Published: 7 April 2021

1. Introduction

Manufacturers of machine parts are increasingly replacing the traditional operation of grinding materials in the hardened state by turning. It allows for constant technological progress, both in the design and technological solutions used in cutting machine tools and in the area of modern cutting materials. In the end, comparatively high machining accuracy is achieved at a lower cost and with a lower environmental impact. This has a direct impact on the high efficiency and productivity of this treatment method [1,2]. However, the problem of replacing the grinding operation with turning lies in the necessity of selecting such machining parameters and such cutting tools for which the machined surface will be characterized by possibly low surface roughness, usually defined by the R_a parameter in the required range of 0.1–2 µm, at the same time with a high material contribution to the roughness profile [3,4].

Finishing turning of toughened materials differs considerably from the turning of "soft" materials, primarily due to the much higher hardness of the machined material (normally higher than 45 HRC). The machining parameters used are significantly reduced due to higher cutting forces compared to traditional turning. Turning of "hard" materials is most often performed without using cooling and lubricating fluids, which results in generated temperatures exceeding 1000 °C in the cutting zone [5].

The cutting blades used in "hard" turning machining applications must be capable of withstanding extremely high mechanical loads and be chemically stable at high temperatures due to the characteristics of the process. Cutting blades based on polycrystalline cubic boron nitride (PCBN) have found the greatest practical use in the machining of iron-based toughened materials. This material is characterized by high chemical stability at temperatures exceeding 1400 °C and also has good impact strength and high thermal shock resistance, while retaining its mechanical properties [6].

The wear of polycrystalline cubic boron nitride (PCBN) tools is one of the most frequently undertaken research topics in the field of machining of hardened materials. Two groups of PCBN-based tool materials can be distinguished [7–10]:

- High-CBN (CBN-H), containing a large (70–95%) volume of CBN in the tool material, most often with a metallic binder;
- Low-CBN (CBN-L), containing a smaller (40–65%) volume of CBN in the tool material, most often with a ceramic binder (e.g., TiC, TiN).

Global cutting tool manufacturers offer both coated and uncoated CBN cutting inserts commercially. The most common commercial coatings on this type of tools include TiN, TiAlN and, less commonly, Ti(C,N), applied using physical vapor deposition (PVD) method Coated PCBN tools, compared to uncoated ones, provide improved cutting performance, like higher resistance to blade wear (especially crater wear) and to the destructive effects of high temperature occurring in the cutting zone [11–13]. This applies specifically to the machining of steel in the toughened state, during which temperatures in the cutting zone exceed 800 °C [14]. According to [15–18], TiN and TiAlN coatings on CBN blades provide a 30% (TiN) to 40% (TiAlN) higher tool life, compared to uncoated CBN tools. TiAlN coating applied on cutting tools, which is a TiN improved by adding aluminum, provides higher, than TiN coating, resistance to oxidation at high temperatures [19]. TiN oxidation process takes place at about 400 °C., while TiAlN oxidizes at temperatures above 600 °C [20,21], which increases the TiAlN-coated tool life.

However, surfaces machined with coated blades are characterized by a relatively constant height of micro-roughnesses [22].

Poulachon et al. [23] examined the effect of the microstructure of the hardened X155CrMoV12 steel on Surface Geometric Structure (SGP) and chip parameters for variable finish turning parameters with CBN-L and CBN-H tools. He found that the highest percentage of M7C3 carbides, which had a hardness of more than twice the hardness of the martensitic matrix of the tested material (800 HV), had the greatest influence on SGP parameters [24]. This affects the formation of the so-called a white layer on the chip surface, an example of which is shown in Figure 1.

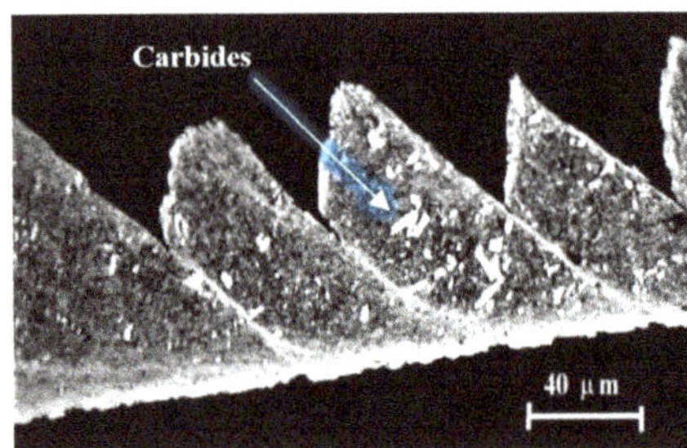

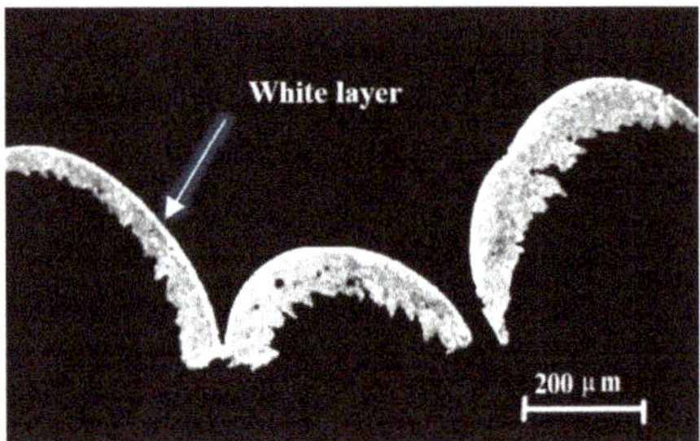

Figure 1. An example of a "white layer" on the chip surface after processing improved steel [21].

The analysis of the obtained results proved that in the case of continuous and semi interrupted processing, greater durability was recorded for the CBN-L tool. CBN-H tools showed better properties in the case of interrupted cuts, which is a direct result of much higher strength, heat resistance and higher ductility of these materials compared to CBN-L. The authors of the cited studies concluded that the microstructure of hardened steel had the main impact on the wear of the tested PCBN tools. These conclusions coincide with those of the current work [25–27].

The above research was continued in many studies [28–30], extending it with a Vc variable cutting speed. CBN 7015 with a TiN coating was used as cutting materials (tested Vc: 150, 250 m/min.) and uncoated CBN 7025 (tested Vc: 150, 195 m/min.). For continuous machining carried out with a lower Vc value, the main wear mechanism of both types of tools was abrasive wear, though was diffusion wear for higher Vc. For interrupted machining, in both cases intense abrasive and diffusion wear was noted, while the flank wear for the CBN 7015 tool was on average 3 times higher than for the 7025.

Another issue analyzed in the area of machining of hardened materials is the influence of protective coatings on the durability and wear of PCBN cutting tools.

Coelho et al. [18] analyzed the effect of anti-wear coatings on PCBN tool wear when machining AISI 4340 (EN40NiCrMo7) hardened steel. Three different TiN and TiC-based coatings were used for the tests. The analysis of the results showed that the tested coatings acted as a thermal barrier between the contact surface of the tool and the grains of the tool material.

The research results presented in the study [31] showed that the coatings on PCBN tools generally increase the chemical stability of the tool material, thus ensuring greater resistance to adhesive wear.

The authors of the work [32] stated, however, that at present the knowledge regarding the direct influence of protective coatings on the durability and wear of PCBN tools is insufficient to determine their usefulness in turning materials in an improved or hardened state.

The authors of the study [33] analyzed the condition of TiN and TiAlN coated cutting tools after finish turning of AISI D2 steel (EN X153CrMoV12). The lowest wear determined by the VB parameter was noted for TiN coated tools containing 50% of CBN in the tool material. On the other hand, the highest values of the tested parameter were obtained for TiAlN-coated tools with a CBN content of 65%.

The performed analysis showed that there is no clearly defined dependence of the influence of the type of tool material and protective coatings on the wear of PCBN cutting tools and the condition of the surface layer of the processed material.

When organizing cutting processes in a manufacturing system, the most complex problems arise when assessing the efficiency and reliability of machining with the use of specific cutting tools, including finish turning tools for improved AISI D2 tool steel (EN X153CrMoV12).

High quality requirements of the machined parts determine technological damage, which is the main object of research in the theory of technological system reliability [18,30,31]. During operation, the technological system is subject to thermal and mechanical effects, which cause damage and change the parameters of its initial state. During the operation of tools, along with the general principles of wear, special ones appear, often leading to their damage. Admittedly, the guides provide the permissible wear values at which the tool should be replaced, but they may differ significantly from those at which the cutting potential is fully used [25,26]. The dominant factor influencing the optimal conditions of tool operation is mainly damage caused by the wear of its cutting surfaces.

The aim of the study was to determine the technological feasibility of replacing the grinding operation of improved AISI D2 tool steel (EN X153CrMoV12) with turning machining using commercially available PCBN cutting blades. The article describes the effect of TiN and TiAlN coatings applied on PCBN-cutting tools on the condition of surface layer and the condition of cutting tools after hard turning. It shows a number of disadvantages of using uncoated tools for turning of a given material, e.g., related to shorter tool life or the occurrence of adverse phenomena on the machined surfaces (chip adhesion, the side flow phenomenon), which negatively affect the shape of the bearing area curves, and thus affected the operational properties of the machined surfaces.

2. Materials and Methods

The research included AISI D2 (EN X153CrMoV12) high-carbon high-chromium cold work tool steel, which is most often used as a material for dies, punches and forming rolls [26]. Its chemical composition is presented in Table 1. Samples in the form of rolls with a diameter of Ø = 50 mm, l = 20 mm were subjected to volume quenching in oi at a temperature of 1020 °C and tempering at a temperature of 150 °C to the assume hardness of 63 ± 2 HRC. The obtained material structure is characterized by the presenc of large primary (1) and secondary (2) carbides formed during tempering (Figure 2), th microhardness of which is even three times higher than the hardness of the obtaine martensitic matrix of the strip-like material [34,35].

Table 1. Chemical composition of AISI D2 steel (EN X153CrMoV12) [36].

Chemical Composition (mas.%)					
C	Si	Mn	Cr	Mo	V
1.5–1.7	0.15–0.4	0.15–0.45	11–13	0.7–1.0	0.6–0.8

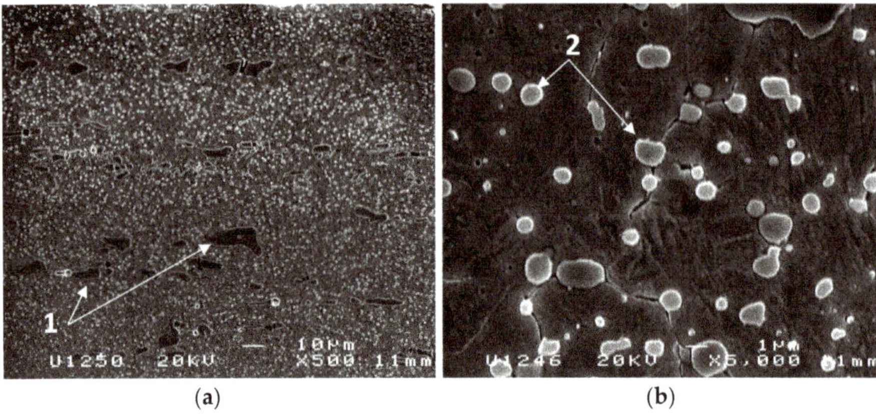

(a) (b)

Figure 2. The structure of the material used in research where: (**a**) magnification 500×, 1—primary carbides, (**b**) magnification 5000×, 2—secondary carbides.

The machining was performed using the CTX 510 machining centre (DMG Mor Nagoya, Japan) with the following cutting parameters: Vc = 160 m/min, ap = 0.2 mn f = 0.1; 0.2; 0.3 mm/rev. The PDJNR2020K11 holder knife was used in the research (κr = 93° α = 6°, γ = −6°) with DNGA 110408 replaceable inserts (re = 0.8 mm). The material of th cutting tools is presented in Table 2.

Table 2. Material of the cutting tools adopted for the test.

Material Type	T1	T2	T3 *	T4 *	T5
Machining type	Continuous and slightly interrupted machining	Continuous machining	Continuous machining	Continuous machining	Continuous machining
PCBN structure	60% CBN in a ceramic binder	50% CBN in a ceramic binder	50% CBN in a ceramic binder	50% CBN in a ceramic binder	65% CBN in a ceramic binder
Coating type	NONE	TiN (PVD)	TiAlN (PVD)	TiAlN (PVD)	TiAlN (PVD)
Coating thickness	–	1–4 µm	2–4 µm	2–4 µm	2–4 µm
Coating hardness	–	2100–2600 HV	2400–2800 HV	2400–2800 HV	2400–2800 HV
Chamfer	BN = 0.1 mm GB = 20°	BN = 0.1 mm GB = 30°	BN = 0.13 mm GB = 25°	BN = 0.13 mm GB = 25°	BN = 0.12 mm GB = 25°
Cutting edge radius **	21.7 µm	17.17 µm	25.26 µm	24.12 µm	22.02 µm

where: BN—chamfer width, GB—chamfer angle; *—different manufacturers; **—average of 3 measurements.

A JEOL JSM-6400 scanning microscope (Tokyo, Japan) was used for metallographic examinations. The condition of the surface layer of machined surfaces (R_a, R_z, RS_m parameters, bearing area curves) and cutting tools was assessed on the three-axis optical measurement system Alicona Infinite SL coupled with the IF-Laboratory Measurement Module (Graz, Austria).

A ZWICK ZHV10 microhardness tester (Wroclaw, Poland) was used to test the depth and degree of material reinforcement. HV0.01 microhardness was measured by keeping the distance between the impressions in each direction of no less than 2.5 of impression diagonal.

Due to the requirements of the tool manufacturers, the machining of "hard" materials with PCBN tools is always performed "dry".

A new cutting tool and a new sample was used for each operation. A lack of roughness measurement results for turning with a T5 blade with a feed f_2 = 0.2 mm/rev. resulted from surface layer damage during turning.

3. Results and Discussion

The analysis of the geometric structure of the machined surfaces was carried out by comparing the values of the key roughness parameters: R_a—arithmetical mean height, R_z—maximum height of profile and RS_m—mean width of the profile elements [36,37].

Figure 3 shows the obtained results of measurements of selected roughness parameters R_a, R_z and RS_m depending on the f feed value and the type of tools.

For feeds f_1 = 0.1 mm/rev. and f_3 = 0.3 mm/rev. the roughness parameter R_a values for all tested cutting tools are similar and have a downward trend (Figure 3a). For the feed f_2 = 0.2 mm/rev. for the only uncoated tool T1 among the tested, a sudden increase in the analyzed parameter was noted. The same dependence can be noticed for the results of the R_z parameter measurement (Figure 3b). The microscopic analysis of the tool T1 condition showed the most intense adhesive wear occurring on the flank surfaces after machining with feeds f_2 = 0.2 mm/rev. and f_3 = 0.3 mm/rev. (Figure 4a,b). For all coated tools, especially T2 and T3, these phenomena occurred with much less intensification (Figure 4c,d).

For feed f_2 = 0.2 mm/rev., the occurrence of numerous chip sticking and the phenomenon of material side flow occurring on the T1 tool-machined surfaces was also noted (Figure 5a,b). Both phenomena could be caused by different tribological properties of the tested tools and different heat distribution in the cutting zone. The uncoated T1 tools, characterized by a higher coefficient of friction than the coated ones, generate a higher temperature in the cutting zone [38]. This, in turn, might have resulted in the plastification of the processed material, leading to the phenomenon of material side flow as well as frequent chip sticking on the processed surface [5], which caused the abnormal increase of roughness parameters (Figure 3). These phenomena did not occur on surfaces machined with coated blades, an example of which is shown in Figure 5c,d.

The largest decrease and, at the same time, the lowest value of all tested parameters of the roughness of the machined surfaces were recorded for the TiAlN coated tools T3 for the feed f_2 = 0.2 mm/rev. These tools were the only ones that recorded a decrease in the value of the RS_m parameter, including for the feed f_2 = 0.2 mm/rev., while for all other tools, the value of this parameter increased. However, the T3 tools were the only ones to record an increase in all tested roughness parameters for the feed f_3 = 0.3 mm/rev., which, according to the tool's manufacturer, is the upper limit of the working range for this type of tool.

The smallest differences in the measurement results of the tested roughness parameters depending on the feed value characterize the T2 tools as the only ones with a TiN coating and show them to characterized by the highest shear angle among the tested tools, i.e., 30°.

In general, it should be assumed that the lowest values were recorded for the material treated with T3 tools, and for the material treated with T1 tools, the highest values of the tested roughness parameters were noted.

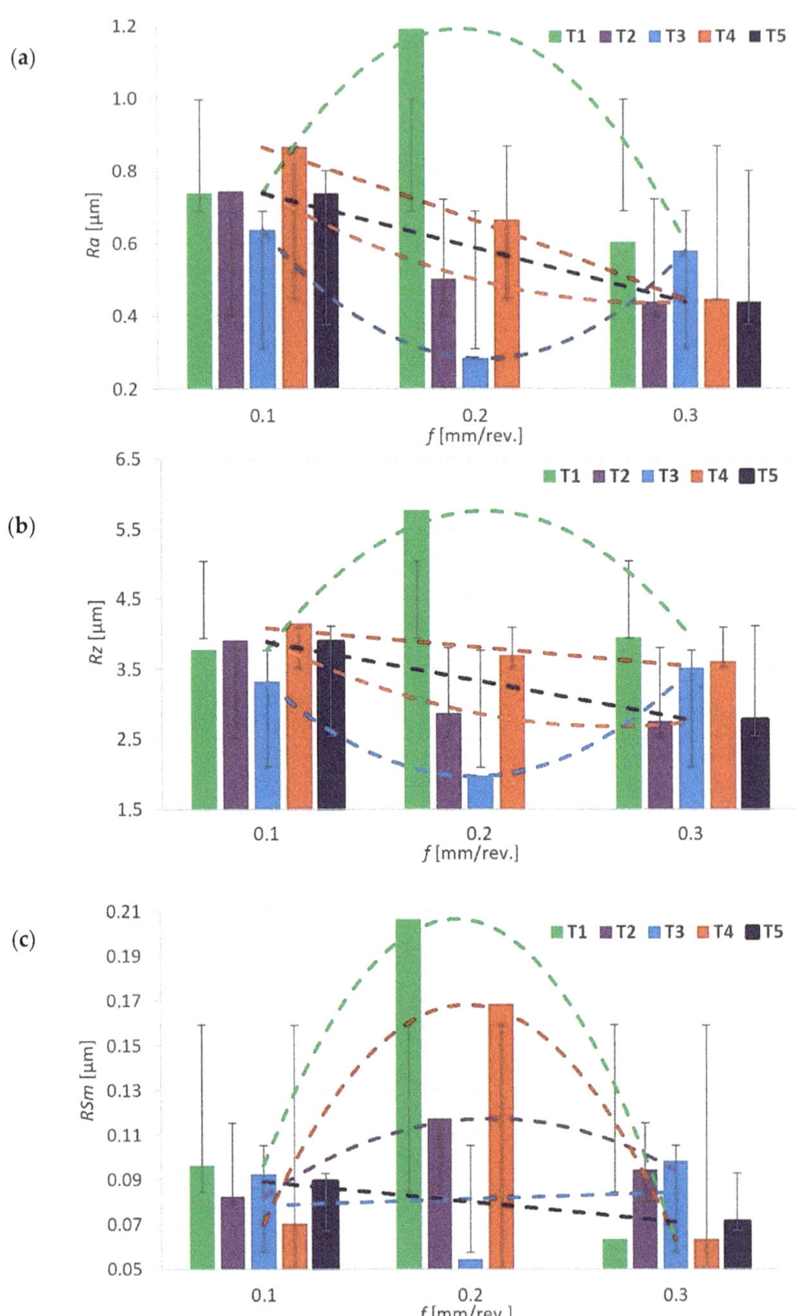

Figure 3. Values of selected parameters of surface roughness depending on the value of f feed and the type of cutting edge where: T1–T5—tool number; (**a**)—R_a (μm); (**b**)—R_z (μm); (**c**)—RS_m (μm).

Figure 4. (**a**,**b**): T1 tool after machining with the feed value f_3 = 0.3 mm/rev., magnification 500×; 1—adhesive wear, 2—abrasive wear; T2 (**c**) and T3 (**d**) tool after machining with the feed value f_3 = 0.3 mm/rev., magnification 200×; 1—adhesive wear.

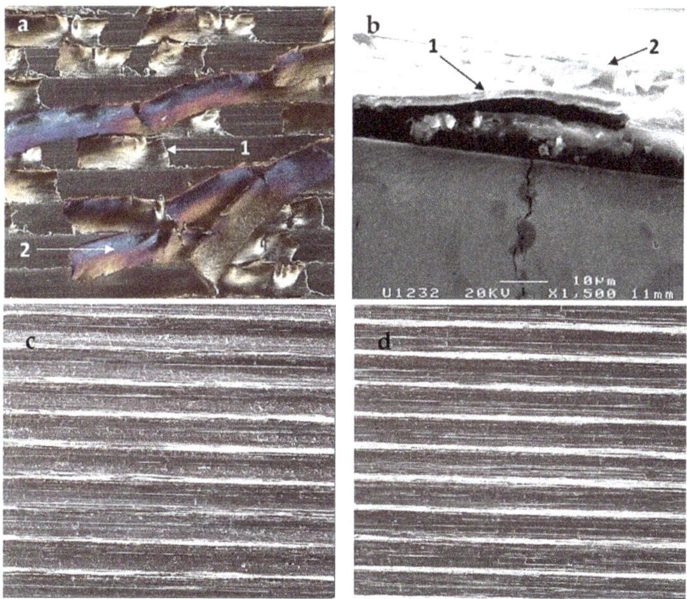

Figure 5. (**a**,**b**): The surface treated with an uncoated T1 tool with the feed value f_2 = 0.2 mm/rev.; 1—side flow of the material, 2—chip glued on the surface; (**c**,**d**): the surfaces treated with coated T2 (**c**) and T3 (**d**) tools with the feed value f_2 = 0.2 mm/rev.; magnification 200×.

All the average measurement results of the roughness parameters R_a and R_z for th f_1 feed significantly exceed the theoretical values given by the manufacturers of indivic ual tools.

Figures 6–8 show the bearing area curve for the surfaces processed with the teste tools and for the entire range of the tested feeds.

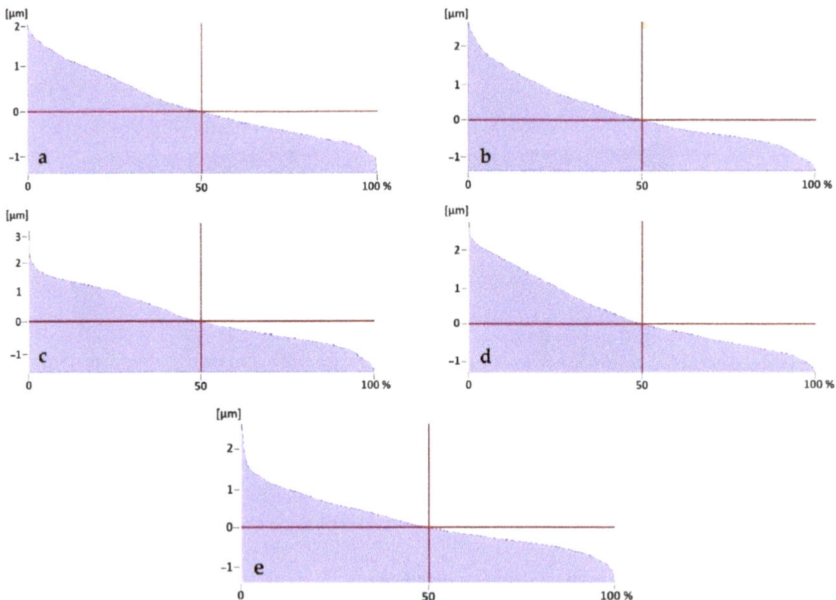

Figure 6. Bearing area curves for the surfaces processed with the T1–T5 tools with feed f_1 = 0.1 mm/rev.: (**a**): T1; (**b**): T2; (**c**): T3; (**d**): T4; (**e**): T5.

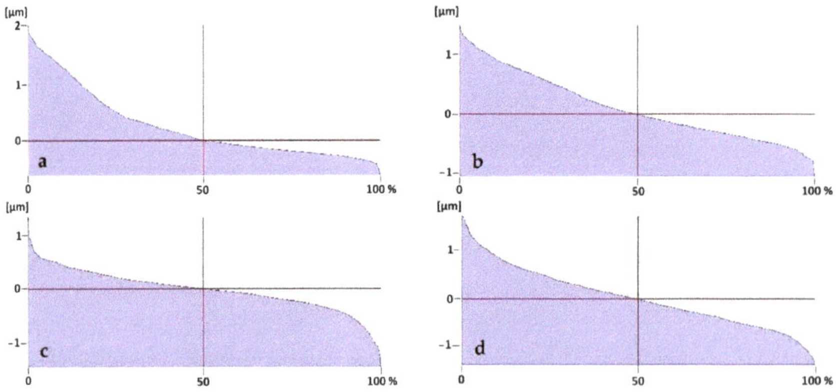

Figure 7. Bearing area curves for the surfaces processed with the T1–T4 tools with feed f_2 = 0.2 mm/rev.: (**a**): T1; (**b**): T2; (**c**): T3; (**d**): T4.

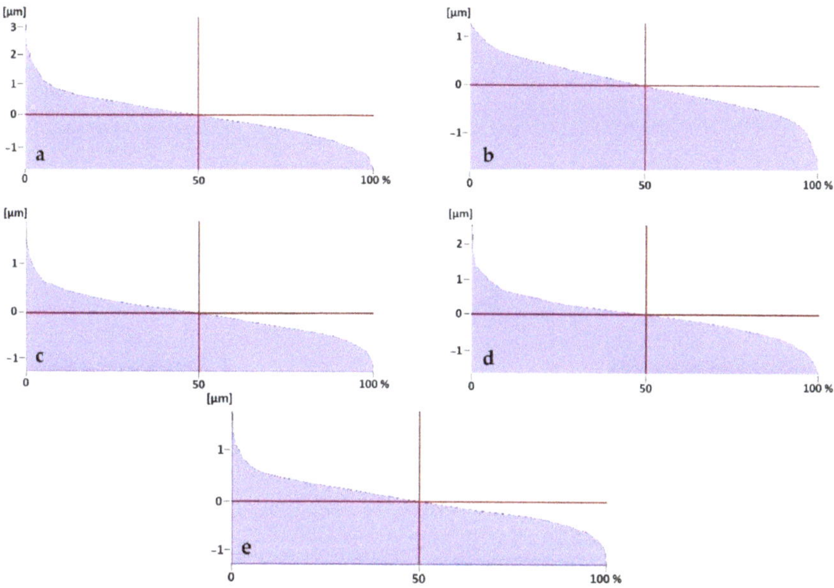

Figure 8. Bearing area curves for the surfaces processed with the T1–T5 tools with feed f_3 = 0.3 mm/rev.: (**a**): T1; (**b**): T2; (**c**): T3; (**d**): T4; (**e**): T5.

The analysis of the bearing area curves shows the advantage of increasing the feed value during the turning process of the tested material. The progressive nature of the curves, indicating good operational properties of the processed material—in particular its high abrasion resistance [39]—was observed for all surfaces processed with a feed of f_3 = 0.3 mm/rev.

The desired shape of the bearing area curve, after turning with a feed of f_2 = 0.2 mm/rev., was only observed for the surface processed with the T3 tool. The remaining curves, including all those obtained after turning with a feed of f_1 = 0.1 mm/rev. have a digressive shape, which possibly indicates insufficient operational properties for these surfaces [40].

Figure 9 shows the results of research on the depth and degree of strengthening of the surface layer of the processed material with the feed value f_3 = 0.3 mm/rev. Figure 10 presents the surface hardening rate of the surface layer after machining with the same feed value.

The greatest changes in the microhardness of the material occur for the surface layer depth of approximately 100 µm. For all tested surfaces, the hardness of the material core was achieved at a depth of 400–500 µm from the outer border of the processed material (Figure 9). The presented test results prove a very similar degree of material tempering for all tested cutting tools (defined as the percentage difference between the microhardness of the surface and the material core—as shown in Figure 9).

The lowest degree of tempering (10.05%) was obtained for the material treated with the T1 uncoated tool. The largest one (14.39%) was the material treated with the T5 TiAlN-coated tool. The reason for this could have been the largest (65%) share of CBN in the composition of the T5 tool material out of all the examined tools, resulting in less heat being transferred to the tool and more heat being passed on to the processed material. This can be confirmed by the sudden decrease of microhardness at the surface layer depth of about 50–60 µm, which may indicate the presence of high temperatures causing material tempering at this depth. However, it should be taken into account that the measurement value could have been influenced by the structure of the material obtained during thermal improvement, characterized by both an uneven primary and secondary distribution of carbides that was much harder than the matrix.

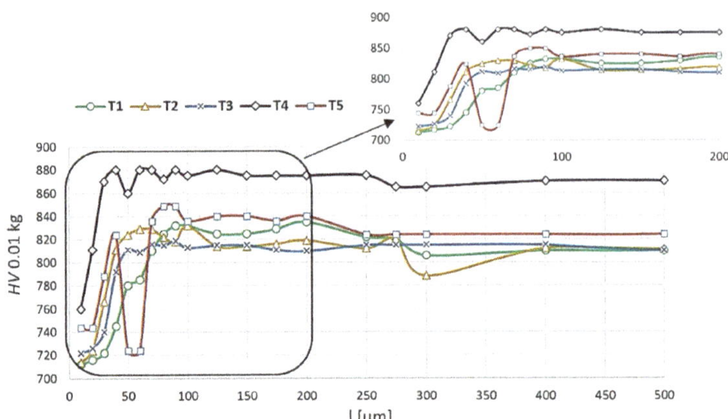

Figure 9. Microhardness of the material processed with the feed f_3 = 0.3 mm/rev. depending on the tested tool where: T1–T5—tool type.

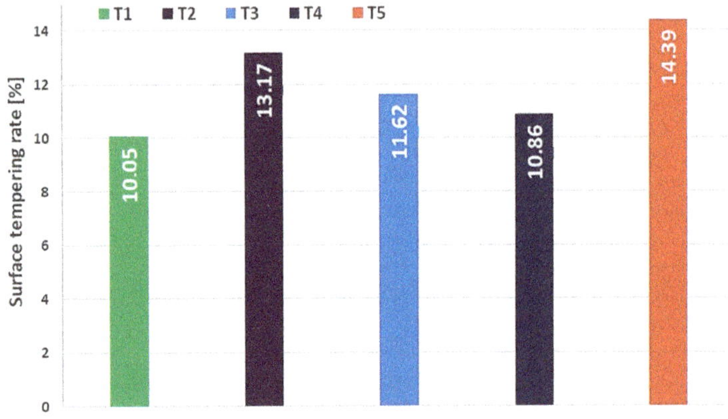

Figure 10. Surface tempering rate of the material processed with the feed f_3 = 0.3 mm/rev. depending on the tested tool where: T1–T5—tool type.

4. Conclusions

In the presented research, the influence of anti-wear coatings applied on PCBN cutting tools on selected parameters characterizing the condition of the surface layer after finish turning of AISI D2 tool steel (EN X153CrMoV12) was analyzed. The performed analysis allows drawing the following cognitive conclusions:

- a positive effect of the use of anti-wear coatings on PBCN tools has been shown to reduce the values of the roughness parameters tested compared to uncoated tools,
- the lowest influence of the feed on the values of all tested roughness parameters was noted for surfaces treated with TiN-coated T2 tools (50 vol.% of CBN) and TiAlN-coated T3 (50 vol.% of CBN) tools,
- the lack of protective coating contributed to the occurrence of intense adhesive wear on the flank surface on the uncoated T1 tools, in the range of the feed values f = 0.2 and 0.3 [mm/rev.],

- the analysis of material surface after treatment with the uncoated T1 tool with the feed f = 0.2 [mm/rev.] showed the occurrence of the phenomenon of lateral material flow and numerous chip deflections,
- for all tested tools, the greatest changes in the microhardness of the structure after the turning process with a feed f = 0.3 [mm/rev.] were obtained for a depth of up to 100 μm, and the hardness of the material core for all tested tools was achieved at a depth of 400–500 μm from the outside material boundary,
- the most optimal parameters of the surface layer, equivalent to parameters after the grinding process, for all analyzed materials were obtained after processing with a feed value of f = 0.3 mm/rev.

Author Contributions: Conceptualization, M.O. and M.J.; methodology, M.J.; software, M.O.; validation, P.K. and M.J.; formal analysis, M.J. and P.K.; investigation, M.O., M.J. and P.K.; resources, M.J. and P.K.; data curation, P.K.; writing—original draft preparation, M.O.; writing—review and editing, P.K.; visualization, M.O.; supervision, M.J.; All authors have read and agreed to the published version of the manuscript.

Funding: The authors gratefully acknowledge the financial support from the program of the Polish Ministry of Science and Higher Education under the name "Regional Initiative of Excellence" in 2019–2022, project No. 003/RID/2018/19, funding amount 11 936 596.10 PLN".

Institutional Review Board Statement: Not applicable.

Informed Consent Statement: Not applicable.

Data Availability Statement: The data presented in this study are available in article.

Conflicts of Interest: The authors declare no conflict of interest.

eferences

Rao, J.; Sharma, A.; Rose, T. Titanium aluminium nitride and titanium boride multilayer coatings designed to combat tool wear. *Coatings* **2017**, *8*, 12. [CrossRef]

Boing, D.; Zilli, L.; Fries, C.E.; Schroeter, R.B. Tool wear rate of the PCBN, mixed ceramic, and coated cemented carbide in the hard turning of the AISI 52100 steel. *Int. J. Adv. Manuf. Technol.* **2019**, *104*, 4697–4704. [CrossRef]

Radha Krishnan, B.; Aravindh, R.; Barathkumar, M.; Gowtham, K.; Hariharan, R. Prediction of surface roughness (AISI 4140 steel) in cylindrical grinding operation by RSM. *Int. J. Res. Dev. Technol.* **2018**, *9*, 702–704.

Puerto, P.; Fernandez, R.; Madariaga, J.; Arana, J.; Gallego, I. Evolution of surface roughness in griding and its relationship with the dressing parameters and the radial wear. *Procedia Eng.* **2013**, *63*, 174–182. [CrossRef]

Ociepa, M.; Jenek, M.; Feldshtein, E.; Leksycki, K. The phonomenon of material side flow during finish turning of EN X153CrMoV12 hardened steel with tools based on polycrystalline cubic boron nitride. *J. Superhard Mater.* **2019**, *41*, 265–271. [CrossRef]

Tillmann, W.; Elrefaey, A.; Wojarski, L. Brazing of cutting tools. In *Advances in Brazing—Science, Technology and Applications*; Elsevier: Amsterdam, The Netherlands, 2013; pp. 423–471.

Bushlya, V.M.; Gutnichenko, O.A.; Zhou, J.M.; Ståhl, J.-E.; Gunnarsson, S. Tool wear and tool life of PCBN, binderless cBN and wBN-cBN tools in continuous finish hard turning of cold work tool steel. *J. Superhard Mater.* **2014**, *36*, 49–60. [CrossRef]

DIN ISO 513:2014–05. *Classification and Application of Hard Cutting Materials for Metal Removal with Defined Cutting Edges—Designation of the Main Groups and Groups of Application (ISO 513:2012)*; Deutsches Institut für Normung: Berlin, Germany, 2014.

Gsander, D. Chancen und Gränzen des high-performance cutting. *Werkzeug Tech.* **2002**, *72*, 35–37.

10. Gordon, S.; Phelan, P.; Lahiff, C. The effect of high speed machining on the crater wear behaviour of PCBN tools in hard turning. *Procedia Manuf.* **2019**, *38*, 1833–1848. [CrossRef]

11. Sveen, S.; Andersson, J.; M'Saoubi, R.; Olsson, M. Scratch adhesion characteristics of PVD TiAlN deposited on high speed steel, cemented carbide and PCBN substrates. *Wear* **2013**, *308*, 133–141. [CrossRef]

12. Sousa, V.F.C.; Silva, F.J.G. Recent advances in turning processes using coated tools—A comprehensive review. *Metals* **2020**, *10*, 170. [CrossRef]

13. Micallef, C.; Zhuk, Y.; Aria, A.I. Recent progress in precision machining and surface finishing of tungsten carbide hard composite coatings. *Coatings* **2020**, *10*, 731. [CrossRef]

14. Caliskan, H.; Panjan, P.; Kurbanoglu, C. Hard Coatings on Cutting Tools and Surface Finish. In *Comprehensive Materials Finishing*; Elsevier: Amsterdam, The Netherlands, 2017; Chapter 3; pp. 230–242. [CrossRef]

15. Taylan, F.; Colak, O.; Kayacan, M. Investigation of TiN coated CBN and CBN cutting tool performance in hard milling appli-cation, Strojniski vestnik. *J. Mech. Eng.* **2011**, *57*, 417–424.

16. Galoppi, G.D.S.; Filho, M.S.; Batalha, G.F. Hard turning of tempered DIN 100Cr6 steel with coated and no coated CBN insert *J. Mater. Process. Technol.* **2006**, *179*, 146–153. [CrossRef]
17. M'Saoubi, R.; Johansson, M.; Andersson, J. Wear mechanisms of PVD-coated PCBN cutting tools. *Wear* **2013**, *302*, 1219–122 [CrossRef]
18. Coelho, R.T.; Ng, E.-G.; Elbestawi, M. Tool wear when turning hardened AISI 4340 with coated PCBN tools using finishin cutting conditions. *Int. J. Mach. Tools Manuf.* **2007**, *47*, 263–272. [CrossRef]
19. Sousa, V.; Da Silva, F.; Pinto, G.; Baptista, A.; Alexandre, R. Characteristics and wear mechanisms of TiAlN-based coatings fo machining applications: A comprehensive review. *Metals* **2021**, *11*, 260. [CrossRef]
20. Kuo, C.-C.; Lin, Y.-T.; Chan, A.; Chang, J.-T. High temperature wear behavior of titanium nitride coating deposited using hig power impulse magnetron sputtering. *Coatings* **2019**, *9*, 555. [CrossRef]
21. Badini, C.; Deambrosis, S.M.; Padovano, E.; Fabrizio, M.; Ostrovskaya, O.; Miorin, E.; D'Amico, G.C.; Montagner, F.; Biamin S.; Zin, V. Thermal shock and oxidation behavior of HiPIMS TiAlN coatings grown on Ti–48Al–2Cr–2Nb intermetallic allo *Materials* **2016**, *9*, 961. [CrossRef] [PubMed]
22. Manokhin, A.S.; Klimenko, S.A.; Klimenko, S.A.; Beresnev, V.M. Promising types of coatings for PCBN tools. *J. Superhard Mate* **2018**, *40*, 424–431. [CrossRef]
23. Poulachon, G.; Bandyopadhyay, B.; Jawahir, I.; Pheulpin, S.; Seguin, E. The influence of the microstructure of hardened tool ste workpiece on the wear of PCBN cutting tools. *Int. J. Mach. Tools Manuf.* **2003**, *43*, 139–144. [CrossRef]
24. Yamamoto, K.; Inthidech, S.; Sasaguri, N.; Matsubara, Y. Influence of Mo and W on high temperature hardness of M7C3 carbid in high chromium white cast iron. *Mater. Trans.* **2014**, *55*, 684–689. [CrossRef]
25. Dziedzic, K.; Józwik, J.; Barszcz, M.; Gauda, K. Wear characteristics of hard facing coatings obtained by tungsten inert gas metho *Adv. Sci. Technol. Res. J.* **2019**, *13*, 8–14. [CrossRef]
26. Pytlak, B. Multicriteria optimization of hard turning operation of the hardened 18HGT steel. *Int. J. Adv. Manuf. Technol.* **2010**, *4* 305–312. [CrossRef]
27. Boing, D.; Schroeter, R.B.; De Oliveira, A.J. Three-dimensional wear parameters and wear mechanisms in turning hardened steel with PCBN tools. *Wear* **2018**, *398–399*, 69–78. [CrossRef]
28. De Godoy, V.A.A.; Diniz, A.E. Turning of interrupted and continuous hardened steel surfaces using ceramic and CBN cuttin tools. *J. Mater. Process. Technol.* **2011**, *211*, 1014–1025. [CrossRef]
29. Grzesik, W. *Advanced Machining Processes of Metallic Materials—Theory, Modeling and Applications*, 2nd ed.; Elsevier: Amsterdan The Netherlands, 2017; p. 578. ISBN 978-0-444-63711-6.
30. Wang, Y.; Cui, X.; Xu, H.; Jiang, K. Cutting force analysis in reaming of ZL102 aluminium cast alloys by PCD reamer. *Int. J. Ad Manuf. Technol.* **2012**, *67*, 1509–1516. [CrossRef]
31. Fuentes, J.A.O.; Keunecke, M. Analysis and application of CBN coated cutting tools. In Proceedings of the 7th Internationa Conference "The Coatings in Manufacturing Engineering", Chalkidiki, Greece, 1–3 October 1998.
32. Uhlmann, E.; Braeuer, G.; Wiemann, E.; Keunecke, M. CBN coatings on cutting tools, production engineering. *Res. Dev. Ger. An Ger. Acad. Soc. Prod. Eng.* **2004**, *11*, 45–48.
33. Ociepa, M.; Jenek, M.; Feldshtein, E. On the wear comparative analysis of cutting tools made of composite materials based o polycrystalline cubic boron nitride when finish turning of AISI D2 (EN X153CrMoV12) steel. *J. Superhard Mater.* **2018**, *40*, 396–40 [CrossRef]
34. Krbat'a, M.; Eckert, M.; Križan, D.; Barényi, I.; Mikušová, I. Hot deformation process analysis and modelling of X153CrMoV1 steel. *Metals* **2019**, *9*, 1125. [CrossRef]
35. Panjan, P.; Drnovšek, A.; Gselman, P.; Čekada, M.; Bončina, T.; Merl, D.K. Influence of growth defects on the corrosion resistanc of sputter-deposited TiAlN hard coatings. *Coatings* **2019**, *9*, 511. [CrossRef]
36. EN ISO 4957:2018. *Tool Steels*; ISO: Geneva, Switzerland, 2018.
37. EN ISO 4287:1997. *Geometrical Product Specifications (GPS)—Surface Texture: Profile Method—Terms, Definitions and Surface Textur Parameters*; ISO: Geneva, Switzerland, 1997.
38. Kumar, C.; Patel, S.K.; Majumder, H.; Khan, A.; Naik, D.K. Numerical analysis of hard machining of AISI 52100 steel usin uncoated and coated Al_2O_3-TiCN based mixed ceramic inserts. In Proceedings of the National Conference on Emerging Trend in Manufacturing & Automation Engineering, NCMAE, Gwalior, India, 6 October 2017.
39. Urban, M.; Monkova, K. Research of tribological properties of 34CrNiMo6 steel in the production of a newly designed sel equalizing thrust bearing. *Metals* **2020**, *10*, 84. [CrossRef]
40. Salcedo, M.C.; Coral, I.B.; Ochoa, G.V. Characterization of surface topography with Abbott Firestone curve. *Contemp. Eng. Sc* **2018**, *11*, 3397–3407. [CrossRef]

Article

The Potential of High-Fluence Ion Irradiation for Processing and Recovery of Diamond Tools

Anatoly M. Borisov [1,2,*], Valery A. Kazakov [3], Eugenia S. Mashkova [4], Mikhail A. Ovchinnikov [4], Sergey N. Grigoriev [2] and Igor V. Suminov [2]

1 Moscow Aviation Institute (National Research University), 125993 Moscow, Russia
2 Moscow State Technological University "STANKIN", 127994 Moscow, Russia; s.grigoriev@stankin.ru (S.N.G.); ist3@mail.ru (I.V.S.)
3 Chuvash State University, 428015 Cheboksary, Russia; cossac@mail.ru
4 Skobeltsyn Institute of Nuclear Physics, Moscow State University, 119991 Moscow, Russia; es_mashkova@mail.ru (E.S.M.); ov.mikhail@gmail.com (M.A.O.)
* Correspondence: anatoly_borisov@mail.ru

Received: 26 November 2020; Accepted: 15 December 2020; Published: 17 December 2020

Abstract: The graphitization and surface growth of synthetic diamonds by high-fluence irradiation with 30 keV argon and carbon ions have been experimentally studied. scanning electron microscope (SEM) and atomic force microscope (AFM) show removal of traces of mechanical polishing. The ion-induced roughness does not exceed 20 nm. Raman spectroscopy and the measurement of electrical conductivity confirm the graphitization of the surface layer when irradiated with argon ions at the temperature of 230 °C and the diamond structure of the synthesized layer when irradiated with carbon ions at the temperature of 650 °C.

Keywords: monocrystalline; high-pressure, high-temperature (HPHT) diamond; chemical vapor deposition (CVD) diamond; high-fluence ion irradiation; Ar^+; C^+; SEM; AFM; Raman spectra; electrical conductivity

1. Introduction

Ionic and neutral atomic and molecular beams are very promising for the synthesis of tool-hardening coatings on an industrial scale [1–9]. Diamond is widely used as an abrasive, as an indenter and cutting tool, and as a heat sink in electronic devices, including as a filler in composite highly heat-conducting materials, and its use in microelectronics is promising. Among the methods for obtaining synthetic diamonds and diamond-like coatings, the method of the ion implantation of carbon has certain advantages. According to [10], the ion implantation of carbon into natural diamond crystals leads to internal epitaxial growth without a visible interface between the grown layer and the crystal surface. The layer synthesized by ion implantation has the same high resistance to acids and oxidation in air at 500 °C as natural diamond. Neither one is polished with aluminum oxide; both behave the same when lapping on a diamond polishing wheel and when scratching with a diamond needle or a diamond indenter.

The influence of high-fluence ion irradiation by 10–30 keV Ar^+, Ne^+, N^+, $N_2{}^+$ and C^+ ions at temperatures from 30 to 720 °C on the conductivity and microstructure of a polycrystalline diamond surface layer was experimentally studied in [11,12]. The increase in diamond temperature during irradiation leads to ion-induced graphitization at $T_{ir} > T_{gr} \approx 200$ °C. The Raman spectra indicate that irradiation with neon and argon ions at temperatures of the diamond more than $T_{ir} > 500$ °C leads to the formation of a nanocrystalline graphite layer that increases the resistivity of the irradiated layer. This effect is not observed under irradiation by nitrogen ions. It was found that dynamic annealing at temperatures above 500 °C leads to the recrystallization of the diamond only in the case of irradiation with carbon ions, and irradiation with impurity ions causes graphitization of the ion-modified diamond layer. Under irradiation with carbon ions, the growth and recrystallization of diamond with a thin (~1 nm) graphite-like layer on the surface occurs.

For the processing and restoration of diamond tools, there is an interest in determining the conditions for the graphitization of the diamond surface to a given thickness and controlled surface growth. A soft graphitized layer on the surface of a diamond can facilitate its mechanical polishing and can be used as a sacrificial layer for planarization. There is an interest in surface build-up for resizing and reshaping a unique diamond tool. In both cases, it is important to reduce or at least maintain the surface roughness during processing.

In this work, graphitization with argon ions, which begins at a temperature of >230 °C, and the growth of diamond under irradiation with carbon ions at a temperature of >500 °C were chosen as the most promising for instrumental application. The effect of high-fluence ion irradiation on the surface morphology was studied by SEM and AFM methods, and the structure of the irradiated layer was studied by Raman spectroscopy and electrical conductivity measurements.

2. Materials and Methods

Samples of synthetic polycrystalline diamond and single crystal Ib diamond were used. A 450 μm-thick polycrystalline diamond plate was prepared by chemical vapor deposition on a silicon substrate 63 mm in diameter [13,14]. A (111) face diamond crystal was obtained by high-pressure, high-temperature growth in Fe–Ni–C liquid from the seed [15,16]. High-fluence irradiation with Ar^+ and C^+ ions was carried out along the normal to the sample surface on a mass monochromator of the Skobeltsyn Institute of Nuclear Physics, Moscow State University [17]. The ion energy was 30 keV. The ion current density reached 0.4 mA/cm^2 at a beam cross section of 0.3 cm^2. The irradiation fluences φt (φ = the ion flux density, and t = the irradiation time) were no less than 10^{18} ions/cm^2 for all the cases of diamond irradiation. The flat resistive furnace in the target holder allows heating the target up to 720 °C. The target temperature was measured with a chromel alumel thermocouple fixed on the irradiated side of the sample outside of the irradiation area. Samples irradiated at T = 30 °C were isochronally annealed in a vacuum with a one-hour thermal treatment as the maximum cycle temperature T was increased sequentially from 100 to 720 °C.

The modified target was examined by SEM with a Lyra 3 (TESCAN, Brno, Czech Republic), AFM with a Dimension V (Veeco, Plainview, NY, USA) and a T64000 Raman spectrometer (Horiba Jobin Yvon, Edison, NJ, USA). Laser radiation with a wavelength of 488 nm for a single diamond crystal and a wavelength of 514 nm for a polycrystalline diamond was used to excite Raman scattering. Electrical measurements of the sheet resistance R_s were used in a four-point probe method at room temperature.

3. Results and Discussion

3.1. Characterization of High-Fluence Ion Irradiation

For the characterization of high-fluence ion irradiation, the depth distributions $\nu(x)$ of the numbers of displacements per atom (dpa) (damage depth profiles) were calculated. The profiles $\nu(x)$ were determined by the depth distributions $\Sigma(x)$ of the average numbers of vacancies formed by one ion per path length, which were derived by the computer simulation of the ion interactions with solids using the SRIM code [18]. At not-too-high fluences $\nu(x) \approx \Phi \cdot \sum(x)/n_0$, where the fluence $\Phi = \varphi \cdot \tau$ is equal to the product of the ion beam flux φ and irradiation time τ, n_0 is the atomic target concentration [19]. At high fluences, dynamic equilibrium conditions are established, in which the profiles of the concentration of implanted particles and radiation damage become stationary [11]. The stationary damage depth profile $\nu_{st}(x)$ is calculated as follows:

$$\nu_{st}(x) = \frac{1}{K}\int_{x}^{Rd} \Sigma(x') \cdot dx', \tag{1}$$

where R_d is the depth of the defect production, $K = Y$ for Ar^+ ion irradiation, $K = |1-Y|$ for C^+ ion irradiation and Y is the sputtering yield.

The calculated stationary profiles $\nu_{st}(x)$ are shown in Figure 1. Based on the calculated profiles of $\nu_{st}(x)$, the thickness t of the modified surface layer was estimated.

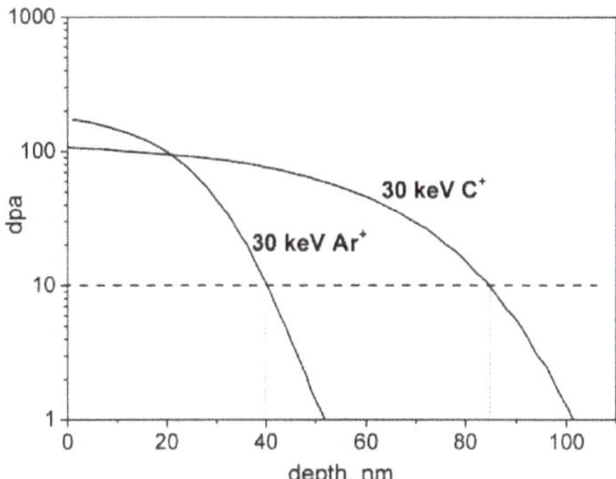

Figure 1. Stationary damage depth profiles under high-fluence ion irradiation.

The value of t was determined from the level of 10 dpa. This value of ν was attained over the depths $x \leq R_d$. Unlike for irradiation by gas ions, for C^+ ion irradiation, the thickness of the modified layer increases due to the implanted carbon ions, since the self-sputtering yield ($Y = 0.21$) is less than 1 [20]. For estimating the thickness t, the thickness of the deposited carbon was added to the modified layer thickness obtained from $\nu_{st}(x)$.

3.2. SEM and AFM

Figure 2 shows SEM images of the (111) face of the diamond before and after irradiation with argon and carbon ions at temperatures of 230 and 650 °C, respectively. The radiation fluences in both cases were equal to 1.3×10^{18} ion/cm^2. Traces of mechanical polishing are clearly visible on the surface before irradiation, which were removed by high-fluence ion irradiation. At the same time, the ion irradiation led to the appearance of etching pits in the form of equilateral triangles with sides of about 20 μm and open micropores. Such micropores are associated, as a rule, with the increased sputtering of areas of clusters of crystal structure defects, which lower the binding energy of surface atoms. At higher magnification, a stochastic nanoglobular relief appears in the SEM images (insets in Figure 2).

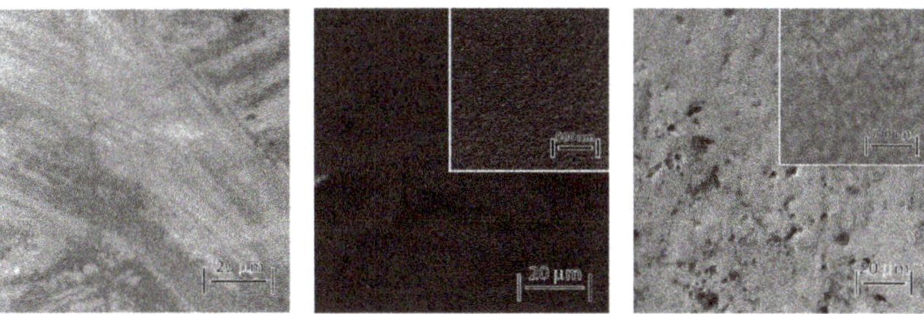

Figure 2. SEM images of the (111) face of the diamond: (**a**) before irradiation, (**b**) after irradiation with 30 keV Ar^+ 230 °C ions and (**c**) after irradiation with 30 keV C^+ 650 °C ions.

The AFM measurements of the relief also show the removal of the traces of mechanical polishing under ion irradiation and the appearance of a stochastic nanorelief (Figure 3). The roughnesses before and after the irradiation of the diamond surface are close; the root-mean-square roughness at a base length of 1 μm is about 20 nm.

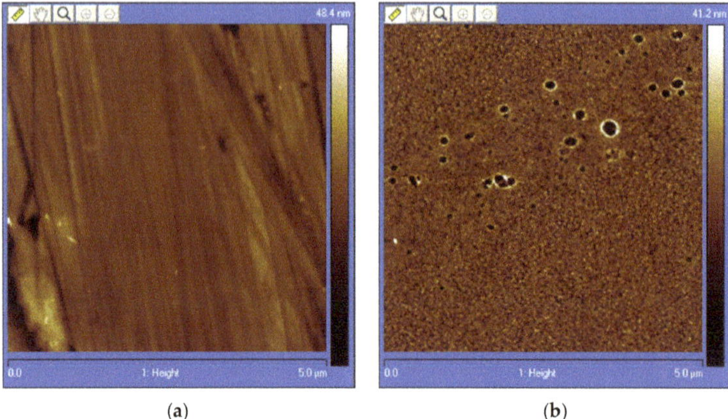

(**a**) (**b**)

Figure 3. AFM images of the (111) face of diamond (**a**) before irradiation and (**b**) after Ar^+ ion irradiation with 30 keV at T = 230 °C.

It should be noted that the described results for the ion-beam treatment of the diamond surface are not common. The high-fluence ion irradiation of most metals and semiconductors leads to a significant development of ordered (in the form of ripples) and stochastic reliefs (in the form of cones, pyramids, ridges, etc.). The ion irradiation of graphite-like carbons also leads to a significant development of the relief. In particular, the irradiation of highly oriented pyrographite, which is the closest in structure to a single crystal of graphite, leads to an extreme development of microrelief in the form of sharp ridges of micron size in a temperature range for the irradiated samples from room temperature to 400 °C [21]. Such a strong effect of the temperature of the irradiated diamond samples on the ion-induced relief was not observed. In particular, the ion-induced relief after irradiation with argon ions with an energy of 30 keV at temperatures of 230 and 400 °C is practically the same [22].

3.3. Raman Spectroscopy

It is known that Raman spectroscopy (RS) is an effective method for studying carbon materials [23]. In RS spectra in the frequency range 1000–1700 cm^{-1}, diamond gives a single narrow peak at 1332 cm^{-1}, and graphite-like materials manifest themselves as characteristic D and G peaks, with frequencies for microcrystalline graphite of 1345 and 1580 cm^{-1}, respectively. The Raman spectra of graphite-like materials can also contain peaks at the frequencies 1200, 1500 and 1620 cm^{-1}. These peaks appear for disordered or nanocrystalline graphite-like materials and are associated with the disorder of the planar structure of crystallites, boundary scattering when crystallites are reduced to nanometer sizes, disorder of translational symmetry, ionic impurities in materials and formation of carbyne chain compounds.

Raman spectroscopy data obtained after the ion irradiation of polycrystalline diamond are shown in Figure 4. Ion-induced graphitization under irradiation with argon ions manifests itself in the RS spectra in the form of broadened D and G peaks. The diamond peak is also observed due to the optical transparency of a thin ~30 nm graphite-like layer. The intensity ratio I_D/I_G and the positions of the D and G peaks correspond to graphite with a high concentration of radiation damage [24].

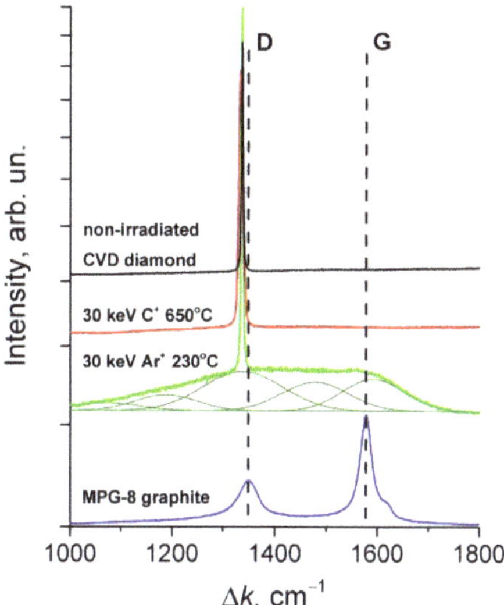

Figure 4. Raman spectra for chemical vapor deposition (CVD) diamond before and after irradiation by 30 keV Ar^+ and C^+ ions. Spectrum for polycrystalline graphite MPG-8 was measured for comparison. The Gaussian analyses of the spectrum are shown by thin lines for diamond irradiated with Ar^+ ions.

The Raman spectrum after irradiation with carbon ions shows that synthesizing a diamond surface with a size of ~0.1 μm practically does not result in a difference from the Raman spectrum of the initial diamond. This indicates a good crystallinity of the synthesized diamond layer upon the implantation of carbon ions. Similar results were obtained for single crystal diamond [12].

3.4. Conductivity of the Ion-Modified Layer

It is known that recrystallization during diamond annealing after ion implantation at temperatures <60 °C occurs only at low irradiation fluences, which do not lead to a vacancy concentration of more than 10^{22} cm^{-3} [25]. The isochronous annealing of the implanted diamond with a final temperature T_a = 1200 °C leads to an almost initial structure. At high irradiation fluences, a disordered irradiated diamond layer undergoes an irreversible transition to graphite-like structures under annealing. The processes of the graphitization of thin layers in diamond under ion irradiation are of interest for the creation of conductors on the surface of diamond and various diamond–graphite heterostructures [26–28].

Due to the dielectric properties of diamond, the analysis of the conductivity of modified layers is a simple and effective method for assessing ion-induced structural changes. This is demonstrated by the dependences of the conductivity $\sigma = (R_s \cdot t)^{-1}$ of the modified diamond layers on the irradiation and annealing temperatures shown in Figure 5. The dotted lines in the figure show the range of conductivity values typical for prepregs of synthetic graphites in the process of graphitization at T_a > 1000 °C. It can be seen that the conductivity of the irradiated-at-30 °C diamond layer, comparable to that of graphitized carbon materials, begins at T_a > 300 °C. The increasing character of the σ (T_a) dependences suggests that an increase in T_a > 700 °C will lead to an even greater increase in the conductivity of the irradiated layer. It should be noted that at $T_a \geq$ 1300 °C, the transformation of diamond into graphite begins. More efficient graphitization occurs with "hot" irradiation (Figure 5) (see [11]). A sharp increase in conductivity begins at a T of about 150 °C; then, after passing the maximum of about 300 °C, a decrease in the conductivity is observed, which is small for irradiation with argon and significant for irradiation with carbon ions.

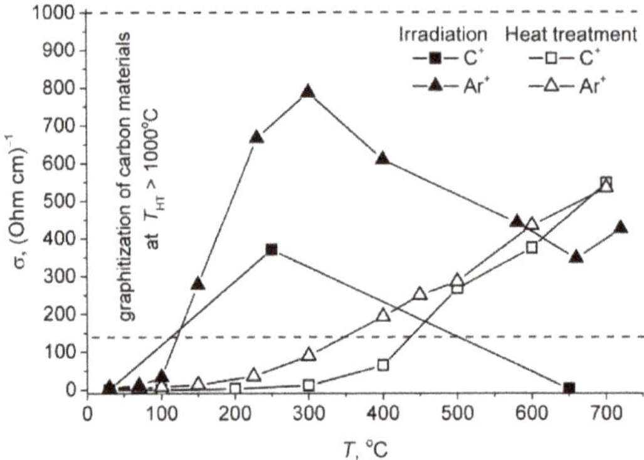

Figure 5. Dependences of the conductivity σ of the irradiated polycrystalline diamond layers on the irradiation temperature T and annealing T_a.

A decrease in conductivity at 400–600 °C is associated with the formation of nanocrystalline graphite, which has a lower conductivity [11,12]. It can be seen from the figure that a graphite-like layer can also be obtained by irradiating diamond with carbon ions at a temperature corresponding to the maximum conductivity of the modified layer. Thus, it can be assumed that irradiation with carbon ions makes it possible to process the diamond either with the formation of a graphite layer or with the growth of the diamond surface. Vacuum arc sources, which have successfully been used for both coating deposition [5–9] and for implantation [29], can be suitable generators of carbon ions.

4. Conclusions

The graphitization and growth of diamond surfaces by the high-fluence (>10^{18} ion/cm^2) irradiation with 30 keV argon and carbon ions of CVD diamond and the (111) face of HPHT diamond were experimentally studied.

To characterize the high-fluence ion irradiation, the depth distributions $\nu(x)$ of the numbers of displacements per atom (dpa) were calculated using the SRIM code. Based on the calculated profiles of $\nu(x)$, the thickness of the modified surface layer was estimated for Ar^+ and C^+ ion irradiation. Unlike for irradiation for argon ions, for carbon ion irradiation, the thickness of the modified layer increased due to the implanted carbon ions.

SEM and AFM showed the removal of traces of mechanical polishing under Ar^+ and C^+ ion irradiation and the appearance of a stochastic microrelief. The roughnesses before and after the ion irradiation of the diamond surface are close—the root-mean-square roughness at a base length of 1 μm is about 20 nm.

Raman spectroscopy before and after the CVD diamond irradiation showed the graphitization of the surface layer when irradiated with argon ions at the temperature of 230 °C and a diamond structure for the synthesized layer when irradiated with carbon ions at the temperature of 650 °C.

The measurement of the conductivity of the modified layer on diamond is a simple and effective method for determining the irradiation or annealing temperature to obtain a graphitized modified layer. Graphite conductivity is obtained at a diamond irradiation temperature of about 300 °C. Annealing a diamond irradiated at room temperature can also lead to a graphitized layer. The measurement of the electrical conductivity of irradiated CVD diamonds showed that the graphitization of the surface layer can also be obtained by irradiation with carbon ions.

Author Contributions: Conceptualization, S.N.G.; methodology, I.V.S.; validation, E.S.M.; investigation, A.M.B., V.A.K., M.A.O. and E.S.M.; writing—original draft preparation, A.M.B.; writing—review and editing, E.S.M. and M.A.O.; project administration, S.N.G. All authors have read and agreed to the published version of the manuscript.

Funding: This research was funded by the Ministry of Science and Higher Education of the Russian Federation, grant number 0707-2020-0025.

Conflicts of Interest: The authors declare no conflict of interest.

Abbreviations

Symbol	Description
AFM	atomic force microscopy
CVD	chemical vapor deposition
G	graphitic carbon band
HPHT	high-pressure, high-temperature
D	disordered carbon band
ID/IG	ratio of the intensities (height) of D and G bands
SEM	scanning electron microscopy

References

1. Grigoriev, S.; Melnik, Y.; Metel, A. Broad fast neutral molecule beam sources for industrial-scale beam-assisted deposition. *Surf. Coat. Technol.* **2002**, *156*, 44–49. [CrossRef]
2. Colligon, J.S.; Vishnyakov, V.; Valizadeh, R.; Donnelly, S.T.; Kumashiro, S. Study of nanocrystalline TiN/Si3N4 thin films deposited using a dual ion beam method. *Thin Solid Film.* **2005**, *485*, 148–154. [CrossRef]
3. Metel, A.; Bolbukov, V.; Volosova, M.; Grigoriev, S.; Melnik, Y. Source of metal atoms and fast gas molecules for coating deposition on complex shaped dielectric products. *Surf. Coat. Technol.* **2013**, *225*, 34–39. [CrossRef]
4. Metel, A.; Bolbukov, V.; Volosova, M.; Grigoriev, S.; Melnik, Y. Equipment for deposition of thin metallic films bombarded by fast argon atoms. *Instruments Exp. Tech.* **2014**, *57*, 345–351. [CrossRef]
5. Sobol', O.V.; Andreev, A.A.; Grigoriev, S.N.; Gorban', V.F.; Volosova, M.A.; Aleshin, S.V.; Stolbovoy, V.A. Physical characteristics, structure and stress state of vacuum-arc tin coating, deposition on the substrate when applying high-voltage pulse during the deposition. *Probl. At. Sci. Technol.* **2011**, *4*, 174–177.
6. Grigoriev, S.N.; Sobol, O.V.; Beresnev, V.M.; Serdyuk, I.V.; Pogrebnyak, A.D.; Kolesnikov, D.A.; Nemchenko, U.S. Tribological characteristics of (TiZrHfVNbTa)N coatings applied using the vacuum arc deposition method. *J. Frict. Wear* **2014**, *35*, 359–364. [CrossRef]
7. Kuzin, V.V.; Grigor'Ev, S.N.; Volosova, M.A. Effect of a tic coating on the stress-strain state of a plate of a high-density nitride ceramic under nonsteady thermoelastic conditions. *Refract. Ind. Ceram.* **2014**, *54*, 376–380. [CrossRef]
8. Volosova, M.A.; Grigor'ev, S.N.; Kuzin, V.V. Effect of titanium nitride coating on stress structural inhomogeneity in oxide-carbide ceramic. Part 4. Action of heat flow. *Refract. Ind. Ceram.* **2015**, *56*, 91–96. [CrossRef]
9. Kuzin, V.V.; Grigoriev, S.N.; Fedorov, M.Y. Role of the thermal factor in the wear mechanism of ceramic tools. Part 2: Microlevel. *J. Frict. Wear* **2015**, *36*, 40–44. [CrossRef]
10. Nelson, R.S.; Hudson, J.A.; Mazey, D.J.; Piller, R.C. Diamond synthesis: Internal growth during C+ ion implantation. *Proc. R. Soc. London Ser. A Math. Phys. Sci.* **1983**, *386*, 211–222. [CrossRef]
11. Borisov, A.M.; Kazakov, V.A.; Mashkova, E.S.; Ovchinnikov, M.A. The regularities of high-fluence ion-induced graphitization of diamond. *Vacuum* **2018**, *148*, 195–200. [CrossRef]
12. Borisov, A.M.; Kazakov, V.A.; Mashkova, E.S.; Ovchinnikov, M.A.; Pitirimova, E.A. On the dynamic annealing of ion-induced radiation damage in diamond under irradiation at elevated temperatures. *J. Surf. Investig. X-ray Synchrotron Neutron Tech.* **2019**, *13*, 306–313. [CrossRef]
13. Borisov, A.M.; Kazakov, V.A.; Mashkova, E.S.; Ovchinnikov, M.A.; Palyanov, Y.N.; Popov, V.P.; Shmytkova, E.A. Optical and electrical properties of synthetic single-crystal diamond under high-fluence ion irradiation. *J. Surf. Investig. X-ray Synchrotron Neutron Tech.* **2017**, *11*, 619–624. [CrossRef]
14. Borisov, A.M.; Kazakov, V.A.; Mashkova, E.S.; Ovchinnikov, M.A.; Shemukhin, A.A.; Sigalaev, S.K. The conductivity of high-fluence noble gas ion irradiated CVD polycrystalline diamond. *Nucl. Instrum. Methods Phys. Res. B.* **2017**, *406*, 676–679. [CrossRef]

15. Andrianova, N.N.; Borisov, A.M.; Kazakov, V.A.; Mashkova, E.S.; Popov, V.P.; Palyanov, Y.N.; Risakhanov, R.N.; Sigalaev, S.K. High-fluence ion-beam modification of a diamond surface at high temperature. *J. Surf. Investig. X-ray Synchrotron Neutron Tech.* **2015**, *9*, 346–349. [CrossRef]
16. Andrianova, N.N.; Borisov, A.M.; Kazakov, V.A.; Mashkova, E.S.; Palyanov, Y.N.; Pitirimova, E.A.; Popov, V.P.; Rizakhanov, R.N.; Sigalaev, S.K. Graphitization of a diamond surface upon high-dose ion bombardment. *Bull. Russ. Acad. Sci. Phys.* **2016**, *80*, 156–160. [CrossRef]
17. Mashkova, E.S.; Molchanov, V.A. *Medium-Energy Ion Reflection from Solids*; Elsevier: Amsterdam, The Netherlands, 1985.
18. Ziegler, J.F.; Biersack, J.P. SRIM. 2013. Available online: www.srim.org (accessed on 18 August 2019).
19. Ehrhart, P.; Schilling, W.; Ullmaier, H. Radiation damage in crystals. *Encycl. Appl. Phys.* **1996**, *15*, 429–457.
20. Roth, J. Chemical sputtering. In *Sputtering by Particle Bombardment II*; Behrisch, R., Ed.; Springer: Berline/Heidelberge, Germany; New York, NY, USA; Tokyo, Japan, 1983; pp. 91–141.
21. Andrianova, N.N.; Borisov, A.M.; Mashkova, E.S.; Shemukhin, A.A.; Shulga, V.I.; Virgiliev, Y.S. Relief evolution of HOPG under high-fluence 30 keV argon ion irradiation. *Nucl. Instrum. Methods Phys. Res. B* **2015**, *354*, 146–150. [CrossRef]
22. Anikin, V.A.; Borisov, A.M.; Kazakov, V.A.; Mashkova, E.S.; Palyanov, Y.N.; Popov, V.P.; Shmytkova, E.A.; Sigalaev, S.K. Diamond single crystal-surface modification under high- fluence ion irradiation. *J. Phys. Conf. Ser.* **2016**, *747*, 12025. [CrossRef]
23. Ferrari, A.C.; Robertson, J. Resonant Raman spectroscopy of disordered, amorphous, and diamondlike carbon. *Phys. Rev. B* **2001**, *64*, 75414. [CrossRef]
24. Niwase, K. Raman spectroscopy for quantitative analysis of point defects and defect clusters in irradiated graphite. *Int. J. Spectrosc.* **2012**, *2012*, 197609. [CrossRef]
25. Kalish, R. Doping diamond by ion-implantation. *Semicond. Semimet.* **2003**, *76*, 145–181. [CrossRef]
26. Popov, V.P.; Safronov, L.N.; Naumova, O.V.; Nikolaev, D.V.; Kupriyanov, I.N.; Palyanov, Y.N. Conductive layers in diamond formed by hydrogen ion implantation and annealing. *Nucl. Instrum. Methods Phys. Res. B.* **2012**, *282*, 100–107. [CrossRef]
27. Philipp, P.; Bischoff, L.; Treske, U.; Schmidt, B.; Fiedler, J.; Hübner, R.; Klein, F.; Koitzsch, A.; Mühl, T. The origin of conductivity in ion-irradiated diamond-like carbon—Phase transformation and atomic ordering. *Carbon* **2014**, *80*, 677–690. [CrossRef]
28. Rubanov, S.; Suvorova, A.; Popov, V.P.; Kalinin, A.A.; Pal'yanov, Y.N. Fabrication of graphitic layers in diamond using FIB implantation and high pressure high temperature annealing. *Diam. Relat. Mater.* **2016**, *63*, 143–147. [CrossRef]
29. Brown, I.G. *The Physics and Technology of Ion Sources*; Lawrence Berkeley Laboratory, University of California: Berkeley, CA, USA, 1989.

Article

Investigation of the Properties of Ti-TiN-(Ti,Cr,Mo,Al)N Multilayered Composite Coating with Wear-Resistant Layer of Nanolayer Structure

Sergey Grigoriev [1], Alexey Vereschaka [2,*], Filipp Milovich [3], Nikolay Sitnikov [4], Nikolay Andreev [3], Jury Bublikov [2], Catherine Sotova [1], Gaik Oganian [1] and Ilya Sadov [1]

1 VTO Department, Moscow State Technological University STANKIN, 127994 Moscow, Russia; s.grigoriev@stankin.ru (S.G.); e.sotova@stankin.ru (C.S.); svartrans88@yandex.ru (G.O.); sadvilj@yandex.ru (I.S.)

2 Institute of Design and Technological Informatics of the Russian Academy of Sciences (IDTI RAS), 127994 Moscow, Russia; yubu@rambler.ru

3 Materials Science and Metallurgy Shared Use Research and Development Center, National University of Science and Technology MISiS, 119049 Moscow, Russia; filippmilovich@mail.ru (F.M.); andreevn.misa@gmail.com (N.A.)

4 Department of Solid State Physics and Nanosystems, National Research Nuclear University MEPhI, 115409 Moscow, Russia; sitnikov_nikolay@mail.ru

* Correspondence: dr.a.veres@yandex.ru

Received: 9 November 2020; Accepted: 13 December 2020; Published: 16 December 2020

Abstract: The article describes the results of an investigation focused on the properties of the Ti-TiN-(Ti,Cr,Mo,Al)N multilayered composite coating with a wear-resistant layer of nanolayer structure. A transmission electron microscope was used to study the coating structure. The examination of the phase composition using selected area diffraction electron pattern has detected the presence of two phases, including c-(Ti,Cr,Mo,Al)N and h-AlN. The cutting properties of the tool with the coating under consideration were studied during the turning of AISI 1045 steel at v_c = 300 m/min, f = 0.25 mm/rev, and a_p = 1.0 mm. After 16 min of cutting, the wear rate for the tool with the Ti-TiN-(Ti,Cr,Mo,Al)N coating was 1.9 times lower compared to the wear rate for the tool with the (Ti,Al)N commercial monolithic coating. As a result of the investigation focused on the fracture pattern on the coating during the cutting, the brittle nature of the fracture has been detected with a noticeable effect of adhesive fatigue mechanisms.

Keywords: nanolayered coating; microparticles; crack formation; tool wear

1. Introduction

Modified coatings developed for various purposes, particularly coatings for metal-cutting tools, are being actively implemented in various areas of manufacturing activity. At the same time, the trends of modern manufacturing suggest toughening requirements for coatings. In particular, an increase in the cutting speed leads to an increase in temperature in the cutting zone, and, accordingly, the crucial feature of the coatings is heat resistance [1–5]. With an increase in temperature, oxidation and diffusion processes become more active and intensify tool wear [6–10]. Accordingly, one of the significant factors is the ability of a coated tool to resist the oxidation and diffusion wear. The coatings of traditional composition, such as TiN, TiC, ZrN, CrN, or (Ti,Al)N, can no longer meet the requirements of the modern manufacturing, and new coatings with enhanced properties are required. One of the ways to develop such coatings is to use a multicomponent composition and nanolayer architecture [11–15].

In particular, the introduction of such elements as Cr and Mo into the coating composition increases its resistance to heat and oxidation and diffusion effects. The coatings based on the (Ti,Al)N system are used rather widely, like nitride coatings containing Cr and Mo. However, fewer studies have considered the coatings based on the multicomponent nitrides, including a complex of these elements.

Regent and Musil [16] investigated the (Ti,Mo)N and (Ti,Cr)N coatings. The hardness of the (Ti,Mo)N coating was 38–40 GPa with the content of molybdenum (Mo) up to 10 at.%), and the δ-TiN phase (111) with miscible Mo dominated in the coating structure. The study of the (Ti,Cr)N coating detected the presence of the solid solution of (Ti,Cr)N (200) and Cr_2N (111), while the phases of TiN (111), (Ti,Mo)N (200), Mo_2N (111), and pure Mo were detected in the (Ti,Mo)N coating [17]. Moreover, the authors suggest that the phase of pure molybdenum in the (Ti,Mo)N coating can significantly reduce the coefficient of friction (COF) in comparison with the (Ti,Cr)N and TiN coatings [17]. The study focused on the properties of the (Ti,Cr)N coating also found that the introduction of chromium (Cr) into the composition of the TiN coating is able to significantly increase the oxidation resistance due to the formation of dense chromium oxide (Cr_2O_3) on the coating surface [18,19]. This oxide is characterised by high heat resistance and does not transform into chromium trioxide (CrO_3) until the temperature reaches 1100 °C [20].

During the studies of the (Cr,Mo)N coating, the phases of fcc-CrN (111) and (200) and amorphous/nanocrystalline Mo_2N were found, but no expected substitutional solid solution of (Cr,Mo)N was detected [21]. Experiments carried out by Kim et al. [22] revealed the substitutional solid solution (Cr,Mo)N with the Mo content of less than 30.4 at.%, and the maximum hardness of the coating (34 GPa) was achieved with the Mo content of 21 at.%. The studies also note that the phases formed in this coating depend on the nitrogen-to-argon ratio in a chamber. With an increase in the nitrogen content, a single fcc solid solution (Cr,Mo)N phase forms instead of the mixture of bcc hexagonal (Cr,Mo)N phases [23]. During the studies of the coating with alternating CrN/Mo_2N layers, the formation of MoO_3 and Cr_2O_3 oxides was detected upon heating to 400–500 °C, and it was molybdenum trioxide (MoO_3) which had a crucial influence on the reduction of friction [24]. In [25–27], the investigation of the (Ti,Al,Mo)N coating revealed the presence of TiN and Mo phases (studied by the X-ray diffraction (XRD) method). A study by the XRD method detected no phase of γ-Mo_2N which can be explained by the fact that in the X-ray diffraction patterns, the reflections of this phase and the phase of MoN coincide with TiN. However, the phase of γ-Mo_2N was detected by the X-ray photoelectron spectroscopy (XPS) method. The hardness of the (Ti,Al,Mo)N coating reaches 40 GPa [28]. The introduction of nickel (Ni) in the composition of the (Ti,Al,Mo)N coating reduced the grain sizes from 40–50 to 10–12 nm. Meanwhile, at temperatures exceeding 500 °C, the Ni-containing coating wore out more intensively, and in [29], that fact was associated with the formation of the $TiNiO_3$ oxide.

In their previous articles, authors of this paper studied the properties of the coatings with wear-resistant layers, including (Ti,Cr,Al)N [30–32], (Zr,Nb,Cr,Al)N [30,32], (Zr,Nb,Ti)N [32], (Zr,Cr,Al)N [32], (Nb,Zr,Ti,Al)N [32], and (Ti,Cr,Al,Si)N [33]. These studies show that the coatings with multi-element compositions often have better performance properties compared to binary and ternary systems.

The studies were focused on the Ti-TiN-(Ti,Cr,Mo,Al)N coating with three-layer architectures according to the recommendations described in in our past works [34,35]. It can be assumed that coating Ti-TiN-(Ti,Cr,Mo,Al)N will have good tribological properties due to the formation of MoO_3 and Cr_2O_3 oxides with high hardness and wear resistance [18–24,29].

2. Materials and Methods

The VIT-2 vacuum plasma unit (IDTI RAS—MSTU STANKIN, Moscow, Russia) [34,36] was used to deposit coatings with the filtered cathodic vacuum arc deposition (FCVAD) technology [34–41]. The VIT-2 unit contained two arc evaporators with a pulsed magnetic field and one arc evaporator with filtering of the vapor-ion flow. Moreover, the complex contains a source of pulsed bias voltage supply to a substrate, a dynamic gas mixing system for reaction gases, systems for automatic chamber pressure control and for process temperature control, and a system for the stepless adjustment of planetary gear rotation.

The VIT-2 unit has three cathode systems, in which cathodes of Al 99.1 at.%, Ti 99.9 at.% and also of Cr-Mo (50–50 at.%) are installed. Cylindrical cathodes with the diameter of 80 mm were used. Three cathodes of Ti-Al (80–20 at.%) were used to deposit the (Ti,Al)N commercial coating.

Coating deposition rate is 100 nm per minute.

The parameters for the process of coating deposition are presented in Table 1.

Table 1. Parameters of stages of the technological process of the deposition of coatings.

Process	p_N (Pa)	U (V)	I_{Ti} (A)	I_{Al} (A)	$I_{Ti\text{-}Al}$ (A)	$I_{Cr\text{-}Mo}$ (A)
Pumping and heating of vacuum chamber	0.06	+20	75	120	75	–
Heating and cleaning of products with gaseous plasma	2.0	100DC/900 AC f = 10 kHz, 2:1	85	80	85	–
Deposition of coating	0.42	−800 DC	75	160	75	120
Cooling of products	0.06	–	–	–	–	–

Note: I_{Ti} = current of titanium cathode, $I_{Ti\text{-}Al}$ = current of Ti-Al cathode, I_{Al} = current of aluminum cathode, $I_{Cr\text{-}Mo}$ = current of Cr-Mo cathode, p_N = gas pressure in chamber, U = voltage on substrate.

The nanoindentation technique and an Instron Wilson Hardness Group Tukon tester at the load of 0.01 N were used to determine the coating microhardness.

During the turning of workpieces made of AISI 1045 steel, a CU 500 MRD lathe (ZMM Sliven, Sliven, Bulgaria) with a ZMM CU500 MRD variable-speed drive (ZMM Bulgaria, Sofia, Bulgaria) was applied. No coolants or lubricants were used during the process of cutting. SNUN ISO 1832:2012 carbide inserts played a role of substrates, with the parameters as follows: $\gamma = -7°$, $\alpha = 7°$, $\lambda = 0$, r = 0.4 mm; cutting mode: f = 0.25 rpm, a_p = 1.0 mm, and v_c = 300 m/min. Four experiments were conducted for each coating, and the obtained values of flank wear were processed to get the polynomial functions exhibited on the curve. The limit wear criterion was assumed as flank wear rate VB_{max} = 0.4 mm. Five tests of cutting properties were carried out, after which the information was statistically processed. Average values were determined for five experiments, these average values were used to plot the graph. Polynomial dependencies were obtained, on the basis of which graphs of the dependence of the flank wear on the cutting time were plotted.

3. Results and Discussion

3.1. Study of the Chemical Composition and Nanostructure of the Ti-TiN-(Ti,Cr,Mo,Al)N Coating

According to the results of 20 conducted measurements, the average hardness of the coating was 42 ± 1.3 GPa, which is fairly high for nitride coatings. The coating structure includes an adhesion layer of Ti with the thickness of about 50 nm, a transition layer of TiN with the thickness of about 600 nm, and a wear-resistant layer of (Ti,Cr,Mo,Al)N with the thickness of about 2700 nm (Figure 1). The thickness of the functional layers of the coating was selected based on their optimal ratio [32,35]. The wear-resistant layer of (Ti,Cr,Mo,Al)N is formed by a 22-nanolayer period with λ of about 120 nm [11,33].

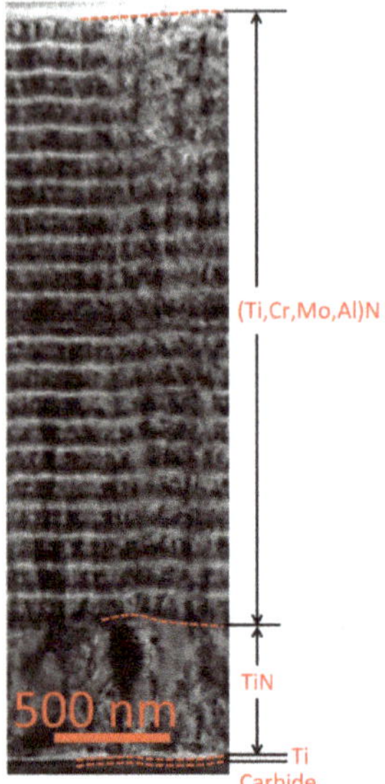

Figure 1. General structure of the Ti-TiN-(Ti,Cr,Mo,Al)N coating (TEM).

The nanostructures of the studied coating Ti-TiN-(Ti,Cr,Mo,Al)N in comparison with the (Ti,Al)N monolayer coating are presented in Figure 2.

Figure 2a illustrates that the nanostructure of Ti-TiN-(Ti,Cr,Mo,Al)N coating includes nanolayers with the high content of Al (lighter bands) and nanolayers with the high content of Cr-Mo and Ti (darker bands). For Ti-TiN-(Ti,Cr,Mo,Al)N coating, the value of nanolayer period λ [11,33] is about 120 nm, and the thicknesses of nanolayers are within a range of 1–8 nm. The results of the studies of the coatings phase compositions using the Selected Area Electron Diffraction (SAED) method are presented in Figure 2c,d.

The analysis of the SAED patterns for coatings Ti-TiN-(Ti,Cr,Mo,Al)N (Figure 2c) detected the presence of two phases. The analysis also found the main cubic phase (Ti,Nb,Zr,Al)N with Fm3m space group. Weak reflections with P6.3mc space group belong to the h-AlN phase. Rings of the h-AlN phase are barely noticeable, and that fact may indicate an extremely insignificant volume of the phase.

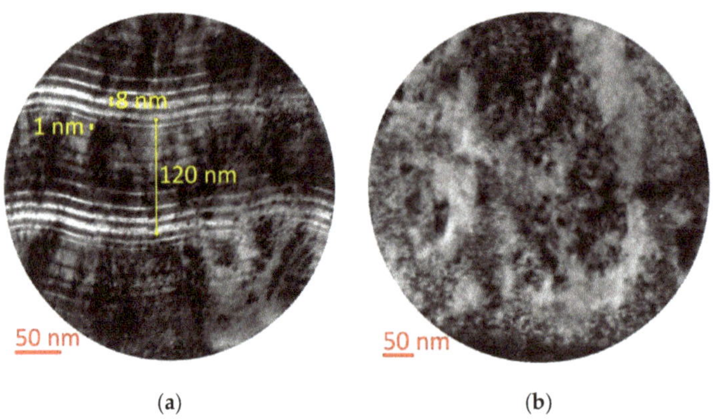

(a) (b)

Figure 2. *Cont.*

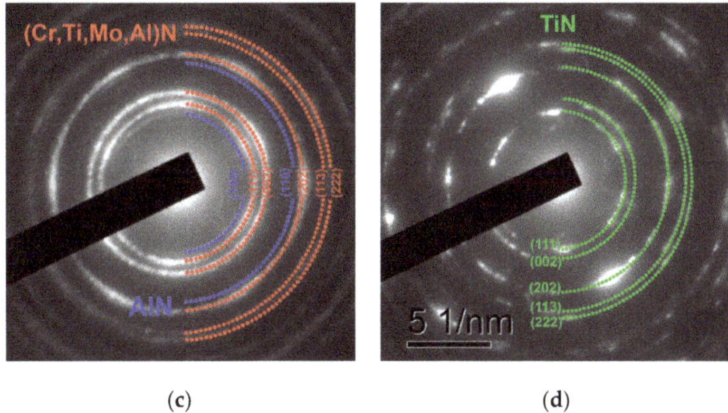

(c) (d)

Figure 2. (**a**,**b**) TEM micrograph of coating and (**c**,**d**) SAED patterns of the coatings: (**a**,**c**) Ti-TiN-(Ti,Cr,Mo,Al)N and (**b**,**d**) (Ti,Al)N monolayer coating (TEM).

The studies of the chemical composition of the coatings (Figure 3a) found the average contents of elements as follows in coating Ti-TiN-(Ti,Cr,Mo,Al)N:

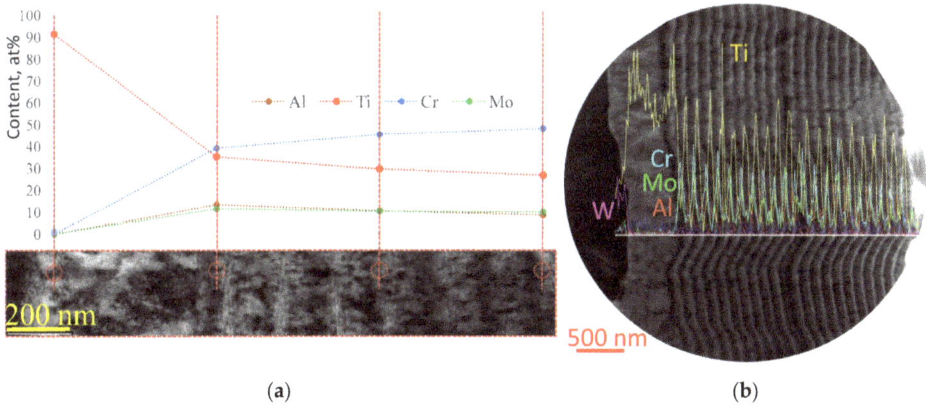

(a) (b)

Figure 3. Distribution of chemical elements over the thickness (**a**) and change in the content of different elements in nanolayers (**b**) of coating Ti-TiN-(Ti,Cr,Mo,Al)N.

Ti—22 at.%, Cr—38 at.%, Al—11 at.%, Mo—10 at.%.

The study of the nature of the distribution of elements in the nanolayer periods (Figure 3b) finds that the content of each element changes significantly within a nanolayer period. In particular, the content of Ti changes from 7 to 60 at.%, and the content of Al—from 3 to 27 at.%. The above ensures a smooth, gradient transition from harder and more wear-resistant layers with the high Al content to more ductile layers with the low Al content.

Let us consider the influence of the nanolayer structure of the coating on its crystalline structure. Earlier, it has been found that the nanolayer structure affects the grain sizes by reducing them [11,33]. At the same time, the grain size of the coating is not always limited by the boundaries of a nanolayer or a nanolayer period [11]. The Ti-TiN-(Ti,Cr,Mo,Al)N coating under study demonstrates columnar crystals with sizes noticeably larger than the value of nanolayer period (Figure 4a), and the grain structure of the coating can be seen more clearly in the reverse contrast image (Figure 4b). The nanolayer structure of this coating does not stop the growth of crystals (Figure 4c). However, as noted earlier [11,33], the presence of the nanolayer structure allows the formation of crystals with significantly smaller sizes than crystals in coatings with monolayer structures. A comparison of SAED patterns on the Ti-TiN-(Ti,Cr,Mo,Al)N nanolayer coating (Figure 2c) and on the (Ti,Al)N monolithic coating (Figure 2d) demonstrates the significantly smaller size of crystals in the Ti-TiN-(Ti,Cr,Mo,Al)N coating.

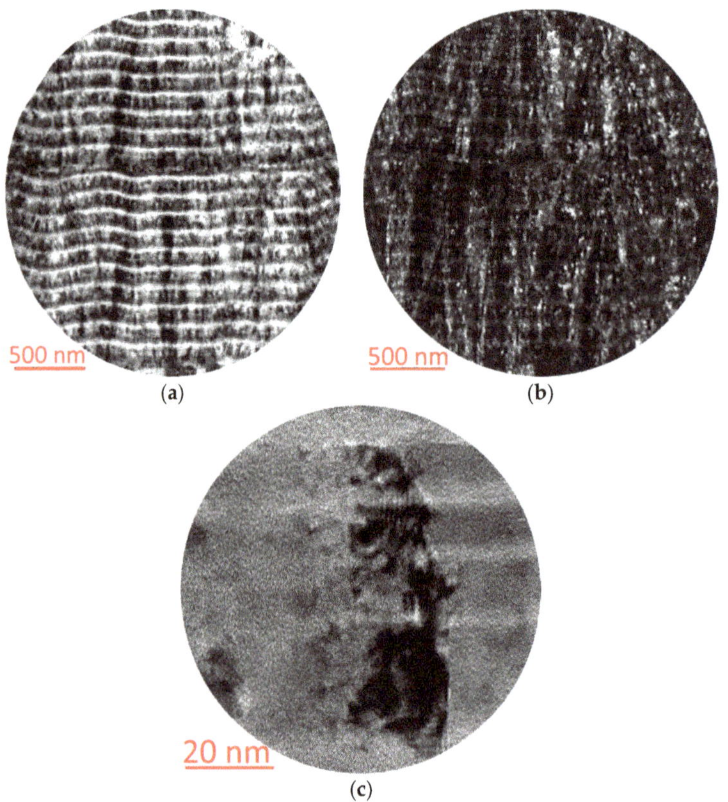

Figure 4. TEM micrograph of Ti-TiN-(Ti,Cr,Mo,Al)N coating and study of the influence of the nanolayer structure of the coating on its crystalline structure. (**a**) nanostructure (**b**) nanostructure in reverse contrast image (**c**) individual grain in a nanostructure.

Figure 5 presents the high-resolution TEM images of coating Ti-TiN-(Ti,Cr,Mo,Al)N, with noticeable crystals with sizes of 5–15 nm. The analysis of interplanar spacing revealed the presence of two phases of h-AlN and c-(Cr,Ti,Mo,Al)N. The obtained results are in line with the SAED pattern presented in Figure 2c.

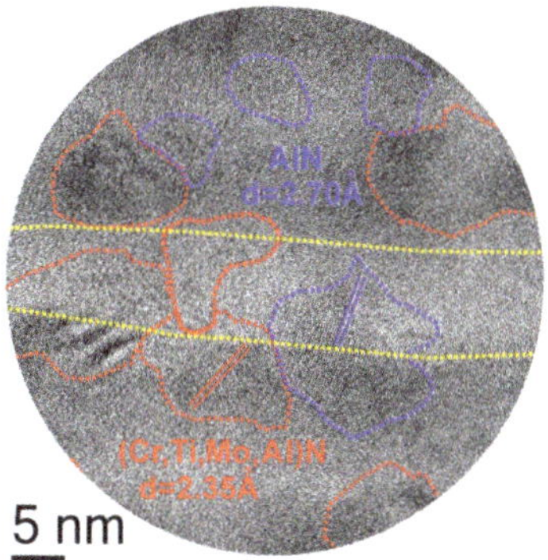

Figure 5. Crystalline structure of coating Ti-TiN-(Ti,Cr,Mo,Al)N (HR TEM).

3.2. Study of the Cutting Properties and the Wear Pattern on Tools with the Ti-TiN-(Ti,Cr,Mo,Al)N Coating

The studies of the cutting properties of the carbide tools found that the use of Ti-TiN-(Ti,Cr,Mo,Al)N coating can significantly reduce the flank wear compared to the tools with the (Ti,Al)N commercial coating. Figure 6 illustrates that after 7 min of operation, the wear of the carbide inserts with the (Ti,Al)N coating increases sharply, while the inserts with Ti-TiN-(Ti,Cr,Mo,Al)N coating demonstrate much lower wear, and the wear rate decreases.

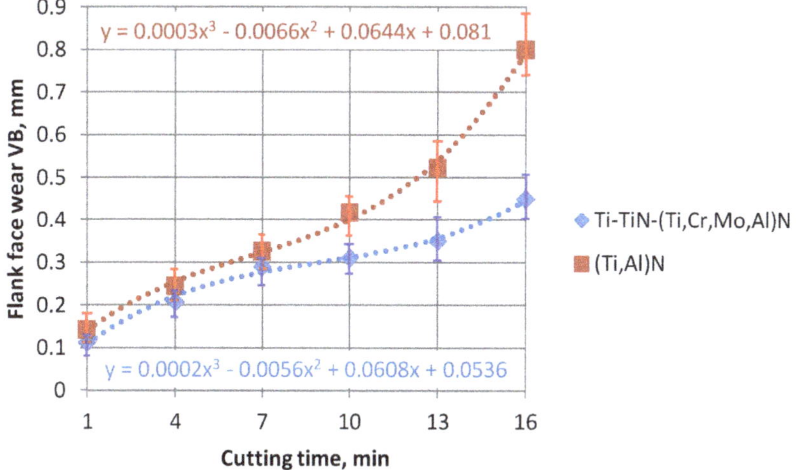

Figure 6. Relation between cutting time and flank wear (VB) on coated carbide tools during the turning of AISI 1045 steel: v_c = 300 m/min, f = 0.25 mm/rev, a_p = 1.0 mm.

The good wear resistance of the Ti-TiN-(Ti,Cr,Mo,Al)N coating can be explained by the formation of tribological oxide films of MoO_3 and Cr_2O_3, which favourably transform the cutting conditions [42–47].

The investigation of wear areas on the rake face of the carbide inserts after 16 min of operation (Figure 7) found that the tool with coating (Ti,Al)N demonstrated the higher rake wear, which manifested itself in the formation of a crater and a notch wear compared to the tool with Ti-TiN-(Ti,Cr,Mo,Al)N coating.

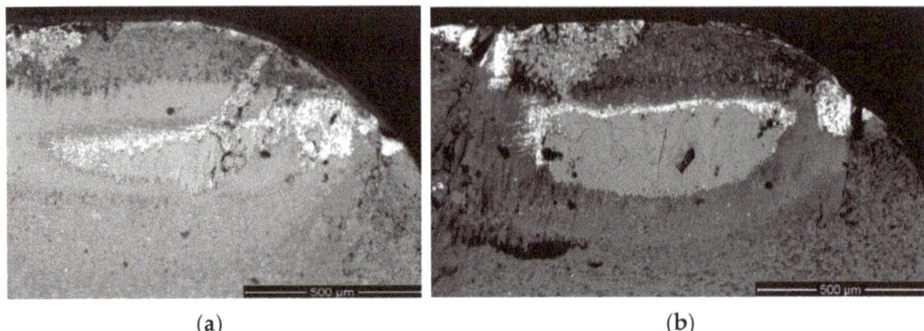

Figure 7. Contact pads on the rake faces of the carbide tools with coatings (**a**) Ti-TiN-(Ti,Cr,Mo,Al)N and (**b**) (Ti,Al)N after 16 min of cutting.

Figures 8 and 9 exhibit the results of the investigation of the fracture pattern on the rake and flank faces of the carbide tools.

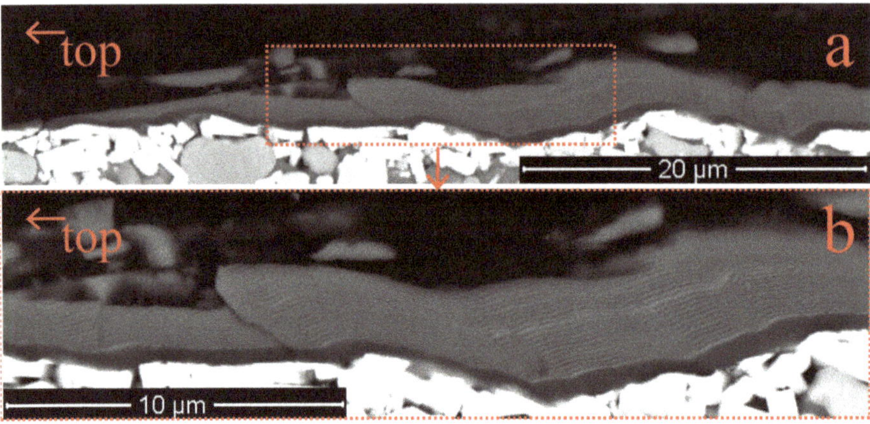

Figure 8. Fracture patterns for Ti-TiN-(Ti,Cr,Mo,Al)N coating (**a**,**b**) on the flank faces of the tools (90° counterclockwise, top left (SEM).

The study of the fracture pattern finds that, given the noticeable microdeformations in the structure of the carbide substrate, Ti-TiN-(Ti,Cr,Mo,Al)N coatings are characterised by sufficient ductility and resistance to brittle fracture.

The study of the fracture patterns on the Ti-TiN-(Ti,Cr,Mo,Al)N coatings on the rake faces of the tools (Figure 9) finds that the fracture process is accompanied by active cracking. The Ti-TiN-(Ti,Cr,Mo,Al)N coating exhibits both inclined and transverse cracks (Figure 9). On the Ti-TiN-(Ti,Cr,Mo,Al)N coating, the brittle fracture accompanied by chipping of some fragments is typical (Figure 9b–e). Figure 9a illustrates that the wear surface of the coating on the boundary of its fracture is quite smooth and there are signs of local chipping of coating fragments.

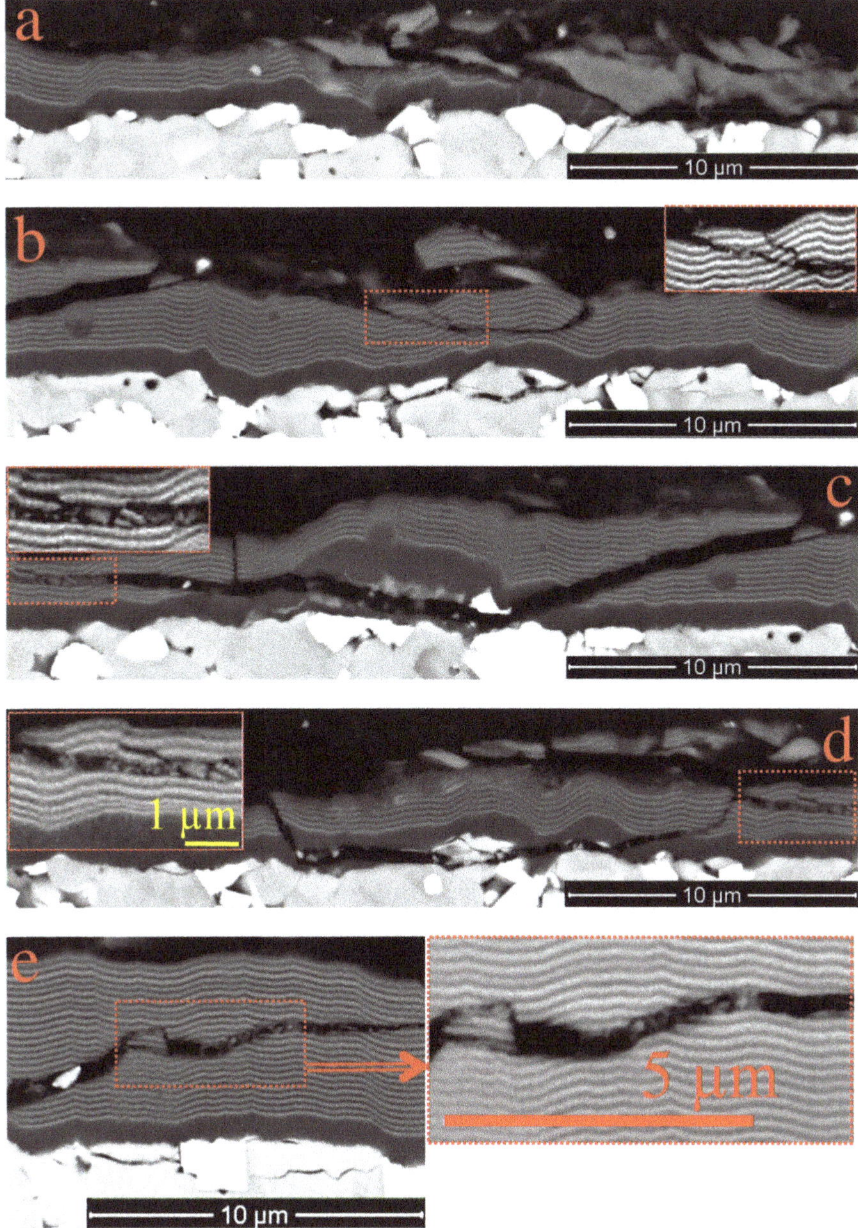

Figure 9. Fracture pattern for coating Ti-TiN-(Ti,Cr,Mo,Al)N on the rake face of the carbide tool (SEM), (**a**) the nature of destruction of the coating at the boundary of the crater of wear, (**b–e**) the nature of cracking in the structure of the coating.

After cutting, delaminations and longitudinal cracks also occur in the structure of the coating on the cutting tool, and the study of them on the rake face of the tool is depicted in Figure 10. The upper part of the image demonstrates a delamination between nanolayers (see Box A for a larger scale).

The delamination occurs not only along the border of two nanolayer periods, but also along the border of individual nanolayers. At the same time, the delamination does not transform into a longitudinal crack, that it, does not cut the nanolayers. As noted earlier [8,30,31,35], such delaminations can reduce the level of internal stresses and can thus play some positive role by slowing down the process of coating fracture. The lower part of the image depicts a delamination transforming into a transverse crack (see Box B for details). The formation of such cracks may be associated with the effect of residual longitudinal compressive stresses. Another reason for the formation of delaminations and longitudinal cracks can be transverse cyclic tensile stresses formed under the influence of adhesive fatigue wear processes arising during cutting [8,30,31,35,48].

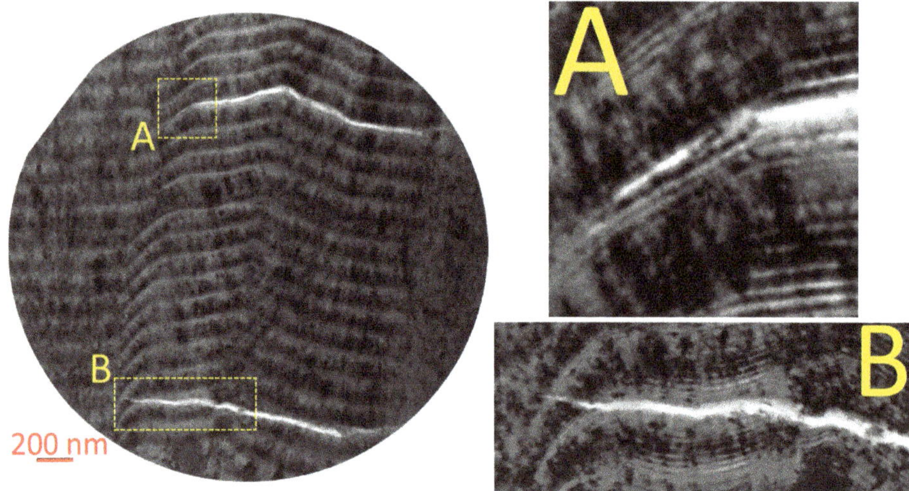

Figure 10. Pattern of the formation of longitudinal cracks and delaminations in the Ti-TiN-(Ti,Cr,Mo,Al)N coating on the rake face of the tool (TEM). (**A**) inter nanolayer delamination (**B**) nanolayer cut by a crack.

4. Conclusions

The properties of the Ti-TiN-(Ti,Cr,Mo,Al)N multilayered composite coating with a wear-resistant layer of nanolayer structure were studied. The conducted studies have found the following:

(1) The average coating hardness was 42 ± 1.3 GPa.

(2) The value of nanolayer period λ is about 120 nm, and the thicknesses of nanolayers are within the range of 1–8 nm.

(3) The studies of the phase composition of the coating have revealed the presence of a main cubic phase of (Ti,Nb,Zr,Al)N with Fm3m space group. Weak reflections with P6.3mc space group belong to the h-AlN phase.

(4) It is found that the grain sizes in the coating under study can significantly exceed the thicknesses of its nanolayers and the value of nanolayer period λ. While nano-sized grains (5–15 nm) are detected, there are also columnar crystals 1–2 μm long.

(5) After 16 min of cutting, the wear rate for the tool with the Ti-TiN-(Ti,Cr,Mo,Al)N coating was 1.9 times lower compared to the wear rate for a tool with the (Ti,Al)N commercial monolithic coating.

(6) The cracking patterns in the coating on the rake and flank faces of the tool demonstrate a considerably brittle nature of fracture, accompanied by the chipping of separate fragments of the coating. At the same time, delaminations between nanolayers of the coating were also detected.

Author Contributions: Conceptualization, A.V. and S.G.; methodology, S.G. and A.V.; formal analysis, F.M., C.S. and I.S.; investigation, F.M., N.S., J.B., N.A., G.O. and I.S.; resources, S.G.; data curation, A.V. and G.O.; writing—original draft preparation, A.V.; writing—review and editing, A.V.; project administration S.G, A.V. and C.S.; funding acquisition, S.G. All authors have read and agreed to the published version of the manuscript.

Funding: This research was funded by Ministry of Science and Higher Education of the Russian Federation, Grant No. 0707-2020-0025.

Conflicts of Interest: The authors declare no conflict of interest.

References

1. Aydin, M.; Karakuzu, C.; Uçar, M.; Cengiz, A.; Çavuşlu, M.A. Prediction of surface roughness and cutting zone temperature in dry turning processes of AISI304 stainless steel using ANFIS with PSO learning. *Int. J. Adv. Manuf. Technol.* **2013**, *67*, 957–967. [CrossRef]
2. Stephenson, D.A. Inverse method for investigating deformation zone temperatures in metal cutting. *J. Eng. Ind. Trans. ASME* **1991**, *113*, 129–136. [CrossRef]
3. Shi, B.; Attia, H.; Tounsi, N. Identification of material constitutive laws for machining—Part I: An analytical model describing the stress, strain, strain rate, and temperature fields in the primary shear zone in orthogonal metal cutting. *J. Manuf. Sci. Eng.* **2010**, *132*, 051008. [CrossRef]
4. Grzesik, W.; Bartoszuk, M.; Nieslony, P. Finite element modelling of temperature distribution in the cutting zone in turning processes with differently coated tools. *J. Mater. Process. Technol.* **2005**, *164–165*, 1204–1211. [CrossRef]
5. Kuzin, V.V.; Grigoriev, S.N.; Fedorov, M.Y. Role of the thermal factor in the wear mechanism of ceramic tools. Part 2: Microlevel. *J. Frict. Wear* **2015**, *36*, 40–44. [CrossRef]
6. Zheng, G.; Xu, R.; Cheng, X.; Zhao, G.; Li, L.; Zhao, J. Effect of cutting parameters on wear behavior of coated tool and surface roughness in high-speed turning of 300M. *Measurement* **2018**, *125*, 99–108. [CrossRef]
7. Vennemann, A.; Stock, H.-R.; Kohlscheen, J.; Rambadt, S.; Erkens, G. Oxidation resistance of titanium-aluminium-silicon nitride coatings. *Surf. Coat. Technol.* **2003**, *174–175*, 408–415. [CrossRef]
8. Vereschaka, A.; Tabakov, V.; Grigoriev, S.; Sitnikov, N.; Milovich, F.; Andreev, N.; Bublikov, J. Investigation of wear mechanisms for the rake face of a cutting tool with a multilayer composite nanostructured Cr–CrN-(Ti,Cr,Al,Si)N coating in high-speed steel turning. *Wear* **2019**, *438–439*, 203069. [CrossRef]
9. Vereschaka, A.; Tabakov, V.; Grigoriev, S.; Sitnikov, N.; Andreev, N.; Milovich, F. Investigation of wear and diffusion processes on rake faces of carbide inserts with Ti-TiN-(Ti,Al,Si)N composite nanostructured coating. *Wear* **2018**, *416–417*, 72–80. [CrossRef]
10. Volosova, M.A.; Grigor'ev, S.N.; Kuzin, V.V. Effect of Titanium Nitride Coating on Stress Structural Inhomogeneity in Oxide-Carbide Ceramic. Part 4. Action of Heat Flow. *Refract. Ind. Ceram.* **2015**, *56*, 91–96. [CrossRef]
11. Grigoriev, S.; Vereschaka, A.; Milovich, F.; Tabakov, V.; Sitnikov, N.; Andreev, N.; Sviridova, T.; Bublikov, J. Investigation of multicomponent nanolayer coatings based on nitrides of Cr, Mo, Zr, Nb, and Al. *Surf. Coat. Technol.* **2020**, *401*, 126258. [CrossRef]
12. Grigoriev, S.N.; Volosova, M.A.; Vereschaka, A.A.; Sitnikov, N.N.; Milovich, F.; Bublikov, J.I.; Fyodorov, S.V.; Seleznev, A.E. Properties of (Cr,Al,Si)N-(DLC-Si) composite coatings deposited on a cutting ceramic substrate. *Ceram. Int.* **2020**, *46*, 18241–18255. [CrossRef]
13. Grigoriev, S.N.; Sobol, O.V.; Beresnev, V.M.; Serdyuk, I.V.; Pogrebnyak, A.D.; Kolesnikov, D.A.; Nemchenko, U.S. Tribological characteristics of (TiZrHfVNbTa)N coatings applied using the vacuum arc deposition method. *J. Frict. Wear* **2014**, *35*, 359–364. [CrossRef]
14. Vereschaka, A.A.; Grigoriev, S.N.; Sitnikov, N.N.; Oganyan, G.V.; Batako, A. Working efficiency of cutting tools with multilayer nano-structured Ti-TiCN-(Ti,Al)CN and Ti-TiCN-(Ti,Al,Cr)CN coatings: Analysis of cutting properties, wear mechanism and diffusion processes. *Surf. Coat. Technol.* **2017**, *332*, 198–213. [CrossRef]
15. Vereschaka, A.S.; Grigoriev, S.N.; Tabakov, V.P.; Sotova, E.S.; Vereschaka, A.A.; Kulikov, M.Y. Improving the efficiency of the cutting tool made of ceramic when machining hardened steel by applying nano-dispersed multi-layered coatings. *Key Eng. Mater.* **2014**, *581*, 68–73. [CrossRef]

16. Regent, F.; Musil, J. Magnetron sputtered Cr-Ni-N and Ti-Mo-N films: Comparison of mechanical properties. *Surf. Coat. Technol.* **2001**, *142–144*, 146–151. [CrossRef]
17. Kuleshov, A.K.; Uglov, V.V.; Chayevski, V.V.; Anishchik, V.M. Properties of coatings based on Cr, Ti, and Mo nitrides with embedded metals deposited on cutting tools. *J. Frict. Wear* **2011**, *32*, 192–198. [CrossRef]
18. Hangwei, C.; Yuan, G.; Lin, Y.; Zhikang, M.; Chenglei, W. High-temperature oxidation behavior of (Ti,Cr)N coating deposited on 4Cr13 stainless steel by multi-arc ion plating. *Rare Metal Mater. Eng.* **2014**, *43*, 1084–1087. [CrossRef]
19. Panjan, P.; Navinsek, B.; Cvelbar, A.; Zalar, A.; Milosev, I. Oxidation of TiN, ZrN, TiZrN, CrN, TiCrN and TiN/CrN multilayer hard coatings reactively sputtered at low temperature. *Thin Solid Films* **1996**, *281–282*, 298–301. [CrossRef]
20. Caplan, D.; Cohen, M. The Volatilization of Chromium Oxide. *J. Electrochem. Soc.* **1961**, *108*, 438–442. [CrossRef]
21. Gu, B.; Tu, J.P.; Zheng, X.H.; Yang, Y.Z.; Peng, S.M. Comparison in mechanical and tribological properties of Cr-W-N and Cr-Mo-N multilayer films deposited by DC reactive magnetron sputtering. *Surf. Coat. Technol.* **2008**, *202*, 2189–2193. [CrossRef]
22. Kim, K.H.; Choi, E.Y.; Hong, S.G.; Park, B.G.; Yoon, J.H.; Yong, J.H. Syntheses and mechanical properties of Cr-Mo-N coatings by a hybrid coating system. *Surf. Coat. Technol.* **2006**, *207*, 4068–4072. [CrossRef]
23. Qi, D.; Lei, H.; Wang, T.; Pei, Z.; Gong, J.; Sun, C. Mechanical, microstructural and tribological properties of reactive magnetron sputtered Cr-Mo-N films. *J. Mater. Sci. Technol.* **2015**, *31*, 55–64. [CrossRef]
24. Koshy, R.A.; Graham, M.E.; Marks, L.D. Temperature activated self-lubrication in CrN/Mo_2N nanolayer coatings. *Surf. Coat. Technol.* **2010**, *204*, 1359–1365. [CrossRef]
25. Sergevnin, V.S.; Blinkov, I.V.; Belov, D.S.; Volkhonskii, A.O.; Skryleva, E.A.; Chernogor, A.V. Phase formation in the Ti–Al–Mo–N system during the growth of adaptive wear-resistant coatings by arc PVD. *Inorg. Mater.* **2016**, *52*, 735–742. [CrossRef]
26. Tomaszewski, L.; Gulbinski, W.; Urbanowicz, A.; Suszko, T.; Lewandowski, A.; Gulbinski, W. TiAlN based wear resistant coatings modified by molybdenum addition. *Vacuum* **2015**, *121*, 223–229. [CrossRef]
27. Yang, K.; Xian, G.; Zhao, H.; Fan, H.; Wang, J.; Wang, H.; Du, H. Effect of Mo content on the structure and mechanical properties of TiAlMoN films deposited on WC–Co cemented carbide substrate by magnetron sputtering. *Int. J. Refract. Met. Hard Mater.* **2015**, *52*, 29–35. [CrossRef]
28. Sergevnin, V.S.; Blinkov, I.V.; Volkhonskii, A.O.; Belov, D.S.; Kuznetsov, D.V.; Gorshenkov, M.V.; Skryleva, E.A. Wear behaviour of wear-resistant adaptive nano-multilayeredTi-Al-Mo-N coatings. *Appl. Surf. Sci.* **2016**, *388*, 13–23. [CrossRef]
29. Sergevnin, V.S.; Blinkov, I.; Volkhonskii, A.; Belov, D.; Chernogor, A. Structure formation of adaptive arc-PVD Ti-Al-Mo-N and Ti-Al-Mo-Ni-N coatings and their wear-resistance under various friction conditions. *Surf. Coat. Technol.* **2019**, *376*, 38–43. [CrossRef]
30. Vereschaka, A.; Tabakov, V.; Grigoriev, S.; Sitnikov, N.; Oganyan, G.; Andreev, N.; Milovich, F. Investigation of wear dynamics for cutting tools with multilayer composite nanostructured coatings in turning constructional steel. *Wear* **2019**, 17–37. [CrossRef]
31. Grigoriev, S.N.; Vereschaka, A.A.; Fyodorov, S.V.; Sitnikov, N.N.; Batako, A.D. Comparative analysis of cutting properties and nature of wear of carbide cutting tools with multi-layerednano-structured and gradient coatings produced by using of various deposition methods. *Int. J. Adv. Manuf. Technol.* **2017**, *90*, 3421–3435. [CrossRef]
32. Vereschaka, A.; Grigoriev, S.N.; Volosova, M.A.; Batako, A.; Vereschaka, A.S.; Sitnikov, N.; Seleznev, A. Nano-scale multi-layered coatings for improved efficiency of ceramic cutting tools. *Int. J. Adv. Manuf. Technol.* **2016**, *90*, 27–43. [CrossRef]
33. Vereschaka, A.; Bublikov, J.I.; Sitnikov, N.N.; Oganyan, G.V.; Sotova, C.S. Influence of nanolayer thickness on the performance properties of multilayer composite nano-structured modified coatings for metal-cutting tools. *Int. J. Adv. Manuf. Technol.* **2017**, *95*, 2625–2640. [CrossRef]
34. Vereschaka, A.A. Development of Assisted Filtered Cathodic Vacuum Arc Deposition of Nano-Dispersed Multi-Layered Composite Coatings on Cutting Tools. *Key Eng. Mater.* **2013**, *581*, 62–67. [CrossRef]
35. Vereschaka, A.; Tabakov, V.; Grigoriev, S.; Aksenenko, A.; Sitnikov, N.; Oganyan, G.; Seleznev, A.; Shevchenko, S. Effect of adhesion and the wear-resistant layer thickness ratio on mechanical and performance properties of ZrN—(Zr,Al,Si)N coatings. *Surf. Coat. Technol.* **2019**, *357*, 218–234. [CrossRef]

36. Metel, A.S.; Bolbukov, V.P.; Volosova, M.A.; Grigoriev, S.N.; Melnik, Y.A. Equipment for deposition of thin metallic films bombarded by fast argon atoms. *Instrum. Exp. Tech.* **2014**, *57*, 345–351. [CrossRef]
37. Sobol', O.V.; Andreev, A.A.; Grigoriev, S.N.; Stolbovoy, V.A.; Volosova, M.A.; Aleshin, S.V.; Gorban', V.F. Physical characteristics, structure and stress state of vacuum-arc tin coating, deposition on the substrate when applying high-voltage pulse during the deposition. *Probl. At. Sci. Technol.* **2011**, *4*, 174–177.
38. Kuzin, V.V.; Grigor'ev, S.N.; Volosova, M.A. Effect of a TiC Coating on the Stress-Strain State of a Plate of a High-Density Nitride Ceramic Under NonsteadyThermoelastic Conditions. *Refract. Ind. Ceram.* **2014**, *54*, 376–380. [CrossRef]
39. Metel, A.S.; Grigoriev, S.N.; Melnik, Y.A.; Bolbukov, V.P. Broad beam sources of fast molecules with segmented cold cathodes and emissive grids. *Instrum. Exp. Tech.* **2012**, *55*, 122–130. [CrossRef]
40. Metel, A.S.; Bolbukov, V.P.; Volosova, M.A.; Grigoriev, S.N.; Melnik, Y.A. Source of metal atoms and fast gas molecules for coating deposition on complex shaped dielectric products. *Surf. Coat. Technol.* **2013**, *225*, 34–39. [CrossRef]
41. Metel, A.S.; Melnik, Y.; Metel, A. Broad fast neutral molecule beam sources for industrial-scale beam-assisted deposition. *Surf. Coat. Technol.* **2002**, *156*, 44–49. [CrossRef]
42. Pogrebnjak, A.; Bagdasaryan, A.; Beresnev, V.; Nyemchenko, U.; Ivashchenko, V.; Kravchenko, Y.; Shaimardanov, Z.; Plotnikov, S.; Maksakova, O. The effects of Cr and Si additions and deposition conditions on the structure and properties of the (Zr-Ti-Nb)N coatings. *Ceram. Int.* **2017**, *43*, 771–782. [CrossRef]
43. Glatz, S.A.; Moraes, V.; Koller, C.M.; Riedl, H.; Bolvardi, H.; Kolozsvári, S.; Mayrhofer, P.H. Effect of Mo on the thermal stability, oxidation resistance, and tribo-mechanical properties of arc evaporated Ti-Al-N coatings. *J. Vac. Sci. Technol. A* **2017**, *35*, 061515. [CrossRef]
44. Lu, Y.-C.; Chen, H.-W.; Chang, C.-C.; Wu, C.-Y.; Duh, J.-G. Tribological properties of nanocomposite Cr-Mo-Si-N coatings at elevated temperature through silicon content modification. *Surf. Coat. Technol.* **2018**, *338*, 69–74. [CrossRef]
45. Voevodin, A.; Muratore, C.; Aouadi, S. Hard coatings with high temperature adaptive lubrication and contact thermal management: Review. *Surf. Coat. Technol.* **2014**, *257*, 247–265. [CrossRef]
46. Hovsepian, P.; Lewis, D.; Luo, Q.; Münz, W.-D.; Mayrhofer, P.; Mitterer, C.; Zhou, Z.; Rainforth, W. TiAlN based nanoscale multilayer coatings designed to adapt their tribological properties at elevated temperatures. *Thin Solid Films* **2005**, *485*, 160–168. [CrossRef]
47. Voevodin, A.A.; Zabinski, J.S. Supertough wear-resistant coatings with 'chameleon' surface adaptation. *Thin Solid Films* **2000**, *370*, 223–231. [CrossRef]
48. Vereschaka, A. Improvement of Working Efficiency of Cutting Tools by Modifying its Surface Properties by Application of Wear-Resistant Complexes. *Adv. Mater. Res.* **2013**, *712*, 347–351. [CrossRef]

Publisher's Note: MDPI stays neutral with regard to jurisdictional claims in published maps and institutional affiliations.

Article

Investigation of the Influence of Microdroplets on the Coatings Nanolayer Structure

Sergey Grigoriev [1], Alexey Vereschaka [2,*], Filipp Milovich [3], Nikolay Sitnikov [4], Nikolay Andreev [3], Jury Bublikov [2], Catherine Sotova [1] and Ilya Sadov [1]

1 VTO Department, Moscow State Technological University STANKIN, 127994 Moscow, Russia; s.grigoriev@stankin.ru (S.G.); e.sotova@stankin.ru (C.S.); sadvilj@yandex.ru (I.S.)
2 IDTI RAS, 127994 Moscow, Russia; yubu@rambler.ru
3 Materials Science and Metallurgy Shared Use Research and Development Center, National University of Science and Technology MISiS, 119049 Moscow, Russia; filippmilovich@mail.ru (F.M.); andreevn.misa@gmail.com (N.A.)
4 Department of Solid State Physics and Nanosystems, National Research Nuclear University MEPhI, 115409 Moscow, Russia; sitnikov_nikolay@mail.ru
* Correspondence: dr.a.veres@yandex.ru

Received: 3 November 2020; Accepted: 8 December 2020; Published: 10 December 2020

Abstract: The paper presents the results of studies focused on the specific features typical for the formation of the shape and structure of microdroplets embedded into the structure of the Ti–TiN–(Ti,Cr,Mo,Al)N and Ti–TiN–(Ti,Al,Nb,Zr)N coatings during their deposition. Three main microdroplet shapes—a sphere, a tear, and a lens—have been considered. The specific features typical for the formation of secondary layered structures on the surface of some microdroplets have also been examined. As a result of the conducted investigations, with the use of scanning and transmission electron microscopy, the influence of microdroplets on the distortion of the nanolayer structure of the coatings was studied. A hypothesis has been proposed concerning a relationship between the microdroplet shape and the presence or absence of secondary structures and the microdroplet sizes and weight, as well as the conditions in the unit chamber during the movement of a microdroplet from a cathode to the deposition surface. Based on the study focused on the shape of the microdroplet core and the specific features typical for the formation of the secondary structure around it, a hypothesis has been proposed, according to which, for some microdroplets, it takes much more time than previously assumed for the movement from a cathode to the deposition surface.

Keywords: nanolayered PVD coating; microdroplets; crack formation; tool wear

1. Introduction

As is known, the formation of microdroplets and their penetration into the structures of modifying coatings are a significant problem, typical of the physical vapour deposition (PVD) method [1–5]. Technologies and equipment—from simple arrays to complex magnetic traps—are being actively developed and used to significantly reduce the number of microdroplets [6–12]. At the same time, the process of the formation of microdroplets and their influence on the coating structure are being studied. In particular, it has been found that the number and mean size of microdroplets increase simultaneously with an increase in the discharge current, the gas pressure, and temperature of the cathode surface [13–16]. In [13,16,17], it is noted that the microdroplet shapes vary, and microdroplets can be classified into three groups, defined as sphere-, lens-, and tear-shaped microdroplets. The number of microdroplets decreases with an increase in the temperature of the substrate [18]. The processes of microdroplet formation and penetration into the coating structure has been studied mainly by using mathematical modeling methods. In several papers, the studies were focused on the mechanism of microdroplet deformation upon hitting a substrate in various deposition conditions. In [19], the cooling of a liquid microdroplet is modeled using the laws of hydrodynamics and assuming an ideal interphase thermal contact. The Lagrangian concept and the hydrodynamic model of Fukai et al. [20] were used to simulate the heat transfer in a microdroplet and a substrate. Wadvogel et al. [21,22] expanded the specified model, taking into account the solidification process and the incomplete interphase thermal contact. Several studies separately consider the role of incomplete thermal contact between a substrate and a microdroplet, which is an essential parameter in the process of heat transfer. In particular, Liu et al. [23] suggest that no ideal thermal contact between a microdroplet and a substrate can be achieved because of the roughness of the substrate surface, the surface tension of the microdroplet, the presence of surface impurities, and gas entrapment. Heat transfer in the real area of contact between the droplet and the substrate surface occurs both due to thermal conductivity and, to some extent, radiation [24]. For microdroplets of molten lead and similar materials used for brazing, the incomplete thermal contact was investigated in [25–27] through experiments. In particular, it was found that the incomplete thermal contact affects the microdroplet shape, and their height relative to the substrate can change by 20% due to changes in the thermal contact resistance. Some values of the interphase heat transfer coefficients for specific pairs of materials were determined both by calculations and experiments in [28]. It was found that the interphase heat transfer coefficient noticeably influences the microdroplet solidification process. In [29], the study focused on the modeling of a liquid microdroplet with the diameter of 80 μm that moved with a velocity of 5 m/s and hit a flat solid surface. Given that the conditions under consideration differ from the coating deposition conditions, this model can be useful, since the general nature of the physical conditions (sizes of the microdroplet, its movement velocity, and penetration of the liquid microdroplet into the solid surface) correlates well with the challenge under consideration. In [29], the investigation also takes into account the possible variance of the heat transfer coefficient over time. In [30], the studies simulate a hollow microdroplet, consisting of a liquid shell enclosing a gas cavity and hitting a solid flat surface. The studies demonstrate that the deformation nature of a hollow microdroplet, upon impact with a solid flat surface, differs significantly from the deformation of a filled microdroplet. In [31], the studies simulate the influence of parameters—such as the microdroplet velocity on impact, the relative distances between two successive microdroplets of molten metal, the substrate temperature, and the microdroplet sizes—on the morphology of the final shape of a microdroplet solidified after the impact. The mechanisms of the microdroplet formation are also considered in detail in [1,32,33]. All the simulation results presented indicate that upon contact of a liquid or highly ductile microdroplet with the deposition surface, the microdroplet is subjected to deformation, and it forms a typical lens-shaped structure, sometimes with a torus-shaped formation along the microdroplet boundaries. All the above papers consider the microdroplet as a liquid or highly ductile substance which does not change during the process of movement from a source to the deposition surface. At the same time, the monitoring proves that a coating structure can include not only lens-shaped microdroplets, but also microdroplets of an almost ideal spherical shape. Such a microdroplet could form only during the cooling until it contacts the deposition surface, otherwise, the microdroplet would have been inevitably deformed.

In [34], Anders considered, in sufficient detail, the issues concerning the formation of microdroplets, their movement from a cathode to a deposition surface, and their effect on the coating structure. In particular, the paper notes that microdroplets can move along trajectories that are different from a simple rectilinear trajectory, and they can repeatedly bounce off the deposition surface. Meanwhile, only the trajectories of large (macro-) droplets of tens of μm can be visually traced. In [34], it is also noted that upon hitting a deposition surface, most of the microdroplets take a donut-like shape; however, some microdroplets (according to the author, the microdroplets with the diameter less than 0.5 μm) solidify before they hit the deposition surface and take a spherical shape. A microdroplet moves in the nitrogen environment, and accordingly, a nitride shell is formed on the microdroplet surface. This is one of the possible reasons for the formation of a solid microdroplet with a regular spherical shape. For example, while moving in the nitrogen environment, a liquid titanium microdroplet can form a shell of TiN, since the melting point of titanium nitride (2930 °C) is much higher than that of pure titanium (1668 °C). As a result, the microdroplet reaching the deposition surface can have a solid nitride shell.

Multicomponent coatings based on nitrides of metals such as Ti, Al, Cr, Mo, Nb, and Zr were analyzed in a large number of studies. In particular, a number of studies have focused on the properties of the (Zr,Nb)N coating, in which a cubic fcc-(Zr,Nb)N mixed crystal phase has been detected as the dominant one [35–38]. The (Ti,Zr)N coating has also been investigated in a number of papers [39–47]. In [39–41], the presence of two cubic phases of fcc-TiN and fcc-ZrN has been detected. Meanwhile, the phase of (Ti,Zr)N (solid solution of Zr in fcc-TiN) is dominating [42–45]. It should also be noted that the hardness and heat resistance of the (Ti,Zr)N coating are significantly higher than those of TiN or ZrN [43,46,47]. The (Ti,Mo)N and (Ti,Cr)N coatings have been studied in [48–52]. In particular, in [48,49], the studies have revealed that the phase of fcc-TiN with Mo dissolved in it dominates in the (Ti,Mo)N coating. An important advantage of the coatings containing Cr and Mo, is their ability to form the dense oxide films (based on Magneli phases) at elevated temperatures, and these films perform tribological and protective functions [50–52]. The studies focused on the (Cr,Mo)N coating have detected the formation of a solid solution of Mo in fcc-CrN [53,54]. In [55], the formation of the tribological films of MoO_3 and Cr_2O_3 has been revealed, with the oxide of MoO_3 having a key effect on the reduction in friction. The properties of four-component coatings based on metals of the indicated elements have also been investigated. In particular, the study focused on the properties of the (Ti,Al,Zr)N coating has found the presence of a multiphase structure with a dominant phase based on fcc-TiN with dissolved Al and Zr, but phases of Zr_3A1N, Ti_3A1N, and A1N have also been detected [56,57]. The coating of (Ti,Al,Mo)N has also been studied in [58–62], MoNAl and Mo in fcc-TiN with the presence of an insignificant amount of MoN phase. The results of the studies performed demonstrate that the multicomponent coatings based on a solid solution of such elements as Cr, Mo, Al, Nb, and Zr in fcc-TiN can significantly increase the hardness, wear and heat resistance.

This study deals with the formation of a microdroplet phase in the coatings of Ti–TiN–(Ti,Cr,Mo,Al)N and Ti–TiN–(Ti,Al,Nb,Zr)N. Earlier, other coatings with similar compositions, including Cr,Mo–(Cr,Mo,Zr,Nb)N–(Cr,Mo,Zr,Nb,Al)N, Cr,Mo–(Cr,Mo)N–(Cr,Mo,Al)N, and Zr,Nb–(Zr,Nb)N–(Zr,Nb,Al)N, had been considered, and they proved to have good mechanical properties, while cutting tools with these coatings demonstrated good cutting properties [60]. Meanwhile, the current study is focused on microdroplets embedded into the structure of the coatings under consideration. It should be noted that the investigation of microdroplets in their dynamic state (while they are moving from a cathode to the deposition surface) is a hard-hitting challenge due to their microscopic sizes, stochastic trajectory of motion, and difficulties in observing in conditions of plasma flow. Accordingly, there are two possible ways to study the above objects, that is, to study solidified microdroplets in the coating structure and to apply methods of mathematical modeling. It should be noted that the development of an adequate model for the motion of a microdroplet is another hard-hitting challenge, taking into account a number of stochastic factors affecting the process.

2. Materials and Methods

To deposit coatings, a VIT-2 unit (VIT—IDTI RAS, Moscow, Russia) was used, which implemented the filtered cathodic vacuum arc deposition (FCVAD) technology [5,63–69].

Three cathode systems were used, in which cathodes of Al 99.1 at.%, Ti 99.9 at.%, and, depending on the type of the coating to be deposited, Cr–Mo (50:50 at.%) or Zr–Nb (50:50 at.%) were installed, respectively. Cathodes of cylindrical shape with a diameter of 80 mm were used.

The parameters for the process of coating deposition are presented in Table 1.

Table 1. Parameters of stages of the technological process of the deposition of coatings.

Process	p_N (Pa)	U (V)	I_{Ti} (A)	I_{Al} (A)	$I_{Zr\text{–}Nb}$ (A)	$I_{Cr\text{–}Mo}$ (A)
Pumping and heating of vacuum chamber	0.06	+20	75	120	65	–
Heating and cleaning of products with gaseous plasma	2.0	100 DC/900 AC f = 10 kHz, 2:1	85	80	–	–
Deposition of coating	0.42	−800 DC	75	160	55	73
Cooling of products	0.06	–	–	–	–	–

Note: I_{Ti} = current of titanium cathode, $I_{Zr\text{–}Nb}$ = current of Zr–Nb cathode, I_{Al} = current of Al cathode, $I_{Cr\text{–}Mo}$ = current of Cr–Mo cathode, p_N = gas pressure in chamber, U = voltage on substrate. Table rotation frequency n = 0.7 rpm.

Due to the fact that the VIT-2 unit uses a micro-droplet reduction system, the droplets in the coating stemming from the Al-cathode are reduced by up to 98% [5,70] and the droplets in the coating stemming from the other cathodes are reduced by up to 85% [71–73]. Microdroplets in the structure of the coatings under study are rather rare. The sizes of the detected microdroplets based on Mo, Cr, Ti, and Zr–Nb usually do not exceed 3 μm (individual microdroplets up to 12 μm in size are rarely observed), because larger microdroplets are separated and do not enter a product (several sources consider the mean microdroplet size of 10–25 μm [1,2]). Samples were subjected to the microstructural studies using an SEM FEI Quanta 600 FEG (SEM 1, Waltham, MA, USA) and a Carl Zeiss EVO 50 with a Bruker energy-dispersive spectrometer (SEM 2, Billerica, MA, USA). A high-resolution transmission electron microscope (HR TEM) JEM 2100 (JEOL, Tokyo, Japan) at an accelerating voltage of 200 kV was used to investigate the coating nanostructure. The analysis of the elemental composition of the coatings was conducted at TEM using the EDX system INCA Energy (OXFORD Instruments, Abingdon, UK) in STEM (scanning transmission electron microscopy) mode. A Strata FIB 205 (FEI) was used to prepare TEM samples.

3. Results and Discussion

The available results from studies of the microstructure of coatings of various compositions and architecture, show that the detected microdroplets can be classified by several parameters (for example, by the shape of a microdroplet and its position in the coating structure) [13,74–76]. Accordingly, the influence of various microdroplet types on the structure of the coating and its fracture pattern also differ. In particular, the stage of the coating deposition process, at which a microdroplet is embedded into the coating structure, has a significant effect. A microdroplet can be embedded at the initial stage of the deposition process, and in such case, it often directly adheres to the substrate surface. A microdroplet can be embedded into the coating structure during the formation of a sequence of nanolayers. Finally, a microdroplet can be formed at the finishing stage of the deposition process. In such case, it penetrates into the coating surface and, since the coating has not yet cooled down and stays considerably ductile, forms a crater around itself and thus, deforms the coating structure. It should be noted that it is very difficult to directly study the process of the formation of a microdroplet, its movement inside the chamber, and penetration into the coating structure because of the small sizes of the microdroplets, their high movement velocity (which varies in the range of 6–120 m/s [1] (according to other data, 100–1000 m/s, depending on the temperature and the evaporated material [16,17]) and the complicated monitoring in the conditions of a plasma flow. Thus, at present, there are only two available ways to study the dynamics of microdroplet development, i.e., mathematical modeling and the study of solidified microdroplets and their effect on the coating structure. Microdroplets can also be classified by their final shape, which may indicate their state at the moment of impact onto the coating. In particular, there are microdroplets of an almost ideal spherical shape, due to which it can be concluded that the microdroplets have been embedded into the coating structure in a solidified state, otherwise their noticeable deformation would be inevitable. At the same time, there are lens-shaped microdroplets.

This shape is typical for a microdroplet hitting the surface in a molten (liquid) state. Finally, there are microdroplets of a transition shape, that is, with signs of noticeable deformation or drop-shaped, which may indicate that at the moment of its penetration into the coating, the microdroplet was no longer liquid, but still was considerably ductile. Let us consider in more detail the structure and shapes of microdroplets of various metals. Since microdroplets of Al are almost completely separated during the deposition process, they could not be detected in the structure of the coatings, despite the fact that Al is contained in the compositions of all the coatings under study.

The structure of a microdroplet embedded into the structure of the Ti–TiN–(Ti,Cr,Mo,Al)N coating was considered. The images reveal (Figure 1) the sections of several clearly pronounced areas with noticeably different crystal sizes, which can indicate the stages of the microdroplet structure formation during the cooling process. Their formation, presumably, occurred during the microdroplet movement from the cathode towards the deposition surface and during the penetration of the microdroplet into the coating structure. The microdroplet core is formed mainly by molybdenum and chromium (Figure 1c). Area I with grains of extremely small sizes (about 3–5 nm) was formed around the microdroplet core. Then areas II and III were formed, and the grain sizes in each new area were noticeably larger than in the previous one (of about 20–40 nm in area III). Presumably, all the three areas form around the core when a microdroplet is moving from the cathode towards the deposition surface, until it hits the surface. The cooling rate of the microdroplet substance is related to the thermal conductivity coefficient (see Table 2). Since aluminium Al and molybdenum Mo are characterized by the highest thermal conductivity, while titanium Ti and zirconium Zr by the lowest thermal conductivity, the probability of the solidification of a microdroplet before its contact with the surface is higher for microdroplets of Al and Mo than for microdroplets of Ti and Zr. The size of a microdroplet and the length of its trajectory from the cathode to the deposition surface are essential. In the microdroplet, the core is formed mainly by Mo and Cr (Figure 1d), while the microdroplet composition also includes Ti and Al. Since the composition of the Mo–Cr cathode does not include Ti and Al, their presence in the microdroplet can be explained by their deposition on the microdroplet surface during its movement from the cathode to the deposition surface.

Table 2. Coefficients of thermal conductivity of metals depending on temperature, W/(m × K) [77].

Chemical Element	127 °C	527 °C	727 °C
Al	240	220	90
Ti	20.4	19.7	20.7
Cr	87.3	71.3	65.3
Mo	134	118	112
Nb	55.2	61.3	64.4
Zr	21.6	21.6	23.7

Let us consider the differences in the selected area electron diffraction (SAED) pattern for the coating structure (Figure 1e) and the microdroplet structure (Figure 1f). The analysis of the SAED pattern for the Ti–TiN–(Ti,Cr,Mo,Al)N coating (Figure 1e) finds the presence of a basic phase of (Cr,Ti,Mo,Al)N with the space group of $Fm\bar{3}m$ and a small amount of AlN (given the intensity of the rings) with the space group of P6.3mc. The SAED pattern for the coating demonstrates the broadening of the rings, related to the (Cr,Ti,Mo,Al)N phase. This broadening apparently arises due to the presence of polycrystalline grains formed by several types of nitrides with the same space group $Fm\bar{3}m$, but with slightly different interplanar spacings for the general interference indices (Miller indices, HKL). Table 3 contains the values of interplanar spacings d (for the first six HKL indices) for the nitrides forming the Ti–TiN–(Ti,Cr,Mo,Al)N coating under study.

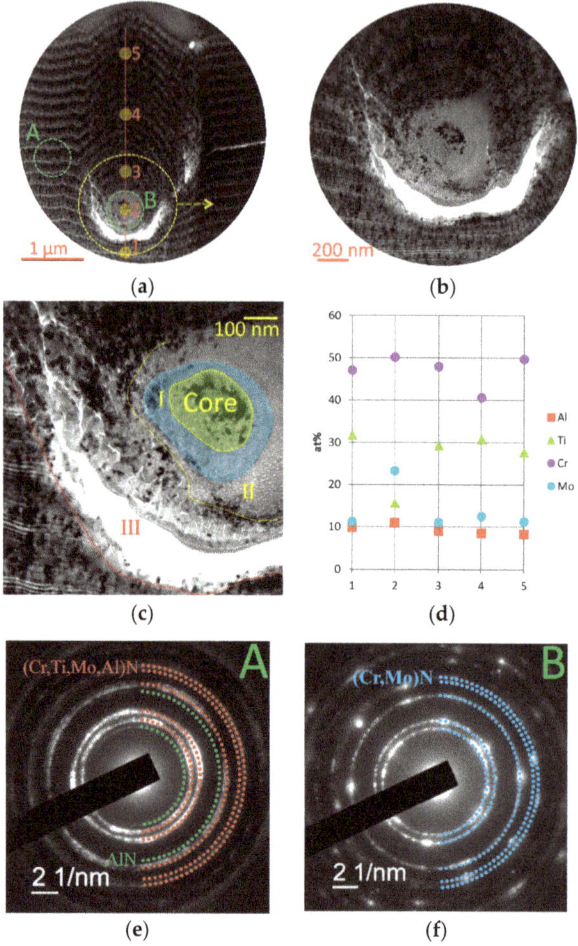

Figure 1. (**a**–**c**) structure and (**d**) chemical composition of the microdroplet in the Ti–TiN–(Ti,Cr,Mo,Al)N coating (TEM), selected area electron diffraction (SAED) pattern for area A (**e**) and area B (**f**). The microdroplet core is surrounded by area I (crystal sizes are 3–5 nm), then by area II (crystal sizes are 8–15 nm), and then by area III (crystal sizes are 20–40 nm).

Table 3. Values of the interplanar spacings d (for the first six HKL indices) for nitrides, forming the Ti–TiN–(Ti,Cr,Mo,Al)N coating.

Phase	CrN	Mo_2N	(Cr,Mo)N	TiN	HKL
Space Group	$Fm\bar{3}m$	$Fm\bar{3}m$	$Fm\bar{3}m$	$Fm\bar{3}m$	
1	2.390 Å	2.408 Å	2.42 Å	2.448 Å	(111)
2	2.070 Å	2.085 Å	2.09 Å	2.120 Å	(002)
3	1.464 Å	1.475 Å	1.48 Å	1.499 Å	(202)
4	1.248 Å	1.257 Å	1.26 Å	1.278 Å	(113)
5	1.195 Å	1.204 Å	1.20 Å	1.224 Å	(222)
6	1.035 Å	1.043 Å	1.04 Å	1.060 Å	(004)

Table 3 also contains the interplanar spacings calculated under the SAED pattern for area B of the microdroplet under the study (Figure 1f). The SAED pattern for the microdroplet (area B) exhibits two differences compared to the SAED pattern for area A of the coating. Firstly, the SAED pattern for the microdroplet includes no AlN phase reflections, and, secondly, there is no noticeable broadening of the SAED rings. No broadening indicates that ether a certain nitride phase dominates in amount, or there are several nitride phases with very close values of d. Table 3 demonstrates that the calculated interplanar spacings correspond most closely to Mo_2N (Fm$\overline{3}$m). During the comparison of the elemental compositions of the coating and of the microdroplet, it can be noticed that in the region of the microdroplet, the amount of titanium decreases, and the amount of molybdenum noticeably increases, while the amount of chromium remains approximately the same. Based on the data obtained during the study of the microdroplet (investigation of its interplanar spacings and elemental composition), it can be assumed that the two phases—chromium nitride and molybdenum nitride—with considerably close interplanar spacings most likely appear in this region.

Thus, the microdroplet consists of chromium and nitrogen nitrides and not of pure metals. The above may confirm the hypothesis about the formation of a hard shell of a microdroplet due to the formation of nitrides that are more refractory than pure metal [34]. However, in this case, not only the shell of the microdroplet, but almost the entire microdroplet consists of nitrides, since the SAED pattern exhibits no rings associated with Cr or Mo.

The study of other examples showing how microdroplets of nearly spherical shape penetrate into the coating structure (Figure 2) reveals certain consistent patterns. Like in the case considered earlier, there is a microdroplet core with a fairly regular spherical shape, and the secondary structures, clearly distinguishable in Figure 2a,b, are formed around the microdroplet core. It should be noted that such a secondary structure is being formed around the microdroplet core not only on the side facing the cathode, but also on the opposite side, which is also noticeable in the earlier considered example (Figure 1). The presence of such structure formed around the spherical core of the microdroplet is difficult to explain if it is assumed that the structure was formed after the microdroplet had hit the deposition surface. In such case, no secondary structure would be formed on the microdroplet side opposite to the source. However, Figure 2b exhibits the microdroplet of Nb surrounded by the secondary structure, consisting of six layers of (Ti,Al,Nb,Zr)N, and the thickness of these layers noticeably increases on the side facing the source (with a thickness of about 250 nm). Meanwhile, such layers are also formed on the reverse side of the microdroplet (with a thickness of about 25 nm). The deposition of the secondary structure layers on the reverse side of the microdroplet core is possible only in cases when a microdroplet did not hit the deposition surface, that is, during the movement of the microdroplet from the cathode to the deposition surface. However, all previously obtained data indicate that the velocity of a microdroplet is very high and that it passes the distance from the cathode to the deposition surface within fractions of a second [1,16,17]. It is clear that the microdroplet cannot solidify during the above time and form a regular sphere (especially taking into account the fact that the coating deposition takes place in a highly discharged medium), and all the more, several layers of the secondary structure cannot form on its surface. Given that the turntable rotation frequency was 0.7 rev/min, it takes 42 s to deposit one layer and 252 s, that is, over 4 min, to deposit six layers. Based on the available methods and models, it can be assumed that free movement of the microdroplet core during such a long time is not possible. However, it should be noted that a moving microdroplet is affected by multidirectional plasma flows and significant magnetic fields. While there are three sources of plasma located at an angle of 90° to each other, a microdroplet is simultaneously affected by three plasma flows, and as it approaches the chamber center, the effect of all three flows can be balanced and at a certain point, such equilibrium effect can keep the microdroplet in a free state (that is, with no contact with the deposition surface or chamber walls). Another source of influence on the microdroplet, is the powerful magnetic field generated by the separation system. The combined effect of the above factors can lead to an effect in which the microdroplet can stay in a free state for a relatively long time, sufficient for its cooling and deposition of a layered secondary structure on the microdroplet core.

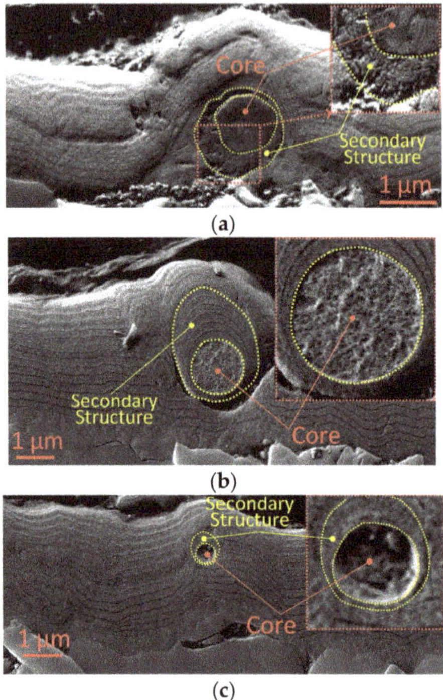

Figure 2. (**a**) the microdroplet of Cr–Mo, embedded into the structure of the Ti–TiN–(Ti,Cr,Mo,Al)N coating, (**b**,**c**) the microdroplets of Nb–Zr, embedded into the structure of the Ti–TiN–(Ti,Al,Nb,Zr)N coating (SEM 2).

Given the small size and weight of the microdroplet core in the cases under study (with microdroplet diameters from 100 to 1000 nm), the above suggests the possibility of its long persistence in a free state. Of course, this hypothesis requires further development, implying thorough mathematical modeling of the process. It should be taken into account that in order to build an adequate model for the movement of a microdroplet, it will be necessary to consider the influence of multidirectional plasma flows (given the presence of at least three plasma sources operating simultaneously, as well as the possibility for the formation of vortex-like plasma flows). The influence of the powerful magnetic field of the plasma flow separation system should also be taken into account, as well as the effect of the internal discharged medium of the chamber, which makes the development of such a model rather complicated.

When a microdroplet has a larger core (3–5 μm), its weight also increases significantly. Accordingly, such microdroplets reach the deposition surface relatively quickly, while there is not enough time for the formation of the secondary structures or an insignificant amount of them are formed only on the side facing the plasma source (Figure 3). However, it is clear that such microdroplets still have enough time to cool down and solidify, although they stay considerably ductile. When microdroplets hit the deposition surface, they undergo plastic deformation, and they become ellipse-shaped (Figure 3b,c) or tear-shaped (Figure 3a). After the penetration of microdroplets into the coating structure, their deposited nanolayers are distorted.

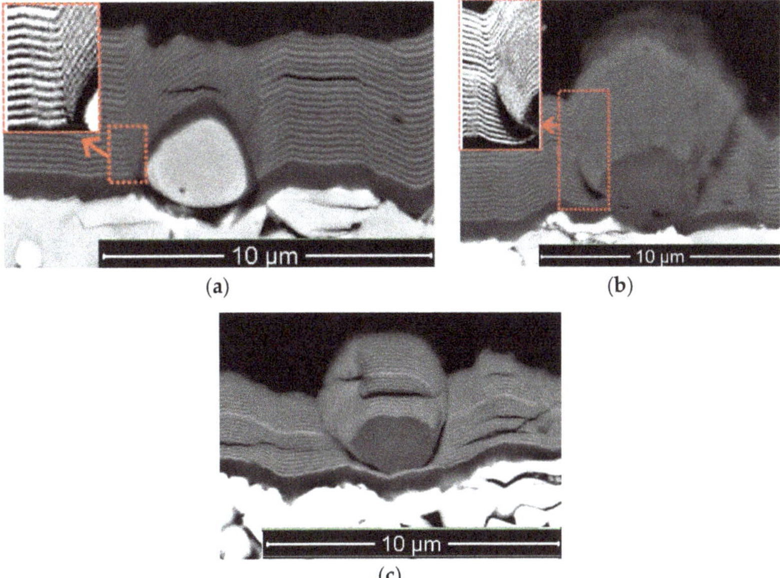

Figure 3. Examples of the influence of microdroplets on the structure of the (a,b) Ti–TiN–(Ti,Cr,Mo,Al)N and (c) Ti–TiN–(Ti,Nb,Zr,Al)N coatings (SEM 1).

Microdroplets of even larger sizes (5–7 µm) also do not have enough time to form a secondary structure in the process of movement. When hitting the coating surface, the above microdroplets can form a crater and noticeably distort the nanolayer structure of the coating (Figure 4). It should be taken into account that immediately after the deposition, during the process of cooling, the coating layers retains their high ductility and when a considerably large object hits the freshly deposited coating layers, it can deform their structure (see Figure 4a, in this image, the very microdroplet was lost, possibly during the manufacturing of a section, but the trace of the microdroplet is clearly visible).

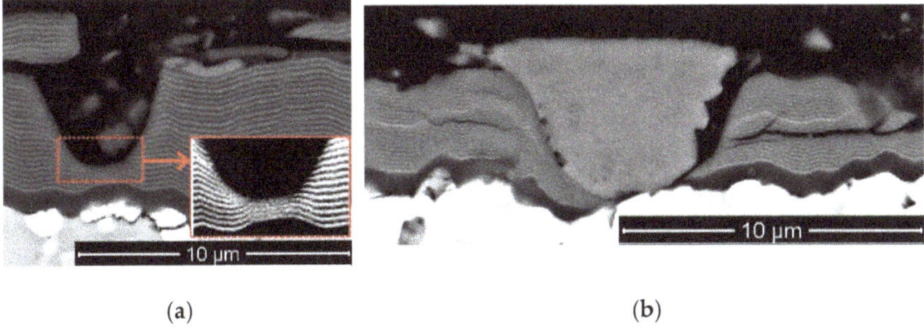

Figure 4. Examples of the influence of microdroplets on the structure of (a) Ti–TiN–(Ti,Cr,Mo,Al)N and (b) Ti–TiN–(Ti,Nb,Zr,Al)N coatings (SEM 1).

Finally, there are also lens-shaped microdroplets embedded into the coating structure, and their shape can be called "correct", because it is fully consistent with the existing models [14–19]. Although most of the microdroplets of the above shape have considerably large sizes and, accordingly, weights (with a diameter of 7–10 µm and more, see Figure 5a,b), there are also microdroplets of smaller sizes, which took the shape of a lens when hitting the deposition surface (the microdroplets of Ti shown in Figure 5c,d). The microdroplets of such a shape have noticeably less effect on the distortion of the nanolayer structure of the coating. It can be assumed that the formation of lens-shaped microdroplets occurs in conditions when the microdroplet weight is large enough and the initial energy of the microdroplet movement allows it to move at a considerably high velocity. There are also cases when the interaction of plasma flows and magnetic fields does not have a significant influence on the velocity of the microdroplet movement or even contributes to the acceleration of such movement. A low thermal conductivity coefficient, which slows down the cooling of a microdroplet, can also play a certain role. In such case, a microdroplet does not have enough time to cool down and no secondary structures manage to form on its surface. Accordingly, when such a microdroplet (in liquid state) hits the deposition surface, it activates the mechanisms described in [14–19].

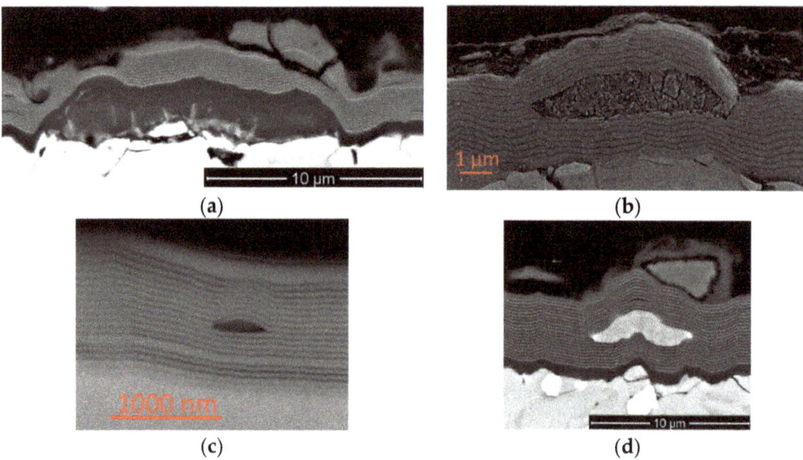

Figure 5. Examples of the influence of microdroplets on the structure of (**a**,**b**) Ti–TiN–(Ti,Nb,Zr,Al)N and (**c**,**d**) Ti–TiN–(Ti,Cr,Mo,Al)N coatings (**a**,**c**,**d**—SEM 1,**b**—SEM 2).

4. Conclusions

Thus, while considering the features typical for the formation of microdroplets embedded into the coating structure, it is possible to make the following conclusions:

There are microdroplets of three main types:

- In the shape of a sphere with a secondary multilayered structure formed around the microdroplet core, and the sizes of such microdroplet cores are within the range of 100–1000 nm. The formation of such microdroplets can also be associated with the high thermal conductivity of the metals forming them;
- In the shape of an ellipse or a tear, when a secondary structure is not typical, and the microdroplet sizes usually stay within 3–5 µm;
- In the shape of a lens, without a surrounding secondary structure, with either large (7–10 µm) or smaller sizes. The formation of such microdroplets can also be associated with the high thermal conductivity of the metals forming them, as well as the formation of nitride compounds that are more refractory compared to pure metal. Due to the above, the surface of the microdroplet solidifies when the microdroplet moves from the cathode to the deposition surface.

While penetrating into a coating, a microdroplet noticeably distorts its nanolayer structure. Such distortion occurs both due to the plastic deformation of nanolayers upon impact of the microdroplet and due to the effect of the microdroplet on the structure of nanolayers deposited after its penetration.

The shape of a microdroplet and the presence or absence of secondary structures, surrounding the microdroplet core, relates both to its size (and, accordingly, weight) and to the conditions affecting the process of the movement of the microdroplet in the unit chamber (the influence of multidirectional plasma flows and magnetic fields). It can be assumed that the movement of some microdroplets with the diameter of 100–1000 nm from the source to the deposition surface can be significantly slowed down (up to tens and even hundreds of seconds) under the influence of multidirectional plasma flows and magnetic fields.

Based on the study focused on the shape of a microdroplet core and the specific features typical for the formation of the secondary structure around it, a hypothesis has been proposed, according to which for some microdroplets, it takes much more time than previously assumed for the movement from a cathode to the deposition surface.

Comprehensive mathematical modeling of the processes under consideration is necessary for a detailed study of the causes and mechanisms typical for the formation of microdroplets of different shapes, as well as the mechanisms of the formation of secondary structures on their surface.

Author Contributions: Conceptualization, A.V. and S.G.; methodology, S.G. and A.V.; formal analysis, F.M., C.S., and I.S.; investigation, A.V., F.M., N.S., J.B., N.A., and I.S.; resources, S.G.; data curation, A.V. and J.B.; writing—original draft preparation, A.V.; writing—review and editing, A.V.; project administration, A.V. and S.G.; funding acquisition, S.G. All authors have read and agreed to the published version of the manuscript.

Funding: This research was funded by Ministry of Science and Higher Education of the Russian Federation, Grant No. 0707-2020-0025.

Conflicts of Interest: The authors declare no conflict of interest.

References

1. Boxman, R.; Goldsmith, S. Macroparticle contamination in cathodic arc coatings: Generation, transport and control. *Surf. Coat. Technol.* **1992**, *52*, 39–50. [CrossRef]
2. Steffens, H.-D.; Mack, M.; Moehwald, K.; Reichel, K. Reduction of droplet emission in random arc technology. *Surf. Coat. Technol.* **1991**, *46*, 65–74. [CrossRef]
3. Akari, K.; Tamagaki, H.; Kumakiri, T.; Tsuji, K.; Koh, E.; Tai, C. Reduction in macroparticles during the deposition of TiN films prepared by arc ion plating. *Surf. Coat. Technol.* **1990**, *1990*, 312–323. [CrossRef]
4. Sanders, D.M. Ion beam self-sputtering using a cathodic arc ion source. *J. Vac. Sci. Technol. A* **1988**, *6*, 1929–1933. [CrossRef]
5. Vereshchaka, A.A.; Mgaloblishvili, O.; Morgan, M.; Batako, A.D.L. Nano-scale multilayered-composite coatings for the cutting tools. *Int. J. Adv. Manuf. Technol.* **2014**, *72*, 303–317. [CrossRef]
6. Michalski, A. Structure and properties of coatings composed of TiN–Ti obtained by the reactive pulse plasma method. *J. Mater. Sci. Lett.* **1984**, *3*, 505–508. [CrossRef]
7. Aksenov, I.I.; Bren′, V.G.; Osipov, V.A.; Padalka, V.G.; Khoroshikh, V.M. Plasma in a stationary vacuum-arc discharge. I. Plasma flux formation. *High Temp.* **1983**, *21*, 160–164.
8. Metel, A.; Bolbukov, V.; Volosova, M.; Grigoriev, S.; Melnik, Y. Equipment for deposition of thin metallic films bombarded by fast argon atoms. *Instrum. Exp. Tech.* **2014**, *57*, 345–351. [CrossRef]
9. Sobol′, O.V.; Andreev, A.A.; Grigoriev, S.N.; Stolbovoy, V.A. See more physical characteristics, structure and stress state of vacuum-arc tin coating, deposition on the substrate when applying high-voltage pulse during the deposition. *Probl. At. Sci. Technol.* **2011**, *4*, 174–177.
10. Metel, A.S.; Grigoriev, S.N.; Melnik, Y.A.; Bolbukov, V.P. Broad beam sources of fast molecules with segmented cold cathodes and emissive grids. *Instrum. Exp. Tech.* **2012**, *55*, 122–130. [CrossRef]
11. Metel, A.S.; Bolbukov, V.P.; Volosova, M.A.; Grigoriev, S.N.; Melnik, Y.A. Source of metal atoms and fast gas molecules for coating deposition on complex shaped dielectric products. *Surf. Coat. Technol.* **2013**, *225*, 34–39. [CrossRef]
12. Metel, A.S.; Melnik, Y.; Metel, A. Broad fast neutral molecule beam sources for industrial-scale beam-assisted deposition. *Surf. Coat. Technol.* **2002**, *156*, 44–49. [CrossRef]

13. Miernik, K.; Walkowicz, J. Spatial distribution of microdroplets generated in the cathode spots of vacuum arcs. *Surf. Coat. Technol.* **2000**, *125*, 161–166. [CrossRef]
14. Anders, S.; Anders, A.; Yu, K.M.; Yao, X.; Brown, I. On the macroparticle flux from vacuum arc cathode spots. *IEEE Trans. Plasma Sci.* **1993**, *21*, 440–446. [CrossRef]
15. Baouchi, A.W.; Perry, A.J. A study of the macroparticle distribution in cathodic-arc-evaporated TiN films. *Surf. Coat. Technol.* **1991**, *41*, 253–257. [CrossRef]
16. Utsumi, T.; English, J.H. Study of electrode products emitted by vacuum arcs in form of molten metal particles. *J. Appl. Phys.* **1975**, *46*, 126. [CrossRef]
17. Vyskočil, J.; Musil, J. Arc evaporation of hard coatings: Process and film properties. *Surf. Coat. Technol.* **1990**, *43–44*, 299–311. [CrossRef]
18. Devia, A.; Benavides, V.; Restrepo, E.; Arias, D.; Ospina, R. Influence substrate temperature on structural properties of TiN/TiC bilayers produced by pulsed arc techniques. *Vacuum* **2006**, *81*, 378–384. [CrossRef]
19. Zhao, Z.; Poulikakos, D.; Fukai, J. Heat transfer and fluid dynamics during the collision of a liquid droplet on a substrate: I—Modeling. *Int. J. Heat Mass Transf.* **1996**, *39*, 2771–2789. [CrossRef]
20. Fukai, J.; Zhao, Z.; Poulikakos, D.; Megaridis, C.M.; Miyatake, O. Modeling of the deformation of a liquid droplet impinging upon a flat surface. *Phys. Fluids A* **1993**, *5*, 2588–2599. [CrossRef]
21. Waldvogel, J.M.; Poulikakos, D.; Wallace, D.B.; Marušák, R. Transport phenomena in picoliter size solder droplet dispension. *J. Heat Transf.* **1996**, *118*, 148–156. [CrossRef]
22. Waldvogel, J.; Poulikakos, D. Solidification phenomena in picoliter size solder droplet deposition on a composite substrate. *Int. J. Heat Mass Transf.* **1997**, *40*, 295–309. [CrossRef]
23. Liu, W.; Wang, G.X.; Matthys, E.F. Determination of the thermal contact coefficient for a molten droplet impinging on a substrate. *Transp. Phenom Mater. Process. Manuf. ASME HTD* **1992**, *196*, 111–118.
24. Attinger, D.; Haferl, S.; Zhao, Z.; Poulikakos, D. Transport phenomena in the impact of a molten droplet on a surface: Macroscopic phenomenology and microscopic considerations. Part II: Heat transferand solidification. *Ann. Rev. Heat Transf.* **2000**, *11*, 65–143.
25. Bennett, T.; Poulikakos, D. Heat transfer aspects of splat-quench solidification: Modelling and experiment. *J. Mater. Sci.* **1994**, *29*, 2025–2039. [CrossRef]
26. Pasandideh-Fard, M.; Mostaghimi, J. On the spreading and solidification of molten particles in a plasma spray process effect of thermal contact resistance. *Plasma Chem. Plasma Process.* **1995**, *16*, S83–S98. [CrossRef]
27. Xiong, B.; Megaridis, C.M.; Poulikakos, D.; Hoang, H. An investigation of key factors affecting solder microdroplet deposition. *J. Heat Transf.* **1998**, *120*, 259–270. [CrossRef]
28. Attinger, D.; Poulikakos, D. On quantifying interfacial thermal and surface energy during molten microdroplet surface deposition. *J. Atom. Spray* **2003**, *13*, 309–319. [CrossRef]
29. Bhardwaj, R.; Longtin, J.P.; Attinger, D. A numerical investigation on the influence of liquid properties and interfacial heat transfer during microdroplet deposition onto a glass substrate. *Int. J. Heat Mass Transf.* **2007**, *50*, 2912–2923. [CrossRef]
30. Kumar, A.; Gu, S. Modelling impingement of hollow metal droplets onto a flat surface International. *Int. J. Heat Fluid Flow* **2012**, *37*, 189–195. [CrossRef]
31. Du, J.; Wei, Z.; Wu, H.; Zhao, G.; Wang, X. Numerical simulation of pileup process in metal microdroplet deposition manufacture. *J. Enhanc. Heat Transf.* **2016**, *23*, 413–430. [CrossRef]
32. Muboyadzhyan, S.A.; Budinovskii, S.A.; Gorlov, D.S.; Doronin, O.N. Structure and microporosity of ion-plasma condensed coatings deposited from a two-phase vacuum-arc discharge plasma flow containing evaporated material microdroplets. *Russ. Met.* **2019**, *2019*, 52–62. [CrossRef]
33. Kashkarov, E.B.; Nikitenkov, N.; Syrtanov, M.S.; Sutygina, A.N.; Gvozdyakov, D.V. Effect of bias potential on the structure and distribution of elements in titanium-nitride coatings obtained by cathodic-arc deposition. *J. Surf. Investig.* **2016**, *10*, 648–651. [CrossRef]
34. Anders, A. *Cathodic Arcs*; Springer Science Business Media, LLC: New York, NY, USA, 2008. [CrossRef]
35. Zhang, S.; Dong, L.; Dong, L.; Gu, H.; Wan, R. Influence of sputtering power on the structure and mechanical properties of Zr–Nb–N nanocomposite coatings prepared by multi-target magnetron Co-sputtering. *Key Eng. Mater.* **2013**, *537*, 307–310. [CrossRef]

36. Klostermann, H.; Fietzke, F.; Labitzke, R.; Modes, T.; Zywitzki, O. Zr–Nb–N hard coatings deposited by high power pulsed sputtering using different pulse modes. *Surf. Coat. Technol.* **2009**, *204*, 1076–1080. [CrossRef]
37. Blinkov, I.V.; Volkhonskii, A.O. The effect of deposition parameters of multilayered nanostructure Ti–Al–N/Zr–Nb–N/Cr–N coatings obtained by the arc-PVD method on their structure and composition. *Russ. J. Non-Ferrous Met.* **2012**, *53*, 163–168. [CrossRef]
38. Makino, Y.; Saito, K.; Murakami, Y.; Asami, K. Phase change of Zr–Al–N and Nb–Al–N films prepared by magnetron sputtering method. *Solid State Phenom.* **2007**, *127*, 195–200. [CrossRef]
39. Yang, G.; Etchessahar, E.; Bars, J.; Portier, R.; Debuigne, J. A miscibility gap in the FCC δ-nitride region of the ternary system titanium-zirconium-nitrogen. *Scr. Met. Mater.* **1994**, *31*, 903–908. [CrossRef]
40. Duwez, P.; Odell, F. Phase Relationships in the Binary Systems of Nitrides and Carbides of Zirconium, Columbium, Titanium, and Vanadium. *J. Electrochem. Soc.* **1950**, *97*, 299. [CrossRef]
41. Sridar, S.; Kumar, R.; Kumar, K.H. Thermodynamic modelling of Ti–Zr–N system. *Calphad* **2017**, *56*, 102–107. [CrossRef]
42. Wang, D.-Y.; Chang, C.-L.; Hsu, C.-H.; Lin, H.-N. Synthesis of (Ti,Zr)N hard coatings by unbalanced magnetron sputtering. *Surf. Coat. Technol.* **2000**, *130*, 64–68. [CrossRef]
43. Uglov, V.V.; Anishchik, V.M.; Zlotski, S.V.; Abadias, G.; Dub, S. Structural and mechanical stability upon annealing of arc-deposited Ti–Zr–N coatings. *Surf. Coat. Technol.* **2008**, *202*, 2394–2398. [CrossRef]
44. Kuo, Y.-L.; Lee, C.; Lin, J.-C.; Yen, Y.-W.; Lee, W.-H. Evaluation of the thermal stability of reactively sputtered (Ti,Zr)N_x nano-thin films as diffusion barriers between Cu and Silicon. *Thin Solid Films* **2005**, *484*, 265–271. [CrossRef]
45. Tonghe, Z.; Yuguang, W.; Zhiyong, Z.; Zhiwei, D. The ternary Ti (Zr, N) phases formation and modification of TiN coatings by Zr^+ MEVVA ion implantation. *Surf. Coat. Technol.* **2000**, *131*, 326–329. [CrossRef]
46. Niu, E.; Li, L.; Lv, G.; Chen, H.; Li, X.; Yang, X.; Yang, S. Characterization of Ti–Zr–N films deposited by cathodic vacuum arc with different substrate bias. *Appl. Surf. Sci.* **2008**, *254*, 3909–3914. [CrossRef]
47. Uglov, V.V.; Anishchik, V.M.; Khodasevich, V.V.; Prikhodko, Z.L.; Zlotski, S.V.; Abadias, G.; Dub, S.N. Structural characterization and mechanical properties of Ti–Zr–N coatings, deposited by vacuum arc. *Surf. Coat. Technol.* **2004**, *180–181*, 519–525. [CrossRef]
48. Regent, F.; Musil, J. Magnetron sputtered Cr–Ni–N and Ti–Mo–N films: Comparison of mechanical properties. *Surf. Coat. Technol.* **2001**, *142–144*, 146–151. [CrossRef]
49. Kuleshov, A.; Uglov, V.V.; Chayevski, V.V.; Anishchik, V.M. Properties of coatings based on Cr, Ti, and Mo nitrides with embedded metals deposited on cutting tools. *J. Frict. Wear* **2011**, *32*, 192–198. [CrossRef]
50. Hangwei, C.; Yuan, G.; Lin, Y.; Zhikang, M.; Chenglei, W. High-temperature oxidation behavior of (Ti,Cr)N coating deposited on 4Cr13 stainless steel by multi-arc ion plating. *Rare Met. Mater. Eng.* **2014**, *43*, 1084–1087. [CrossRef]
51. Panjan, P.; Navinsek, B.; Cvelbar, A.; Zalar, A.; Milosev, I. Oxidation of TiN, ZrN, TiZrN, CrN, TiCrN and TiN/CrN multilayer hard coatings reactively sputtered at low temperature. *Thin Solid Films* **1996**, *281*, 298–301. [CrossRef]
52. Caplan, D.; Cohen, M. The volatilization of chromium oxide. *J. Electrochem. Soc.* **1961**, *108*, 438. [CrossRef]
53. Kim, K.H.; Choi, E.Y.; Hong, S.G.; Park, B.G.; Yoon, J.H.; Yong, J.H. Syntheses and mechanical properties of Cr–Mo–N coatings by a hybrid coating system. *Surf. Coat. Technol.* **2006**, *207*, 4068–4072. [CrossRef]
54. Qi, D.; Lei, H.; Wang, T.; Pei, Z.; Gong, J.; Sun, C. Mechanical, microstructural and tribological properties of reactive magnetron sputtered Cr–Mo–N films. *J. Mater. Sci. Technol.* **2015**, *31*, 55–64. [CrossRef]
55. Koshy, R.A.; Graham, M.E.; Marks, L.D. Temperature activated self-lubrication in CrN/Mo_2N nanolayer coatings. *Surf. Coat. Technol.* **2010**, *204*, 1359–1365. [CrossRef]
56. Liu, W.; Li, A.; Wu, H.; He, R.; Huang, J.; Long, Y.; Deng, X.; Wang, Q.; Wang, C.; Wu, S. Effects of bias voltage on microstructure, mechanical properties, and wear mechanism of novel quaternary (Ti, Al, Zr)N coating on the surface of silicon nitride ceramic cutting tool. *Ceram. Int.* **2016**, *42*, 17693–17697. [CrossRef]
57. Liu, W.; Li, A.; Wu, H.; Long, Y.; Huang, J.; Deng, X.; Wang, C.; Wang, Q.; Wu, S. Effects of gas pressure on microstructure and performance of (Ti, Al, Zr)N coatings produced by physical vapor deposition. *Ceram. Int.* **2016**, *42*, 17436–17441. [CrossRef]

58. Sergevnin, V.S.; Blinkov, I.V.; Belov, D.; Volkhonskii, A.O.; Skryleva, E.A.; Chernogor, A.V. Phase formation in the Ti–Al–Mo–N system during the growth of adaptive wear-resistant coatings by arc PVD. *Inorg. Mater.* **2016**, *52*, 735–742. [CrossRef]
59. Tomaszewski, Ł.; Gulbiński, W.; Urbanowicz, A.; Suszko, T.; Lewandowski, A.; Gulbiński, W. TiAlN based wear resistant coatings modified by molybdenum addition. *Vacuum* **2015**, *121*, 223–229. [CrossRef]
60. Yang, K.; Xian, G.; Zhao, H.; Fan, H.; Wang, J.; Wang, H.; Du, H. Effect of Mo content on the structure and mechanical properties of TiAlMoN films deposited on WC–Co cemented carbide substrate by magnetron sputtering. *Int. J. Refract. Met. Hard Mater.* **2015**, *52*, 29–35. [CrossRef]
61. Sergevnin, V.S.; Anikin, V.N.; Volkhonskii, A.; Belov, D.; Kuznetsov, D.; Gorshenkov, M.; Skryleva, E. Wear behaviour of wear-resistant adaptive nano-multilayered Ti–Al–Mo–N coatings. *Appl. Surf. Sci.* **2016**, *388*, 13–23. [CrossRef]
62. Grigoriev, S.; Vereschaka, A.; Milovich, F.; Tabakov, V.; Sitnikov, N.; Andreev, N.; Sviridova, T.; Bublikov, J. Investigation of multicomponent nanolayer coatings based on nitrides of Cr, Mo, Zr, Nb, and Al. *Surf. Coat. Technol.* **2020**, *401*, 126258. [CrossRef]
63. Vereschaka, A.A. Development of assisted filtered cathodic vacuum arc deposition of nano-dispersed multi-layered composite coatings on cutting tools. *Key Eng. Mater.* **2013**, *581*, 62–67. [CrossRef]
64. Vereschaka, A.A.; Grigoriev, S.N.; Sitnikov, N.N.; Oganyan, G.V.; Batako, A. Working efficiency of cutting tools with multilayer nano-structured Ti–TiCN–(Ti,Al)CN and Ti–TiCN–(Ti,Al,Cr)CN coatings: Analysis of cutting properties, wear mechanism and diffusion processes. *Surf. Coat. Technol.* **2017**, *332*, 198–213. [CrossRef]
65. Grigoriev, S.N.; Sobol, O.V.; Beresnev, V.M.; Serdyuk, I.V.; Pogrebnyak, A.D.; Kolesnikov, D.A.; Nemchenko, U.S. Tribological characteristics of (TiZrHfVNbTa)N coatings applied using the vacuum arc deposition method. *J. Frict. Wear* **2014**, *35*, 359–364. [CrossRef]
66. Vereschaka, A.; Tabakov, V.; Grigoriev, S.; Aksenenko, A.; Sitnikov, N.; Oganyan, G.; Seleznev, A.; Shevchenko, S. Effect of adhesion and the wear-resistant layer thickness ratio on mechanical and performance properties of ZrN–(Zr,Al,Si)N coatings. *Surf. Coat. Technol.* **2019**, *357*, 218–234. [CrossRef]
67. Vereschaka, A.; Tabakov, V.; Grigoriev, S.; Sitnikov, N.; Milovich, F.; Andreev, N.; Sotova, C.; Kutina, N. Investigation of the influence of the thickness of nanolayers in wear-resistant layers of Ti–TiN–(Ti,Cr,Al)N coating on destruction in the cutting and wear of carbide cutting tools. *Surf. Coat. Technol.* **2020**, *385*, 125402. [CrossRef]
68. Vereschaka, A.; Aksenenko, A.; Sitnikov, N.; Migranov, M.; Shevchenko, S.; Sotova, C.; Batako, A.D.L.; Andreev, N. Effect of adhesion and tribological properties of modified composite nano-structured multi-layer nitride coatings on WC–Co tools life. *Tribol. Int.* **2018**, *128*, 313–327. [CrossRef]
69. Vereschaka, A. Improvement of working efficiency of cutting tools by modifying its surface properties by application of wear-resistant complexes. *Adv. Mater. Res.* **2013**, *712*, 347–351. [CrossRef]
70. Volosova, M.A.; Grigor'ev, S.N.; Kuzin, V.V. Effect of titanium nitride coating on stress structural inhomogeneity in oxide-carbide ceramic. Part 4. Action of heat flow. *Refract. Ind. Ceram.* **2015**, *56*, 91–96. [CrossRef]
71. Kuzin, V.V.; Grigor'ev, S.N.; Volosova, M.A. Effect of a TiC coating on the stress-strain state of a plate of a high-density nitride ceramic under nonsteady thermoelastic conditions. *Refract. Ind. Ceram.* **2014**, *54*, 376–380. [CrossRef]
72. Kuzin, V.; Grigoriev, S.N.; Fedorov, M.Y. Role of the thermal factor in the wear mechanism of ceramic tools. Part 2: Microlevel. *J. Frict. Wear* **2015**, *36*, 40–44. [CrossRef]
73. Vereschaka, A.A.; Grigoriev, S.N.; Volosova, M.A.; Batako, A.; Vereschaka, A.S.; Sitnikov, N.; Seleznev, A. Nano-scale multi-layered coatings for improved efficiency of ceramic cutting tools. *Int. J. Adv. Manuf. Technol.* **2016**, *90*, 27–43. [CrossRef]
74. Vereschaka, A.; Tabakov, V.; Grigoriev, S.; Sitnikov, N.; Oganyan, G.; Andreev, N.; Milovich, F. Investigation of wear dynamics for cutting tools with multilayer composite nanostructured coatings in turning constructional steel. *Wear* **2019**, *420–421*, 17–37. [CrossRef]
75. Vereschaka, A.; Grigoriev, S.; Sitnikov, N.; Milovich, F.; Aksenenko, A.; Andreev, N. Investigation of performance and cutting properties of carbide tool with nanostructured multilayer Zr–ZrN–(Zr0.5,Cr0.3,Al0.2)N coating. *Int. J. Adv. Manuf. Technol.* **2019**, *102*, 2953–2965. [CrossRef]

76. Vereschaka, A.S.; Grigoriev, S.N.; Tabakov, V.P.; Sotova, E.S.; Vereschaka, A.A.; Kulikov, M.Y. Improving the efficiency of the cutting tool made of ceramic when machining hardened steel by applying nano-dispersed multi-layered coatings. *Key Eng. Mater.* **2014**, *581*, 68–73. [CrossRef]
77. Engineering ToolBox. Thermal Conductivity of Metals, Metallic Elements and Alloys. 2005. Available online: https://www.engineeringtoolbox.com/thermal-conductivity-metals-d_858.html (accessed on 1 November 2020).

Article

Mathematical Modeling of Carbon Flux Parameters for Low-Pressure Vacuum Carburizing with Medium-High Alloy Steel

Haojie Wang [1], Jing Liu [2,*], Yong Tian [2], Zhaodong Wang [2,*] and Xiaoxue An [2]

1 School of Mechanical Engineering, Shenyang University of Technology, Shenyang 110870, China; wanghaojie8219@126.com

2 State Key Laboratory of Rolling and Automation, Northeastern University, Shenyang 110819, China; tianyong@ral.neu.edu.cn (Y.T.); 15140110204@163.com (X.A.)

* Correspondence: liujingxf@126.com (J.L.); zhaodongwang@263.net (Z.W.); Tel.: +86-24-8105-1555 (J.L.); +86-24-8105-1555 (Z.W.)

Received: 9 October 2020; Accepted: 5 November 2020; Published: 9 November 2020

Abstract: Low-pressure vacuum carburizing adopts a pulse process mode to improve the carburizing efficiency and reduces gas and energy consumption. Carbon flux is the key to accurately control the time of strong infiltration and diffusion in each pulse. In order to obtain the carbon fluxes with various materials under diffident carburizing process conditions, an evenly segmented carbon flux method is proposed. A systematic study with each model using different materials (12Cr2Ni4A, 16Cr3NiWMoVNbE, and 18Cr2Ni4WA represent different initial carbon concentrations and different alloy compositions), carburizing temperatures, and carburizing pressures to determine the effect of these conditions on carbon flux is conducted. Compared with traditional segmented carbon flux method, an evenly segmented carbon flux method can predict the actual carbon flux more precisely and effectively in order to finely control the pulse carburization process. The paper also indicates that carbon fluxes increase with the increase of pressure. The optimal carburization pressure for low-pressure vacuum carburization is 300 Pa. Raising the carburization temperature to 980 °C instead of 920 °C can increase effective carbon flux by more than 30%. Among the material compositions, alloy content has the biggest impact over the carbon, initial carbon concentration the second, and saturated carbon concentration the third biggest impact.

Keywords: modeling; carbon flux; low-pressure vacuum carburizing; medium-high alloy steel

---

1. Introduction

Low-pressure carburization especially combined with high-pressure gas quenching is a new technology that allows accurate control over the carbide morphology, size, distribution, surface carbon content, and layer depth [1–3]. It also offers energy savings and environmental protection and is being applied to various gear steels. However, for high alloy steel, it is easy to exceed the standard of carburized layer carbide after carburizing. The main reason is the accuracy of carbon flux in the process of formulation. Carbon flux represents the mass of material passing through a unit area perpendicular to diffusion direction per unit time. The value of carbon flux is related to factors such as the material composition, temperature, carburizing pressure, and catalyst [4,5]. During carburization, the gas contacting the steel surface undergoes a chemical reaction, but the growth in mass cannot be measured in real time, rendering it difficult to measure the carbon flux. Hwang et al. [6,7] studied the effect of material composition and surface oxidation on carbon flux, and Kula et al. [8] studied the effect of carbides on carbon flux at the surface. Furthermore, Karabelchtchikova and Sisson [9] proposed an improved integration method to determine the carbon flux in the gas boundary layer in austenite, stating that the carbon flux changes with both time and carbon concentration on the workpiece surface. Zajusz et al. [10] optimized the carbon transfer coefficient in their research. They analyzed the effects of pressure, reaction rate, and alloy elements through extensive experiments, thereby obtaining the effect of process conditions on carbon flux, which achieved good results when applied to the model of carburizing. Lowell [11] studied the effect of workpiece surface roughness and the type of contaminant on the carbon flux of low-pressure vacuum carburizing surface. Gorockiewicz et al. [12,13] found that in the boost stage of carburization, carbon atoms are first adsorbed by the workpiece surface to form crystals, and then diffused as a carbon source. Su [14] proposed a saturated carburizing model for carbide migration and calculated the variation of average carbon flux with carburizing time. However, these studies mainly focus on low carbon steel and low alloy steel. For high alloy steel, it is difficult to control the surface carbides. The carbon flux is affected by many factors, so it is hard to determine the carbon flux accurately [15,16].

For low-pressure vacuum carburizing, the carbon potential of the atmosphere cannot be measured or controlled. The measurement of carbon flux is also difficult and there is no suitable method for on-line continuous testing. Philippe et al. [17] experimentally measured the change in carbon flux using a thin saturated carburizing iron foil with a carburizing atmosphere on one side and a decarburizing atmosphere on the other side. Therein, the carbon flux was determined by measuring the carbon flow rate in a decarburizing atmosphere. Owing to the short duration of the boost stage, the actual measurement precision and response time were difficult to control. Therefore, the application of this method to low-pressure vacuum carburization in real-world applications requires further research.

Oriental Furnace (Japan) and Dowa Holdings proposed a method to determine the carbon concentration using a high-temperature hydrogen probe [18] to analyze the hydrogen content in a gas mixture. Khan et al. [19] studied the cracking of propane during low-pressure vacuum carburization for carburization control. Furthermore, Yada et al. [20] and Makino et al. [21] numerically simulated and tested the carburization process and detected the decomposition of acetylene using the volume fraction of hydrogen in the gas mixture to calculate the carbon potential in the furnace. Thus far, this technique has not been successfully applied in industrial applications, and the values of relevant parameters remain confidential.

Currently, majority carbon flux determination methods utilize an experimentally obtained average value, without adequately considering the effects of process parameters such as carburizing temperature and pressure and material composition; this results in a relatively large experimental error. In practice, carbon flux continuously changes in the carburizing process; hence, most carbon flux values exhibit large deviation in real-life applications, particularly for high alloy materials. To the best of our knowledge, there has been no report on mathematical modeling related to this issue.

In the present study, a comparison of the carburized layer organizations obtained by the average and segmented average carbon flux models has been given. It is found that carbide network is more easily to form on the surface-carburized layer with an average carbon flux model. To accurately control the carburizing process, a systematic study with each model using different materials (12Cr2Ni4A, 16Cr3NiWMoVNbE, and 18Cr2Ni4WA represent different initial carbon concentrations and different alloy compositions), carburizing temperatures, and carburizing pressures is conducted to determine the effect of these conditions on carbon flux.

Nomenclature			
J	average carbon flux, g/(mm^2·s)	p	pressure of the carburized gas, Pa
J_e	segmented average carbon flux, g/(mm^2·s)	T	carburizing temperature, °C
Δm	mass difference, mg	R	gas constant, J/(mol·K)
S	surface area, mm^2	k	Pressure and tem/perature dependent constant
t	time, s	A	Pressure and temperature dependent constant
Δt	segmented carburizing time, s		

2. Mathematical Modeling of Carbon Flux

Carbon flux is the main parameter that influences model calculations and determines the control precision of the carburizing process. The transfer process of carbon atoms from the atmosphere to the workpiece surface can be defined by three types of boundary conditions. In typical atmospheric carburizing, the carbon potential control method is used, i.e., the third type of boundary condition. The difference between the carbon concentration of the atmosphere and the carbon concentration of the metal surface acts as the driving force for carbon to enter into the steel from the gas phase. During carburization, an oxygen probe is used to measure the carbon concentration in the atmosphere, and the steel foil method is applied to reverse calculate the transfer coefficient, thus achieving the detection and control of the carburization process [8].

Owing to the vacuum environment of low-pressure vacuum carburizing, the carburizing gas concentration is low, and the carbon concentration in the atmosphere cannot be measured using an oxygen probe. In this scenario, the second type of boundary condition is used, i.e., the carbon flux. Different from atmospheric carburizing, low-pressure vacuum carburizing has a low carburizing pressure. The number of carburizing gas molecules is extremely small and the molecules can be quickly decomposed and absorbed onto the steel surface, instantly boosting the surface carbon flux to the maximum. Therefore, the surface carbon flux model normally used for atmospheric carburizing is no longer applicable [22]. The value of carbon flux is related to the carburizing gas type, pressure, temperature, alloy type, and surface state. To study low-pressure vacuum carburizing, the overall average carbon flux method is generally used [23], as shown in Equation (1). This method obtains the carbon flux by experimentally measuring the mass increment before and after a certain period of carburization.

$$J = \frac{\Delta m}{S \cdot t}. \tag{1}$$

Average Carbon Flux vs. Segmented Average Carbon Flux

The value of carbon flux is directly related to carburizing pressure, carburizing temperature, and material properties [24]. In process design, the average carbon flux (mass increment detection after carburizing for 120 s or longer) is measured. Herein, the cylindrical samples are carburized for 30, 60, 90, and 120 s, and are subsequently weighed. Different from typical carbon flux measurement, the carbon flux of the samples at different times of carburization is calculated separately and defined as the segmented average carbon flux (J_e), as shown in Equation (2).

$$J_e = \frac{\Delta m}{S \cdot \Delta t}. \tag{2}$$

High alloy materials are more likely to absorb carbon atoms. To better determine the accuracy of the original carbon flux model, a carburizing experiment was performed on 16Cr3NiWMoVNbE steel at a carburizing temperature of 920 °C and carburizing pressure of 300 Pa. The average carbon flux over 120 s and the segmented average carbon flux were measured. The obtained carbon flux over carburizing time is shown in Figure 1.

Figure 1 shows that the average carbon flux and segmented average carbon flux both decrease as carburizing time increases. During carburization, the carbon concentration gradient at the carburized layer gradually decreases; hence, the number of carbon atoms entering the sample surface per unit time decreases. In the initial stage of carburization from 0 to 60 s, the carbon concentration gradient at the gas–solid interface is relatively large. After 30 s, the carbon flux obtained using the segmented average method is 2.5 times that of the carbon flux obtained by the overall average method, which result in the obtained value exceeding the carbide standard in the carburized layer.

Herein, two carbon flux models using the average carbon flux and segmented average carbon flux are used to calculate the carburizing process; the corresponding results are shown in Figure 2. The number of carburizing pulses is 19 according to both models. As carburizing proceeds, the carburizing time gradually decreases, and the diffusion time gradually increases. Carburizing stops once the surface carbon concentration reaches 1.3%, and then restarts once the concentration reaches 1.1% through diffusion. For the last pulse, the surface carbon concentration decreases to 0.9%, and the carbon concentration at 1.1 mm of the carburized layer increases to 0.42%. The carburizing process is then completed. The total carburizing time is 18,500 s.

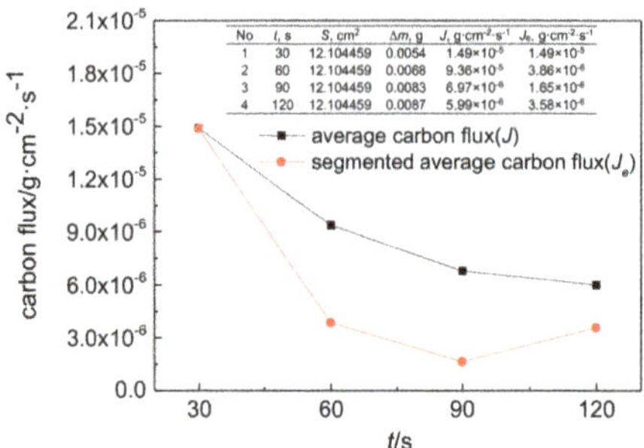

Figure 1. Variation in carbon flux with carburizing time.

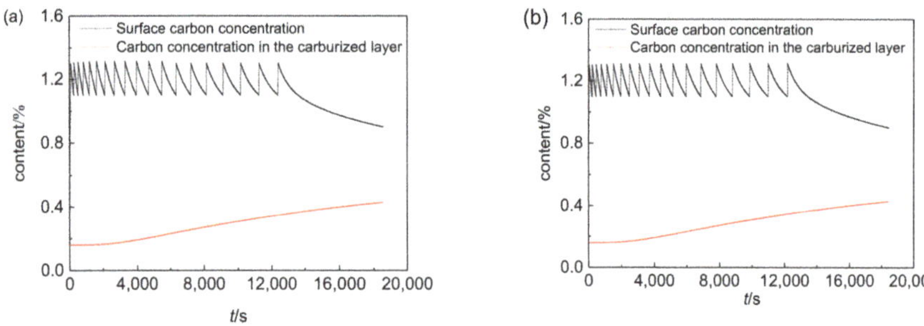

Figure 2. Process curves of (**a**) the average carbon flux method and (**b**) the segmented average carbon flux method.

The changes in surface carbon concentration over carburizing time and boost time along with the number of pulses obtained from the simulations using average carbon flux and segmented average carbon flux are shown in Figures 3 and 4, respectively.

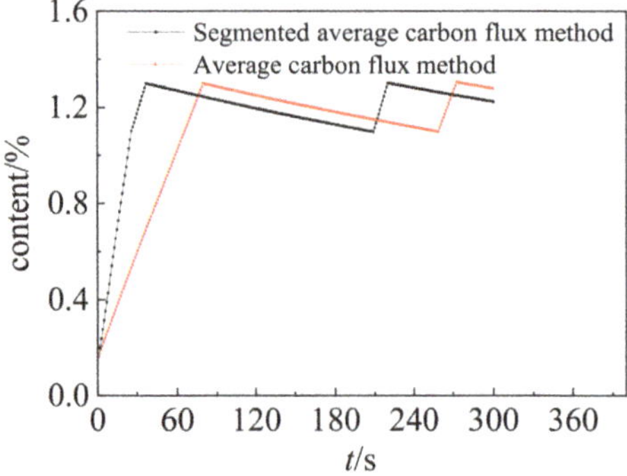

Figure 3. First three pulse curves obtained via different carbon flux measurement methods.

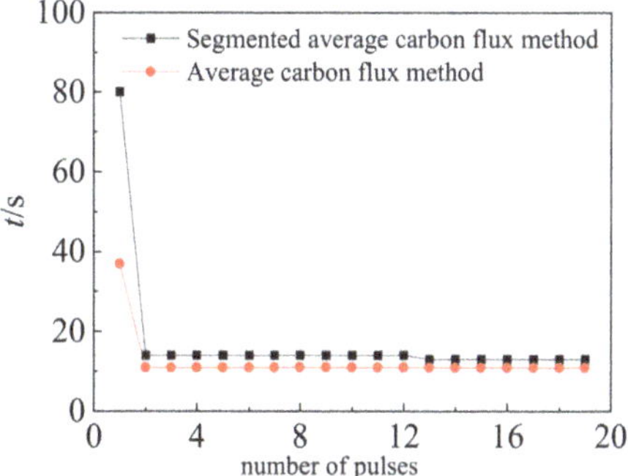

Figure 4. Carburizing time of each pulse obtained via different carbon flux measurement methods.

Using the average carbon flux model, the pulse time of the first boost stage is 80 s, and the pulse time of boost stages thereafter (2nd–19th) is 14 s. However, the data obtained using the segmented average carbon flux method shows that the first boost had a pulse time of 37 s, and the boost stages thereafter (2nd–19th) had a pulse time of 11 s. The carburizing time of the first boost pulse obtained using the average carbon flux method is 2.2 times longer than the segmented average carbon method. For subsequent pulse times, the first boost pulse obtained using the average carbon flux method is 1.3 times longer than the segmented average carbon method.

The variation in the different boost times will cause differences in the carbide morphology of the carburized layer. The results of carburized layer organization after the different carburizing processes (carburizing and quenching) are shown in Figure 5. For the same diffusion time, the surface organization of the carburized layer obtained using the average carbon flux model appears to have a discontinuous carbide network and some continuous carbide network, whereas the surface organization obtained using the segmented carbon flux model has more diffused carbides, which verifies the efficiency of the proposed model and enables a more accurate and efficient characterization of the actual carbon flux value, effectively avoiding the effects of the first pulse and achieving more precise control over the time of subsequent pulses.

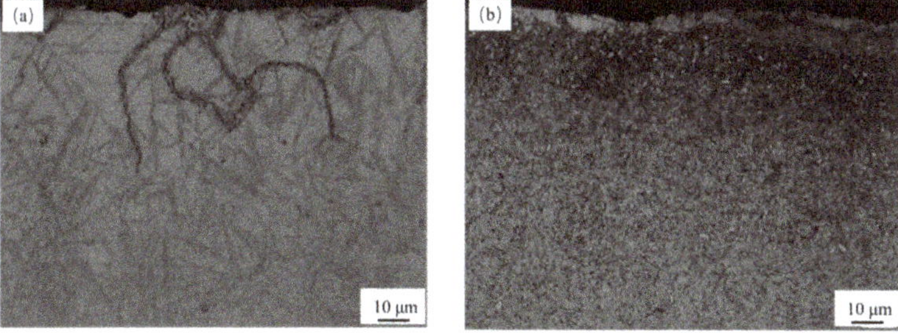

Figure 5. Surface organization of carburized layer of 16Cr3NiWMoVNbE steel under (**a**) the average carbon flux model and (**b**) the segmented average carbon flux model after carburizing and quenching.

Based on this influencing factor, a systematic study on the carbon flux model with different materials, carburizing temperature, and carburizing pressure conditions was conducted to determine the effect of the different process conditions on carbon flux and to build a mathematical model for carbon flux.

3. Materials and Methods

3.1. Experimental Materials

To study the overall pattern of carbon flux and consider the contingency of the experiment, three alloy steel materials were selected for carburization: 12Cr2Ni4A, 16Cr3NiWMoVNbE, and 18Cr2Ni4WA. Table 1 shows the chemical composition of the studied alloy steels. These materials enable the study of carbon flux values with different initial carbon contents and different alloying elements under different carburizing conditions. For this experiment, the samples were processed to have an outer diameter of 24 mm, a small hole with a diameter of 3 mm, and a thickness of 3 mm. As the mass increment is calculated in milligrams, the samples were specially cleaned to maintain the validity of the experiment. The vacuum furnace had a degreasing effect. The samples were first cleaned with alcohol and then dried, followed by heating in a vacuum furnace to 920 °C. They were kept warm for 30 min and then left to cool in the vacuum cooling chamber. The samples were weighed before and after carburization using a LE204E 1/10000 scale (Dongbo Thermal Technology Co., Ltd., Shenyang, China), and the average of the three experiments was considered the measurement value. The experiments were carried out in a dual-chamber vacuum carburizing furnace.

Table 1. Chemical compositions of the alloy steels (wt.%).

Grade	C	Cr	Ni	Mn	Si	W	V	Mo	Nb	Ce	Fe
12Cr2Ni4A	0.12	1.5	3.5	0.5	0.3	-	-	-	-	-	Bal.
16Cr3NiWMoVNbE	0.16	2.8	1.4	0.5	0.8	1.2	0.45	0.5	0.15	0.1	Bal.
18Cr2Ni4WA	0.18	1.5	3.9	0.4	0.23	0.9	-	-	-	-	Bal.

3.2. Experimental Procedures of Carbon Flux Measurement

The low-pressure vacuum carburizing furnaces were heated to 920 °C, 950 °C, and 980 °C. The pretreated samples were transferred into a hot chamber from a cold chamber and kept warm for 20 min to ensure they were fully heated. Then, the carburizing process was performed. The pressure in the carburizing furnaces was rapidly increased to 100, 200, and 300 Pa with a large flow setting. The closed-loop control of the carburizing pressure was realized by controlling the pumping speed and air intake volume of the root pump to maintain the pressure within the range of ±10 Pa. After stabilizing the pressure for 30, 60, 90, 120, and 150 s, the carburizing gas was quickly extracted, the samples were taken out of the furnaces, and then immediately quenched and weighed after cleaning. The schematic of the pressure maintaining process in the boost stage is shown in Figure 6.

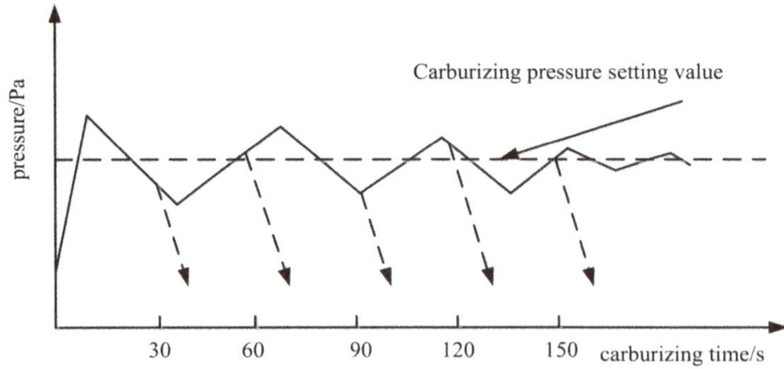

Figure 6. Schematic of the boost stages of the carburizing process.

4. Results and Discussion

4.1. Carbon Flux Measurement

4.1.1. Effect of Carburizing Pressure on Surface Carbon Flux

The surface mass increment, average carbon flux changes, and segmented average carbon flux changes were investigated at a carburizing temperature of 950 °C, and carburizing pressures of 100, 200, and 300 Pa to determine the effect of carburizing pressure on the surface carbon flux. Figure 7 shows the surface mass increment, average carbon flux, and segmented average carbon flux of 12Cr2Ni4A steel over time.

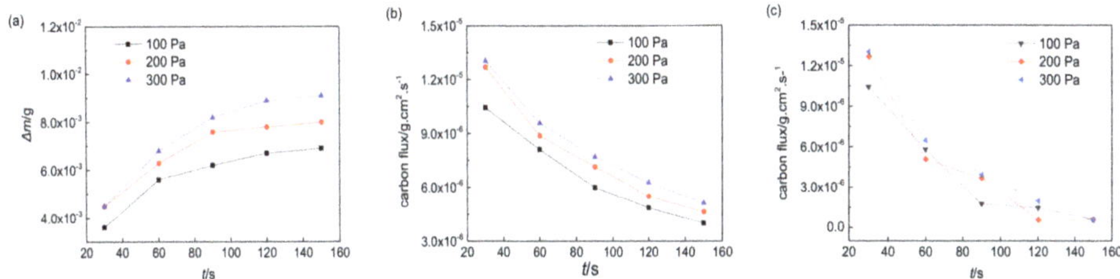

Figure 7. Effect of carburizing pressure on (**a**) mass increment, (**b**) average carbon flux, and (**c**) segmented average carbon flux for 12Cr2Ni4A steel at 950 °C.

During the boost stage of 12Cr2Ni4A steel, the mass linearly increased quickly from 0 to 60 s. From 60 to 120 s, the mass gradually decreased, and approached zero after 120 s. The segmented average carbon flux value increased from 1.04×10^{-5} to 1.36×10^{-5} at 30 s, and from 8.11×10^{-6} to 9.58×10^{-6} at 60 s. The segmented average carbon flux values are 5.98×10^{-6} and 7.7×10^{-6} at 90 s, respectively. From 120 to 150 s, the segmented average carbon flux remained nearly identical at the three pressures mentioned above, exhibiting relatively small flux values.

Figure 8 shows the surface mass increment, average carbon flux, and segmented average carbon flux for 16Cr3NiWMoVNbE steel over time.

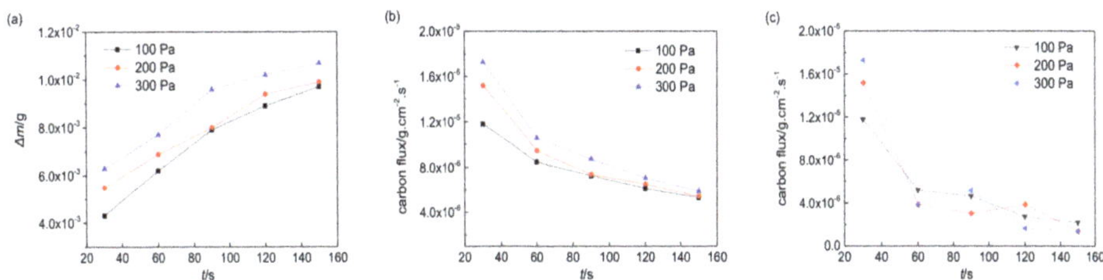

Figure 8. Effect of carburizing pressure on (**a**) mass increment, (**b**) average carbon flux, and (**c**) segmented average carbon flux for 16Cr3NiWMoVNbE steel at 950 °C.

During the boost stage for 16Cr3NiWMoVNbE steel, the mass linearly increased relatively quickly with increasing pressure. After 90 s, the mass increment gradually decreased and approached zero after 120 s. The surface carbon flux value increased from 1.17×10^{-5} to 1.72×10^{-5} at 30 s and from 5.2×10^{-6} to 3.8×10^{-6} at 60 s. The effect of pressure on the carbon flux was small after 90 s.

Figure 9 shows the surface mass increment, average carbon flux, and segmented average carbon flux of 12Cr2Ni4A steel over carburizing time. During the boost stage, the mass linearly increased relatively quickly from 0 to 90 s. After 90 s, the mass increment gradually decreased and approached zero after 120 s. The surface carbon flux value increased from 6.5×10^{-6} to 1.3×10^{-5} at 30 s and from

5.7×10^{-6} to 4.5×10^{-6} at 60 s. After 90 s, the effect of pressure on the carbon flux was small, and the values of the flux were small.

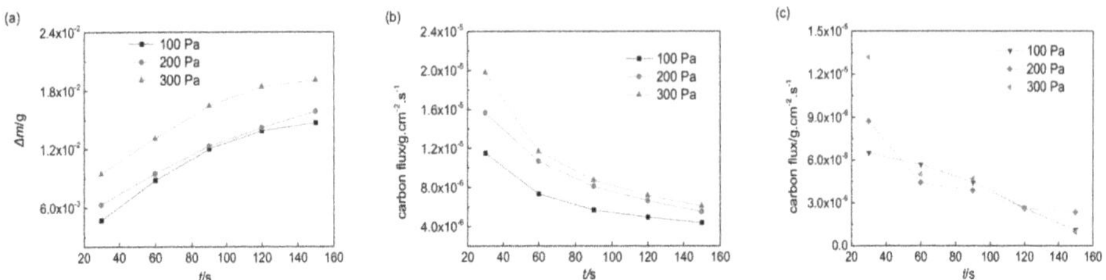

Figure 9. Effect of carburizing pressure on (**a**) mass increment, (**b**) average carbon flux, and (**c**) segmented average carbon flux of 18Cr2Ni4WA steel.

The relationship between carburizing pressure and carbon flux of the above three materials shows that under the same carburizing temperature of 950 °C and carburizing pressures of 100, 200, and 300 Pa, the mass increment, average carbon flux, and segmented carbon flux of the materials exhibit the same trend. The mass of the samples increased between 30 and 90 s. From 90 to 150 s, the rate of mass increase gradually reduced, the average carbon flux tended to stabilize, and the segmented average carbon flux approached zero. With increasing pressure, the overall carbon flux tended to increase. Under the carburizing pressures of 200 and 300 Pa, the carbon flux values exhibited small differences. As the pressure increased, the time required to saturate the carbon concentration on the sample surface decreased.

The carbon flux value was microscopically characterized as the flow rate of acetylene in contact with the workpiece surface, which is microscopically reflected as the number of activated carbon atoms and the time of contact between the activated carbon atoms and the workpiece. During the boost stage of carburization, the acetylene reaction produced hydrogen gas. As carburization proceeded, the gas in the furnace gradually transformed from acetylene to a mixture of acetylene and hydrogen. The flow rate of the acetylene on the sample surface was influenced by the speed of the acetylene on the workpiece surface and the relative effective density (density of acetylene in the gas mixture in the furnace), both of which are pressure dependent.

The process of low-pressure vacuum carburizing requires stable carburizing pressure and carburizing temperature. When the carburizing temperature, acetylene inlet gas flow, and furnace volume were fixed to maintain the dynamic balance of gas pressure in the furnace, adjusting the outlet pumping speed of the furnace was necessary. At a lower carburizing pressure, the outlet speed was at its highest, the gas exchange efficiency in the furnace was high, the relative effective density of acetylene increased, and the probability of active gas flowing through the surface of the workpiece increased; however, the effective density of active gas in the furnace was low. Considering the two factors, if the outlet pumping speed is too large, the contact time between the effective carbon atoms and the workpiece will be shortened, causing the decomposed active carbon atoms to be pumped out of the furnace without contact with the workpiece. This would further decrease the flow rate of acetylene on the surface of the sample, resulting in a decrease in the carbon flux. With the increase in carburizing pressure in the furnace, the outlet pumping speed gradually decreased, and the density of the carburizing atmosphere in the furnace increased. The number of active carbon atoms increased, and the carbon flux gradually increased. When the carburizing pressure reached a certain value, the gas flow in the furnace decreased, and although the density of the workpiece surface increased, the hydrogen content gradually increased per unit time, and the density of the effective active gas gradually decreased, additionally decreasing the value of carbon flux.

For the three representative materials discussed herein, at a carburizing temperature of 950 °C, the value of carbon flux was the smallest at a carburizing pressure of 100 Pa. When the pressure was low and the gas exchange rate was high, part of the decomposed active carbon atoms may have been pumped out before contacting the surface. When the carburizing pressure increased to 200 Pa, the relative density plays a dominant role in effective carbon flux, the pressure increased, the pumping rate decreased, the impact of the flow rate was small, and the effect of the gas exchange rate was small. The time for which acetylene molecules remained on the surface of the workpiece and the relative density increased. When the pressure increased to 300 Pa, the value of carbon flux was close to that under 200 Pa, indicating that the flow rate and the relative effective density attained equilibrium. Therefore, in actual carburization, a pressure of 300 Pa will have relatively little effect on the process.

4.1.2. Effect of Carburizing Temperature on Surface Carbon Flux

Based on the effect of carburizing pressure in the previous section, the mass increment, average carbon flux, and segmented average carbon flux variations at different carburizing temperatures were investigated.

Figure 10 shows the surface mass increment, average carbon flux, and segmented average carbon flux with carburizing time for 12Cr2Ni4A steel under a carburizing pressure of 300 Pa at carburizing temperatures of 920, 950, and 980 °C.

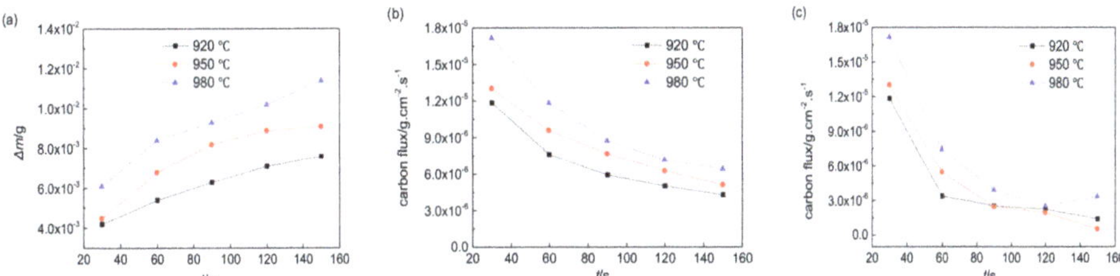

Figure 10. Effect of carburizing temperature on (**a**) mass increment, (**b**) average carbon flux, and (**c**) segmented average carbon flux of 12Cr2Ni4 steel under 300 Pa.

During the boost stage for all carburizing temperatures, the mass linearly increased relatively quickly for 12Cr2Ni4A steel from 0 to 60 s. From 60 to 120 s, the mass increment gradually decreased and approached zero after 120 s. The average carbon flux value decreased by 50%–60%, and the segmented average carbon flux value decreased by ~80%. On increasing the carburizing temperature from 920 °C to 980 °C, the segmented average carbon flux value of the surface also increased from 1.18×10^{-5} to 1.72×10^{-5} at 30 s and from 3.38×10^{-6} to 7.48×10^{-6} at 60 s. At 90 s, the segmented average carbon flux values at 920 and 950 °C were 2.53×10^{-6} and 2.48×10^{-6}, respectively. The carbon flux corresponding to 980 °C was 3.94×10^{-6}. From 120 to 150 s, the segmented average carbon flux remained nearly identical at all three pressures, exhibiting relatively small flux values.

Figure 11 shows the surface mass increment, average carbon flux, and segmented average carbon flux with carburizing time for 16Cr3NiWMoVNbE steel under a carburizing pressure of 300 Pa at the carburizing temperatures of 920, 950, and 980 °C.

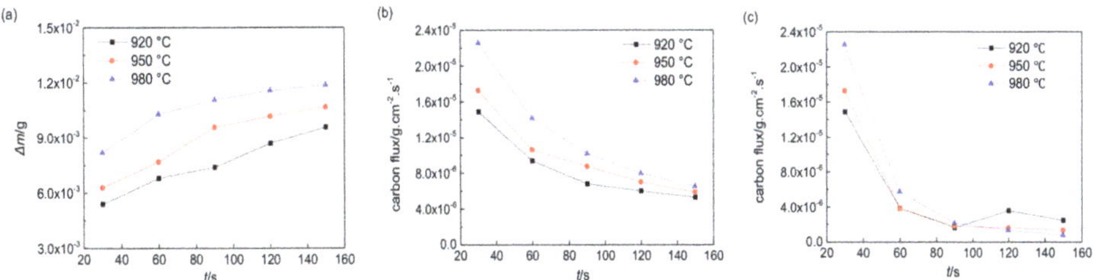

Figure 11. Effect of carburizing temperature on (**a**) mass increment, (**b**) average carbon flux, and (**c**) segmented average carbon flux of 16Cr3NiWMoVNbE steel under 300 Pa.

The mass linearly increased relatively quickly for 16Cr3NiWMoVNbE steel from 0 to 30 s. From 60 to 90 s, the mass increment gradually decreased and approached zero after 90 s. On increasing the carburizing temperature from 920 to 980 °C, the segmented average carbon flux value of the surface also increased from 1.49×10^{-5} to 2.26×10^{-5} at 30 s and from 3.86×10^{-6} to 5.78×10^{-6} at 60 s, respectively. At 90 s, the segmented average carbon flux values at 920 and 950 °C were 1.6×10^{-6} and 1.8×10^{-6}, respectively, i.e., they were practically identical. The carbon flux at 980 °C had a minimum of 1.38×10^{-6}. From 120 to 150 s, the segmented average carbon flux exhibited relatively small flux values under all temperatures.

Figure 12 shows the surface mass increment, average carbon flux, and segmented average carbon flux with carburizing time for 18Cr2Ni4WA steel under a carburizing pressure of 300 Pa at different carburizing temperatures. For 16Cr3NiWMoVNbE steel, the mass linearly increased relatively quickly from 0 to 60 s. From 60 to 150 s, the mass increment gradually decreased and approached zero after 150 s. On increasing the carburizing temperature from 920 to 980 °C, the segmented average carbon flux value of the surface also increased: from 1.19×10^{-5} to 1.98×10^{-5} at 30 s, and from 3.88×10^{-6} to 4.6×10^{-6} at 60 s. From 90 to 120 s, the segmented average carbon flux value at all three temperatures was relatively small and consistent.

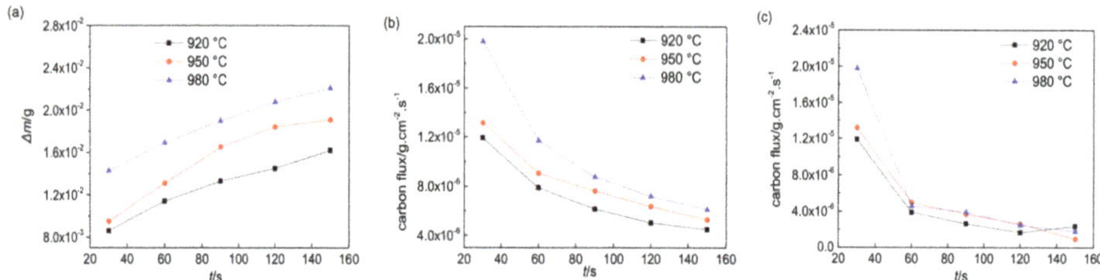

Figure 12. Effect of carburizing temperature on (**a**) mass increment, (**b**) average carbon flux, and (**c**) segmented average carbon flux of 18Cr2Ni4WA steel under 300 Pa.

As can be observed in Figures 10–12, the average carbon flux and segmented carbon flux show the same trend for all three materials at the same holding temperature. Under the same holding pressure, with an increase of carburization temperature, the average carbon flux, and the segmented average carbon flux both gradually increased.

As carburizing proceeded, carbon atoms gradually accumulated on the steel surface until saturation was attained, and the mass growth became gradually slower until equilibrium was reached. The main driving force for carburizing is the carbon concentration gradient at the gas–solid interface, where the carbon concentration in the atmosphere is in a state of supersaturation. During carburization, the carbon concentration on the steel surface gradually increased, causing the concentration gradient to decrease slowly, thereby reducing the driving force as well as the effective carbon flux value.

As the carburizing temperature increased, the maximum solid solubility of carbon in austenite gradually increased, along with the carbon concentration gradient on the surface of the carburized layer. The steel surface needed more carbon atoms to reach saturation, and thus the carburizing time required increased. In addition, the diffusion coefficient of carbon in steel became larger, resulting in the number of surface carbon atoms entering inside increasing. Under the joint effect of these two factors, both the overall average carbon flux and segmented average carbon flux values of steel samples increased with carburizing temperature. The effective carbon flux increased by more than 30% on increasing the temperature from 920 to 980 °C. Furthermore, the time required for the carbon concentration on the sample surface to reach saturation also increased with the carburization temperature, which is consistent with the results of this experiment.

4.1.3. Effect of Material on Surface Carbon Flux

The mass increment, average carbon flux, and segmented average carbon flux changes of three different materials were studied at the carburizing temperatures of 920, 950, and 980 °C and carburizing pressures of 100, 200, and 300 Pa to determine the effect of the three carburizing materials on surface carbon flux. Figures 13 and 14 show the effect of carburizing pressure and carburizing temperature on the average carbon flux within 30 s for the three steel materials. 12Cr2Ni4A is indexed as steel #1, 16Cr3NiWMoVNbE as steel #2, and 18Cr2Ni4WA as steel #3.

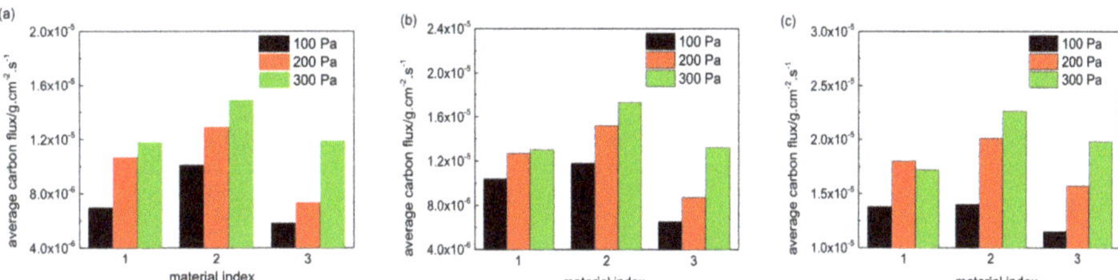

Figure 13. Effect of carburizing pressure on average carbon flux of three materials (**a**) 100 Pa; (**b**) 200 Pa; (**c**) 300 Pa.

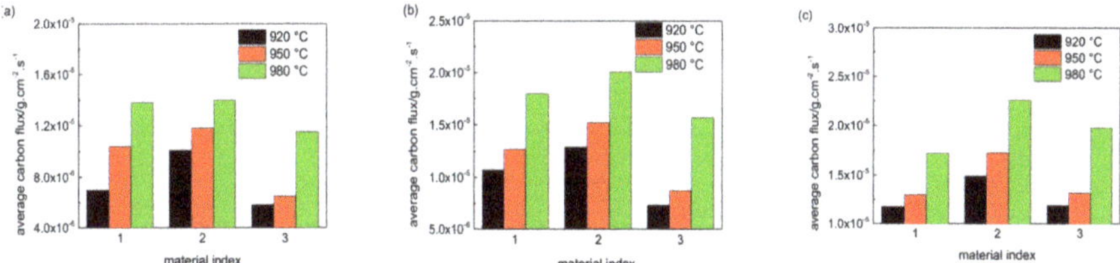

Figure 14. Effect of carburizing temperature on average carbon flux of three materials in 30s (**a**) 920 °C; (**b**) 950 °C; (**c**) 980 °C.

From the experimental data at the same carburization temperature and pressure, the order of the carbon flux values at 30 s was 12Cr2Ni4A, 18Cr2Ni4WA, and then 16Cr3NiWMoVNbE (without considering diffusion). The average carbon flux is affected by three factors: original carbon content, saturated carbon concentration, and alloy content. The carbon content of 12Cr2Ni4A, 16Cr3NiWMoVNbE, and 18Cr2Ni4WA was 0.12%, 0.16%, and 0.18%, respectively. In terms of alloy elements, 18Cr2Ni4WA has 1% more W than 12Cr2Ni4A, and 16Cr3NiWMoVNbE steel has a more complex alloy system, which contains a large amount of W, Mo, V, and Nb. In the alloy composition of the three materials, Cr, W, Mo, V, and Nb are strong carbide forming elements, whereas Si and Ni are non-carbide forming elements. The alloy content had the greatest effect on carbon flux, followed by the initial carbon concentration, and then saturated carbon concentration. Therefore, 16Cr3NiWMoVNbE steel is more receptive to active carbon atoms.

4.2. Carbon Flux Modeling for the Carburization Process

The amount of carbon atoms transferred at the metal surface is defined as carbon flux, which is determined by temperature and pressure, according to physicochemical principles. In practical, the carbon transfer from the atmosphere to the steel surface is proportional to the adsorption capacity of carburizing gas medium, so it follows the empirically derived Freundlich equation [25]:

$$J = kp^{\frac{RT}{A}} \tag{3}$$

As can be seen from the above analysis, in the first carburizing pulse, which occurs generally within the first 30 s, the carbon concentration gradient and the carbon flux are both relatively large. When the carbon concentration on the workpiece surface reaches a certain level (an intermediate carbon concentration), the carbon concentration gradient decreases by more than 80%. The driving force for carbon atoms to enter the sample has decreased by a relatively large amount, and the carbon flux also shows an order-of-magnitude decrease, which is a process maintained in the subsequent 30 s to 90 s. After this, the carbon flux value of the pulse takes the average carbon flux value at 90 s owing to its small carbon concentration gradient. In the carburization process, the initial carburization of the first pulse plays a dominant role in carburization. Therefore, a regression analysis on the adsorption amount of three materials under different carburization temperature and pressure conditions within the first 30 s was conducted. The carbon flux values of the three materials in 30 s are shown in Tables 2–4.

Table 2. Effect of carburizing temperature and pressure on average carbon flux of 12Cr2Ni4A steel.

Pressure/Pa	Temperature/°C	Average Carbon Flux in 30 s/g·cm^{-2}·s^{-1}
100	920	6.96×10^{-6}
	950	1.04×10^{-5}
	980	1.38×10^{-5}
200	920	1.07×10^{-5}
	950	1.27×10^{-5}
	980	1.80×10^{-5}
300	920	1.18×10^{-5}
	950	1.30×10^{-5}
	980	1.72×10^{-5}

Table 3. Effect of carburizing temperature and pressure on average carbon flux of 16Cr3NiWMoVNbE steel.

Pressure/Pa	Temperature/°C	Average Carbon Flux in 30 s/g·cm^{-2}·s^{-1}
100	920	1.01×10^{-5}
	950	1.18×10^{-5}
	980	1.40×10^{-5}
200	920	1.29×10^{-5}
	950	1.52×10^{-5}
	980	2.01×10^{-5}
300	920	1.49×10^{-5}
	950	1.73×10^{-5}
	980	2.26×10^{-5}

Table 4. Effect of carburizing temperature and pressure on average carbon flux of 18Cr2Ni4WA steel.

Pressure/Pa	Temperature/°C	Average Carbon Flux in 30 s/g·cm^{-2}·s^{-1}
100	920	5.82×10^{-6}
	950	6.51×10^{-6}
	980	1.15×10^{-5}
200	920	7.34×10^{-6}
	950	8.73×10^{-6}
	980	1.57×10^{-5}
300	920	1.19×10^{-5}
	950	1.32×10^{-5}
	980	1.98×10^{-5}

Equation (3) is converted to obtain Equation (4):

$$\ln J = \ln k + \frac{RT}{A}\ln p = \frac{R}{A}\ln p \cdot T + \ln k, \tag{4}$$

$$\frac{R}{A}\ln p = M, \tag{5}$$

$$\ln k = N. \tag{6}$$

p is taken as a constant. In this way, linear regression can be performed on the data of the three different materials in Tables 2–4.

Figure 15 shows the regression curves of carburization temperature at different pressure.

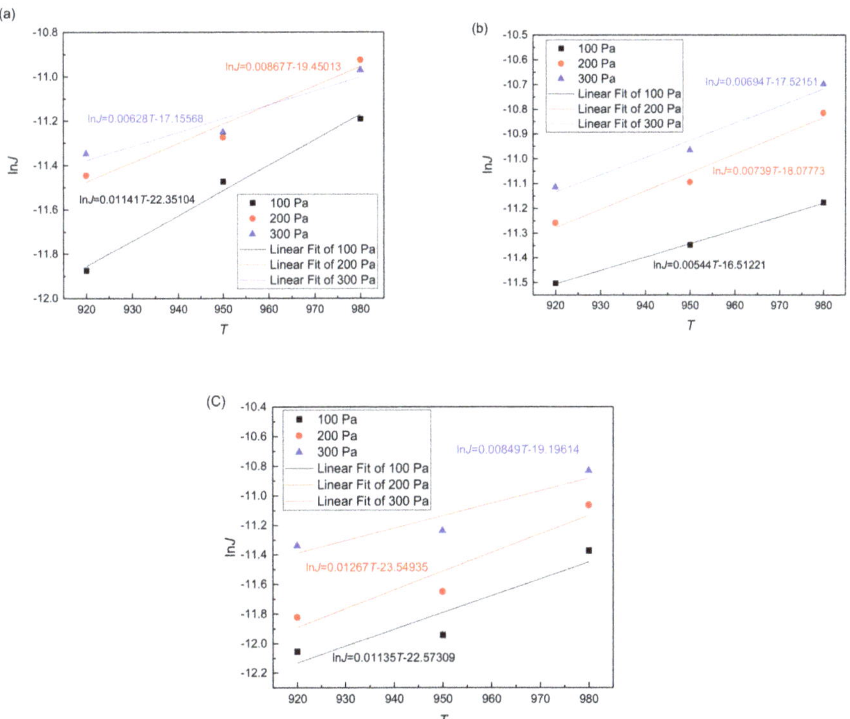

Figure 15. Regression curves of carburization temperature and ln$J$: (**a**) 12Cr2Ni4A steel, (**b**) 16Cr3NiW MoVNbE steel, and (**c**) 18Cr2Ni4WA steel.

The regression results are shown in Equations (7)–(9).

For 12Cr2Ni4A steel:

$$J = 3.26 \times 10^{-9} \cdot p^{\frac{8.314T}{4412}}. \tag{7}$$

For 16Cr3NiWMoVNbE steel:

$$J = 1.14 \times 10^{-8} \cdot p^{\frac{8.314T}{5512}}. \tag{8}$$

For 18Cr2Ni4WA steel:

$$J = 2.35 \times 10^{-10} \cdot p^{\frac{8.314T}{3366}}. \tag{9}$$

5. Conclusions

The engineering modeling of key process parameters, including carburizing temperature, saturated carbon concentration in austenite, diffusion coefficient, and carbon flux, were investigated to meet the requirements of the process parameter model and database for low-pressure vacuum carburization conditions. The following conclusions have been drawn.

1. The use of segmented average carbon flux enables more accurate and efficient characterization of the actual carbon flux required, not only effectively avoiding the effect of the first pulse but also enabling the fine control of the timing of the subsequent pulses. Therefore, the segmented average carbon flux method should be selected when formulating the vacuum carburizing process.
2. For the 16Cr3NiWMoVNE steel carburizing 30 s, the carbon flux obtained using the segmented average method is 2.5 times than that of the carbon flux obtained by the overall average method, which result in the obtained value exceeding the carbide standard in the carburized layer;
3. For the 16Cr3NiWMoVNE steel, the first pulse carburizing time and subsequent pulse time obtained by using the overall average method are 2.2 and 1.3 times longer than those obtained by using the segmented average method;
4. Carbon flux increases with pressure. The optimum carburizing pressure is 300 Pa for low-pressure vacuum carburizing. On increasing the carburizing temperature from 920 °C to 980 °C, the effective carbon flux value increased by more than 30%. Furthermore, with the increase in carburizing temperature, the time required for the surface carbon concentration of the sample to reach saturation also increases;
5. In terms of the degree of influence of material composition on the value of carbon flux, the alloy content has the greatest influence, followed by the initial carbon concentration, and then the saturated carbon concentration;
6. Future work will focus on the effect of different alloy elements and content on carbon flux in detail, and establish a more general mathematical model of carbon flux under the optimal carburizing temperature and pressure, which is verified by experiments.

Author Contributions: All authors contributed equally to this paper and research. Conceptualization, H.W. and Z.W.; data curation, X.A.; formal analysis, Y.T. and J.L.; supervision, Z.W.; writing—review and editing, H.W. All authors have read and agreed to the published version of the manuscript.

Funding: This research was funded by the National Key Research and Development Program of China (2017YFB0305300).

Acknowledgments: The authors would like to express their thanks to teachers and students in the State Key Lab of Rolling and Automation, Northeastern University, for their continuous help and support.

Conflicts of Interest: The authors declare no conflict of interest.

References

1. Sawicki, J.; Krupanek, K.; Stachurski, W.; Buzalski, V. Algorithm Scheme to Simulate the Distortions during Gas Quenching in a Single-Piece Flow Technology. *Coatings* **2020**, *10*, 694. [CrossRef]
2. Derevyanov, M.Y.; Livshits, M.Y.; Yakubovich, E.A. Vacuum Carburizing Process: Identification of Mathematical Model and Optimization. *IOP Conf. Ser. Mater. Sci. Eng.* **2018**, *327*, 022022. [CrossRef]
3. Li, W.; Li, A.; Liang, Y.; Zhang, Z.; Cui, J. Effect of vacuum carburizing on surface properties and microstructure of a tungsten heavy alloy. *Mater. Res. Express* **2020**, *7*, 016558. [CrossRef]
4. Yuan, Z.X.; Yu, Z.S.; Tan, P.; Song, S.H. Effect of rare earths on the carburization of steel. *Mater. Sci. Eng. A* **1999**, *267*, 162–166. [CrossRef]
5. Yan, M.F.; Liu, Z.R. Influence of rare earths on carbon diffusion and transfer coefficients of carburizing process. *Chin. J. Rare Earths* **2001**, *19*, 9–11.
6. Hwang, J.I.; Melville, A.T.; Jhee, T.G.; Kim, Y.K. Equilibration of plain carbon and alloy steels with endothermic carburizing atmospheres: Part I. Activity of carbon in plain carbon steels. *Met. Mater. Int.* **2010**, *15*, 159–173. [CrossRef]

7. Hwang, J.I.; Jhee, T.G.; Kim, Y.K.; Hwang, T.Y. Equilibration of plain carbon and alloy steels with endothermic carburizing atmospheres: Part II. Oxidation of plain carbon and armco steels during carburizing at 1000 °C and 1038 °C. *Met. Mater. Int.* **2010**, *16*, 693–699. [CrossRef]
8. Kula, P.; Pietrasik, R.; Dybowski, K. Vacuum carburizing-process optimization. *J. Mater. Process. Technol.* **2005**, *164*, 876–881. [CrossRef]
9. Karabelchtchikova, O.; Sisson, R.D. Carbon diffusion in steels: A numerical analysis based on direct integration of the flux. *J. Phase Equil. Diff.* **2006**, *27*, 598–604. [CrossRef]
10. Zajusz, M.; Tkacz, S.K.; Danielewski, M. Modeling of vacuum pulse carburizing of steel. *Surf. Coat. Technol.* **2014**, *258*, 646–651. [CrossRef]
11. Lowell, J. The Effects of Contamination and Cleaning on AISI 9310 Vacuum Carburized Steel. Master's Thesis, Worcester Polytechnic Institute, Worcester, MA, USA, 2009.
12. Gorockiewicz, R.; Lapiński, A. Structure of the carbon layer deposited on the steel surface after low pressure carburizing. *Vacuum* **2010**, *85*, 429–433. [CrossRef]
13. Gorockiewicz, R. The kinetics of low pressure carburizing of alloy steels. *Vacuum* **2011**, *86*, 448–451. [CrossRef]
14. Su, Y. *Measurement of Critical Parameters in Vacuum Carburizing and Development of Simulation Software*; Shanghai Jiaotong University: Shanghai, China, 2009.
15. Yin, L.; Wang, T.; Ma, X.; Fu, Z.; Hao, G.; Li, L.; Wang, L. Pre-Coated Fe–Ni Film to Promote Low-Pressure Carburizing of 14Cr14Co13Mo4 Steel. *Coatings* **2019**, *9*, 304. [CrossRef]
16. Yin, L.; Ma, X.; Tang, G.; Fu, Z.; Yang, S.; Wang, T.; Wang, L.; Li, L. Characterization of carburized 14cr14co13mo4 stainless steel by low pressure carburizing. *Surf. Coat. Technol.* **2019**, *358*, 654–660. [CrossRef]
17. Jacquet, P.; Rousse, D.R.; Bernard, G.; Lambertin, M. A novel technique to monitor carburizing processes. *Mater. Chem. Phys.* **2003**, *77*, 542–551. [CrossRef]
18. Zhang, J.G.; Cong, P.W. Overview and development of vacuum carburizing technology. *Metal. Heat Treat.* **2003**, *28*, 52–55.
19. Khan, R.U.; Bajohr, S.; Buchholz, D.; Reimert, R.; Minh, H.D.; Norinaga, K.; Janardhanan, V.M.; Tischer, S.; Deutschmann, O. Pyrolysis of propane under vacuum carburizing conditions: An experimental and modeling study. *J. Anal. Appl. Pyrolysis* **2007**, *81*, 148–156. [CrossRef]
20. Yada, K.; Watanabe, O. Reactive flow simulation of vacuum carburizing by acetylene gas. *Comput. Fluids* **2013**, *34*, 26. [CrossRef]
21. Makino, S.; Inagaki, M.; Ikehata, H.; Tanaka, K.; Inagaki, K. Multi-step simulation of vacuum-carburizing reactor based on kinetics approach. *ISIJ Int.* **2020**, *60*, 2000–2006. [CrossRef]
22. Wang, H.J.; Wang, B.; Tian, Y.; Misra, R.D.K. Optimizing the low-pressure carburizing process of 16Cr3NiWMoVNbE gear steel. *J. Mater. Sci. Technol.* **2019**, *35*, 1218–1227. [CrossRef]
23. Wang, X.L.; Zurecki, Z.; Sisson, R.D. Development of nitrogen-hydrocarbon atmospheric carburizing and process control methods. *J. Mater. Eng. Perform.* **2013**, *22*, 1879–1885. [CrossRef]
24. Tian, Y.; Wang, H.J.; An, X.; Wang, Z. Experimental study on carbon flux in vacuum carburizing. *Mater. Res. Exp.* **2019**, *6*, 096516. [CrossRef]
25. Kanô, F.; Abe, I.; Kamaya, H.; Ueda, I. Fractal model for adsorption on activated carbon surfaces: Langmuir and freundlich adsorption. *Surf. Sci.* **2000**, *467*, 131–138. [CrossRef]

Article

Influence of DLC Coatings Deposited by PECVD Technology on the Wear Resistance of Carbide End Mills and Surface Roughness of AlCuMg2 and 41Cr4 Workpieces

Sergey N. Grigoriev, Marina A. Volosova *, Sergey V. Fedorov and Mikhail Mosyanov

Department of High-Efficiency Machining Technologies, Moscow State University of Technology STANKIN, Vadkovskiy per. 3A, 127055 Moscow, Russia; s.grigoriev@stankin.ru (S.N.G.); sv.fedorov@stankin.ru (S.V.F.); mmosyanov@yandex.ru (M.M.)

* Correspondence: m.volosova@stankin.ru; Tel.: +7-916-308-49-00

Received: 25 September 2020; Accepted: 26 October 2020; Published: 28 October 2020

Abstract: The primary purpose of this work was to study the effectiveness of using diamond-like coatings (DLC) to increase the wear resistance of carbide end mills and improve the surface quality of the processed part when milling aluminum alloy and low-carbon steel. The functional role of forming an adhesive sublayer based on (CrAlSi)N immediately before the application of the external DLC film by plasma-enhanced chemical vapor deposition (PECVD) technology in the composition of a multicomponent gas mixture containing tetramethylsilane was established in the article. The article shows the degree of influence of the adhesive sublayer on important physical, mechanical, and structural characteristics of DLCs (hardness, modulus of elasticity, index of plasticity, and others). A quantitative assessment of the effect of single-layer DLCs and double-layer (CrAlSi)N/DLCs on the wear rate of end mills during operation and the surface roughness of machined parts made of aluminum alloy AlCuMg2 and low-carbon steel 41Cr4 was performed.

Keywords: diamond-like coatings; nitride sublayer; index of plasticity; adhesive bond strength; end mills; hard alloy; wear resistance; milling of aluminum alloys; milling of structural steels; surface quality

1. Introduction

The carbide end mills are the most popular and versatile tools for processing a wide range of metals, alloys, and non-metallic materials [1–5]. The outstanding capabilities of modern carbide mills are provided by a variety of original design solutions, the correct selection of tool geometry, and the use of a wide range of modern wear-resistant coatings [6–8]. The application of wear-resistant coatings with a thickness of 3–7 μm to the working surfaces of end mills gives them the characteristics necessary for specific processing conditions and to ensure that the coating effectively complements the physical and mechanical properties of the hard alloy substrate and together, they will have increased wear resistance during cutting [9–11]. Among the numerous coatings, thin-film diamond-like coatings (DLC) represent a separate group of particular interest. Due to their excellent anti-friction properties and good resistance to abrasive wear, these coatings are now successfully used in mechanical engineering and metalworking as surface protection of machine parts operating under conditions of increased friction with mating parts and cutting tools made of high-speed steels and hard alloys, processing non-ferrous metals and alloys, composite materials, and others [12–16].

For the formation of DLCs of various structures and properties, research organizations and manufacturing enterprises are currently well equipped and have technological equipment based on the principles of physical vapor deposition (PVD), chemical vapor deposition (CVD), as well as their combination [17–22]. However, whatever method of production we use, the structure of the formed DLC assumes the simultaneous presence of various forms of carbon in them, which can exist in different hybridizations. Modern technologies for the deposition of DLCs provide the formation of films with a different ratio between sp^3 (diamond) and sp^2 (graphite-like) hybridizations combined in an amorphous structure (Figure 1). The properties of the formed coating depend on the type of bonds that hold the carbon atoms.

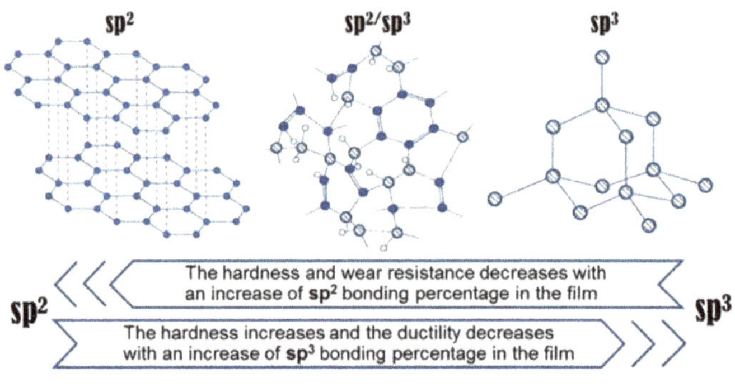

● - sp² carbon atom, ⊘ - sp³ carbon atom, ○ - hydrogen atom

Figure 1. Different hybridizations of carbon in diamond-like coatings (DLC) films.

The higher the proportion of sp^3 bonds, the closer the properties of such material is to diamond—increase in its hardness and resistance to abrasive wear. In turn, the higher the content of sp^2 bonds, the closer the properties of the material are to graphite—its hardness and wear resistance are reduced at temperatures that occur during cutting. When creating DLCs, the main task of technologists is to get a coating with a structure in which the number of sp^3 links is increased compared to sp^2 links. Furthermore, of great importance for the characteristics of coatings is the amount of hydrogen and other components that are always present in coatings deposited by industrial technologies [23–27].

DLCs have specific features: with an increase in film thickness, the level of residual stresses increases noticeably, the strength of its adhesive bond with the substrate decreases, and, as a result, the coating does not effectively resist the loads acting on it during the cutting of materials (especially cyclic loads during interrupted cutting, for example, milling).

According to the known physical laws, a film of thickness h will peel off when the elastic energy per volume unit due to stress σ exceeds the energy required for the formation of two new surfaces by exfoliation [27,28]:

$$h < \frac{4\gamma E}{\sigma^2} \tag{1}$$

where γ is surface energy (J/m^2), σ is stress (Pa), h is the coating thickness (μm), and E is the modulus of elasticity (Pa).

This dependence sets the upper limit of the film thickness, above which it peels off spontaneously. When a tool with a coating is exposed to power loads, the stresses from the external load are additionally superimposed on the residual stresses in the coating, and accordingly, the destruction occurs at significantly lower loads.

Undoubtedly, the stress state of a DLC is influenced by the essential physical and mechanical characteristics—hardness (H), modulus of elasticity (E), as well as their ratio (H/E), called "index of plasticity". There is one more feature that must be taken into account when developing coatings for a hard alloy tool—due to an abrupt change in the modulus of elasticity of the coating layer and the surface layer of the tool material, significant tangential shear stresses arise at their interface, which can lead to peeling of the coating during cutting. Therefore, to reduce technological stresses at the interface and ensure increased strength of the adhesive bond between the coating and the tool, it is necessary to provide, as close as possible, the characteristics of the modulus of elasticity of these materials [29–34].

To minimize the problems described above, various technological approaches are used when creating DLCs: they are doped with different elements and compounds (silicon, tungsten, and others), and can form adhesive sublayers (intermediate coatings) [35–40]. A universal technical solution cannot be found. When developing the architecture of a DLC and the technology of its deposition, it is necessary to be guided by the specific operating conditions of the coated product, and it is necessary to evaluate the effectiveness of the chosen approach under the conditions of the action of real operating loads or as close as possible to them.

In this work, the goal was to comprehensively study the effect of the formation of an adhesive sublayer based on (CrAlSi)N on carbide end mills before applying the outer DLC film using the PECVD method in the presence of a multicomponent gas mixture containing tetramethylsilane, in comparison with applying only a single-layer DLC, and determining the degree of influence of the adhesive sublayer on the critical physical, mechanical, and structural characteristics of the DLC. Furthermore, we quantified the influence of single-layer (without sublayer) and double-layer (with sublayer) DLCs on the wear rate of end mills during operation and the surface roughness of machined parts made of aluminum alloy AlCuMg2 and low-carbon steel 41Cr4. For evaluating the effectiveness of the DLC under different heat and power loads, we specifically selected two fundamentally different processed materials for research. Compared to other structural materials, aluminum alloys are well amenable to machining, relatively low power and temperature loads on the cutting tool arise. At the same time, they have specific features—repeated adhesion of aluminum particles to the cutting edge of the tool during the cutting process, followed by their tearing out from the working surfaces of the tool, which intensifies tool wear and reduces the quality of the processed surface [41–44]. When machining low-alloy structural steels by end mills, the thermal load is noticeably higher, and the main problem is the wear on the flank of the tool as a result of abrasion caused by hard components in the work material. Furthermore, adhesion wear of the working surfaces of the end mill is observed due to the local grasp of the processed and tool materials, followed by separation of the smallest particles from the tool, which are carried away by the descending chips [45,46].

A separate comment is required to avoid doubt about the validity of the authors' choice of the DLC for the tool processing ferrous steel. According to classical concepts, at elevated temperatures, carbon is intensely dissolved in iron, and graphitization of diamond crystals occurs. However, in this case, the authors do not use a diamond tool but study the behavior of a multicomponent DLC film containing various modifications of carbon, and compounds based on silicon, which is a component of the gas mixture during condensation of the DLC film. We purposefully selected iron-containing steel as the material to be processed to study the functioning of the adhesive sublayer and the outer DLC layer under the loads typical for milling this material. We should add that the authors have research experience in using DLCs for turning hardened bearing steels, and in their previous works, a specific effect was observed from the use of coatings for ceramic tools [21,47–49]. This makes it possible to use a carbide tool as an object of research in this work.

The choice of (CrAlSi)N as an adhesive sublayer material for the functioning of the DLC is not accidental and is explained by the great potential of using this compound for the needs of tool production [49–51]. This nitride coating compares favorably with traditional nitride films such as (TiAl)N, (TiCr)N, and (TiNbAl)N since it is a nanocomposite, its deposition does not entirely mix the components but forms two phases. Its structure consists of AlCrN nanocrystals embedded in an amorphous SiN matrix, which provides a high level of strength of interatomic bonds between atoms of the amorphous and crystalline phases [52,53]. The introduction of silicon into the coating reduces the internal residual stress at the "hard alloy-coating" interface, and the formation of oxides by the coating components at high temperatures slows down the tribochemical reactions on the contact surfaces of the cutting tool. These features, and previously obtained experimental data [49,54,55], allow us to count on the effectiveness of the composite compound (CrAlSi)N for carbide end mills as an adhesive sublayer before applying the external DLC.

2. Materials and Methods

2.1. Processed Materials, Cutting Tools, and Operational Testing Methods

As the processed material, two types of materials widely used in mechanical engineering were chosen—aluminum alloy AlCuMg2 and structural low-alloy steel 41Cr4, the chemical composition of which is given in Tables 1 and 2.

Table 1. Chemical composition of the AlCuMg2 alloy used for testing.

Element	Al	Cu	Mg	Mn	Fe	Si	Zn	Ti	Cr	Impurities
Content (%)	92.0	4.3	1.6	0.7	0.4	0.5	0.25	0.15	0.1	0.15

Table 2. Chemical composition of the 41Cr4 steel used for testing.

Element	Fe	Cr	Mn	C	Si	S	Cu	P
Content (%)	97.0	1.2	0.9	0.44	0.37	0.034	0.032	0.024

The processed materials used in work differ significantly in their physical and mechanical characteristics—the aluminum alloy AlCuMg2 has a tensile strength of 245 MPa with a hardness of 105 HB, and 41Cr4 steel, 635 MPa and 210 HB, respectively. Such differences predetermine a significant difference in the heat and power loads that act on the cutter's cutting edge during cutting. When machining 41Cr4 steel, more than a twofold increase in the component of the cutting force Pz and the temperature in the cutting zone is observed.

As cutting tools for carrying out a set of experimental studies we used 3-flute (for processing AlCuMg2 alloy) and 5-flute (for processing 41Cr4 steel) end mills with a diameter of 6 mm, made of tungsten-cobalt hard alloy 6WH10F (WC, 90%; Co, 10%) with hardness HRA 92.1 and density 14.50 g/cm^3. Figure 2 shows a general view of the design of the end mills, and Table 3 shows their design and geometric parameters.

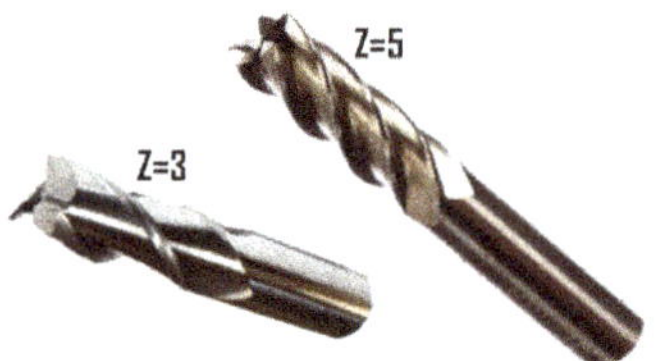

Figure 2. Design of end mills used in research.

Table 3. Design and geometric parameters of end mills.

Tool Parameters	Processed Material	
	AlCuMg2	41Cr4
Mill type	end	
Shank type	cylindrical	
Material	6WH10F	
External diameter D, mm	6	
Cutter length L, mm	36	40
Working unit length l, mm	16	
Number of teeth Z	3	5
Shank diameter d, mm	6	
Inclination angle of the chip grooves ω	30°	

To perform a complex of metallographic and metallophysical studies, square plates with dimensions of 12.0 mm × 12.0 mm × 5.0 mm were manufactured of 6WH10F hard alloy.

Operational tests of the end mills were performed at the CTX beta 1250 TC turning/milling centre (DMG MORI Co., Ltd., Bielefeld, Germany). An ER collet chuck made according to DIN 69893-1 HSK-A63 was used to clamp the end mills. The workpieces were milled in rods with a diameter of 65 mm of aluminum alloy AlCuMg2 and structural low-alloy steel 41Cr4. In the experiments, the coated end mills were tested without coolant to also provide better observation of wear, and that correlates with standard tool tests. Tool tests were carried out using a program written in the SINUMERIK 840D SL system (Siemens AG, Regensburg, Germany) while executing the strategy of milling the plane of the workpiece end face in the following cutting modes (Figure 3):

- for AlCuMg2 alloy—cutting speed Vc = 188.4 m/min, feed rate F = 2000 mm/min, feed per tooth Fz = 0.068 mm/tooth, milling width B = 2 mm, and milling depth t = 0.5 mm;
- for 41Cr4 steel—cutting speed Vc = 188.4 m/min, feed rate F = 750 mm/min, feed per tooth Fz = 0.015 mm/tooth, milling width B = 2 mm, and milling depth t = 0.2 mm.

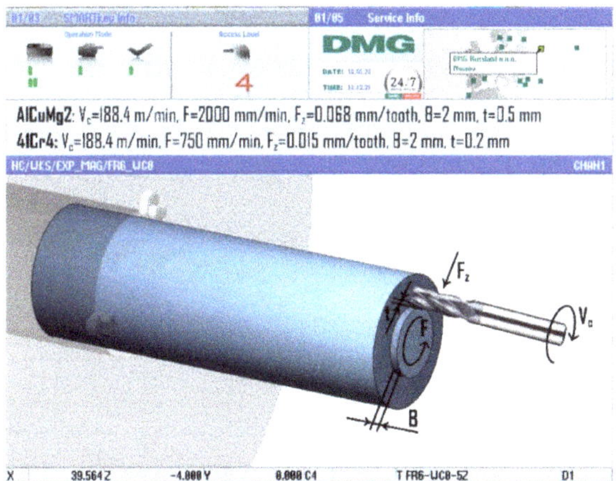

Figure 3. Processing strategy and cutting conditions for machining AlCuMg2 and 41Cr4 steel workpieces with end mills.

The limit size of the wear chamfer on the flank surface (0.4 mm) was taken as the criterion for loss of performance (failure) of the end mills. Tool wear resistance was defined as the cutting time until the cutter reached the limit of wear. To quantify wear, we used a metallographic optical microscope Stereo Discovery V12 (Carl Zeiss Microscopy GmbH, Jena, Germany); the milling cutters were placed in a particular device at an angle of 45° on its table. Each tooth of the cutter was measured, and the largest amount of wear was detected, and this was taken into account when processing the results of experiments and constructing curves for the dependence of wear on the flank surface of the cutting time, based on which the resistance value was calculated.

When processing 41Cr4, one pass of the cutter was 3.5 min, and after each pass, the wear chamfer was measured. When processing AlCuMg2, one pass of the cutter was 3 min, and taking into account that the flank surface wear is less intense when processing aluminum alloys, wear was measured every three passes. After each cycle of experiments, a disk was cut off from the workpiece, and its end surface, previously treated with an end mill, was examined on a Hommel Tester T8000 stylus profiler by JENOPTIK Industrial Metrology (VS-Schwenningen, Germany) to quantify the surface roughness.

2.2. Technology and Equipment for the Deposition of DLC Coatings on End Mills

To apply DLCs to the surface of the end mills, they were placed on the turntable of the vacuum chamber of a hybrid technological unit (Figure 4), the design of which allows the sequential application of coatings in a single cycle by various methods: vacuum-arc deposition by evaporation of cathode materials (Arc-PVD); plasma-chemical gas-phase deposition in a glow discharge plasma by chemical reaction and decomposition of gas mixture components-plasma-enhanced chemical vapor deposition (PECVD). Two types of coatings were applied: (1) double-layer coatings consisting of an adhesive sublayer (CrAlSi)N and the outer DLC layer; (2) single-layer DLCs.

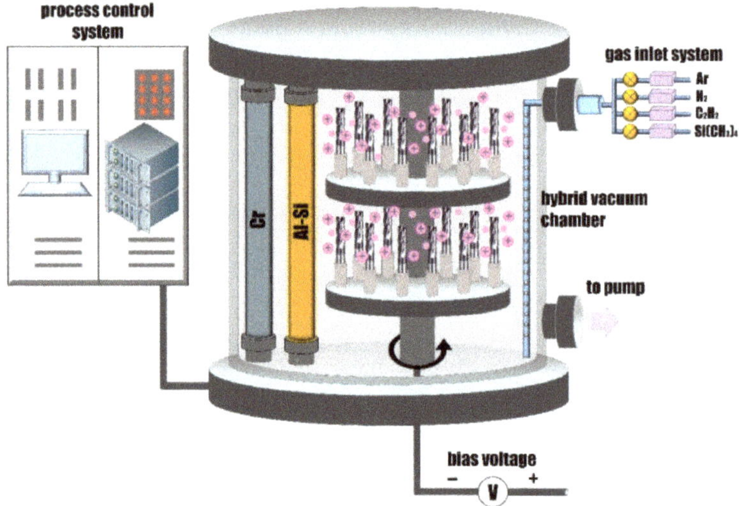

Figure 4. The principal application of DLC-coatings and (CrAlSi)N/DLC-coatings on end mills with the use of hybrid technology unit.

We used the original PI311 series technological unit developed by PLATIT AG (Selzach, Switzerland). The unit (Figure 4) is equipped with a gas filtration device and a multi-channel gas inlet system, a vacuum pumping system, a source of bias voltage to the turntable, and two cylindrical rotating cathodes made of chromium (Cr) and silumin (AlSi). For minimizing micro drops in the formed coating (CrAlSi)N, the unit implements the original lateral rotating cathodes (LARC) technical solution. This technology does not involve the use of bulky magnetic separators to minimize micro drops but uses rotating cathodes that are placed on the periphery of the chamber, and virtual gates that work without any mechanical elements. Following the generally accepted classification and terminology, the DLCs deposited in this work refer to hydrogenated diamond-like a-C: H films.

Preliminary cleaning of the cutters, which is necessary to achieve adequate strength of the adhesive bond of the deposited coating, was carried out with argon ions with an energy of 500 eV at a pressure of 1 Pa using a non-self-sustaining gas discharge ignited between the cathodes. At the same time, a negative voltage of 400 V was applied to the rotary table with the processed samples. The electron flow between the cylindrical cathodes created a high-density plasma, by which the samples were efficiently cleaned of impurities and oxides for 15 min before the deposition of the coating. After the cleaning was completed at a pressure of 1.5 Pa, gradually decreasing to 0.7 Pa, an adhesion sublayer (CrAlSi)N was formed by two Cr and AlSi cathodes when a gas mixture of nitrogen (volume fraction 90%) and argon (volume fraction 10%) and supplying negative voltage to the turntable of 500 V. The time of deposition of the (CrAlSi)N layer was 25 min, which provided a thickness of about 1.5 μm. Then the outer diamond-like layer was formed by plasma-chemical gas-phase deposition in a glow discharge plasma by starting a chemical reaction and decomposing the components of the gas mixture supplied to the chamber: acetylene C_2H_2 (volume fraction 78%), argon Ar (volume fraction 7%), and tetramethylsilane $Si(CH_3)_4$ (volume fraction 5%). The deposition time of the diamond-like layer was 180 min, which provided a thickness of about 2.5 μm. For the application of a single-layer DLC, the process step associated with the formation of an adhesive sublayer was excluded, and after cleaning the cutters, the DLC layer was deposited directly. The DLC layer was deposited at a pressure of 4.0 Pa. This value's choice is because a decrease in this parameter leads to a decrease in the productivity of the process when an increase leads to an excessive increase in the structure of the graphite-like component's coating.

A Tescan VEGA3 LMH scanning electron microscope was used to analyze the structure of DLCs deposited using the technology described above.

2.3. *X-ray Photoelectron and* Diffraction *Analysis of DLCs*

X-ray photoelectron spectroscopy (XPS) was used to study the chemical and electronic state of carbon atoms on the surface of samples with DLCs. The analysis was performed using Thermo Scientific's K-ALPHA X-ray photoelectron spectrometer (Thermo Fisher Scientific Inc., Bremen, Germany).

The samples were both shot in the initial state of the coating surface and after surface treatment for 6 min with an Ar^+ ion beam to remove the adsorbed layers with high carbon and oxygen content. High-resolution photography was performed using monochromatic and polychromatic Mg Kα and Al Kα radiation at a power of 200–300 W and a voltage of 14 kV, pressure in the chamber 7×10^{-8} Pa.

XPS analysis was performed for various peaks of the DLC layer within the prominent C1s peak: sp^2 and sp^3 peaks, characterizing carbon hybridization, and C=O and C–O peaks, characterizing various compounds of the components with oxygen and other elements. Besides, we analyzed the peaks of the phases containing silicon compounds in the coating. For each component of the DLC layer, the following characteristics were measured using XPS: start binding energy (Start BE), binding energy (BE), end binding energy (End BE), full width at half maximum (FWHM), and atomic percent (Atomic %).

A Panalytical Empyrean X-ray diffractometer (PANalytical, Almelo, The Netherlands) with monochromatic Cu Kα-radiation was used to study the stresses. The stresses were determined by the classical $\sin^2\psi$ method with a grazing beam at a fixed angle of incidence of the beam ψ_0 and scanning along the 2θ axis.

2.4. *Microhardness, Modulus of Elasticity, and Adhesive Bond Strength of DLCs*

We used the method of nanoindentation with a Berkovich diamond indenter on CSEM Nano hardness tester (CSM Instruments, Needham, MA, USA) equipped with specialized software (version 4.0) based on the algorithm of Oliver and Pharr [26,56]. This method, in contrast to the Vickers pyramid microhardness estimation scheme, is more reliable, since, if necessary, it allows excluding the influence of the hardness of the underlying layers on the results of measurements of the characteristics of the outer DLC layer. Before starting the measurements, the equipment was calibrated on reference samples with the known modulus of elasticity and hardness.

In experiments, the value of the applied load and the corresponding depth of indenter insertion were selected based on the condition of not more than 15% of the thickness of the DLC. The measurements were performed at loads of 1.0 and 4.0 mN, and the corresponding maximum indenter insertion depths were 0.09 and 0.38 μm, respectively. The tests were carried out as follows: after the load on the coated sample reached the maximum value, unloading began, and the load acting on the indenter gradually decreased to zero. The duration of the load-unload cycle was 50 s. According to the measurement results, typical experimental curves of continuous indentation were obtained—"the dependence of the load on the depth of indentation". The first curve corresponds to loading and reflects the resistance of the material to the penetration of the indenter. The second describes the return of deformation after the removal of the external load and characterizes the elastic properties of the material. In this case, the deformation response (indentation depth) is simultaneously recorded in nanometer resolution.

The obtained experimental data allowed us to not only judge the nano hardness (H) but also the modulus of elasticity (E) of the DLC. If we obtain a quantitative value of the H/E ratio (index of plasticity) [57–59], we can approximate the viscosity of the coating and its ability to resist possible deformation and destruction during cutting. For reducing possible measurement errors, each sample was processed five times, resulting in an average hardness value.

To quantify the strength of the adhesive bond of DLCs to carbide substrates, we used the CSEM micro scratch tester system. It implements the method of scratching with a diamond cone indenter (apex angle of 120° and apex radius of 100 μm) with a variable load from 1 to 30 N. The spectra of the acoustic emission signal were constantly recorded. Their bursts make it possible to reliably judge the initial stage of cracking and subsequent peeling of the coatings. Hard-alloy samples with coatings moved at a constant speed, and the recording and processing of acoustic emission spectra were carried out using specialized software (version 4.0). Thus, when conducting scratch tests, is solved the main problem of determining the critical load, at which there is an abnormal change in the indentation depth of the indenter and the separation of the coating.

In addition to the method described above, a qualitative assessment of the adhesive bond strength of the coating to the substrate was performed by pressing the Rockwell indenter at a load of 15 N/mm^2 on a Wilson Hardness R574T device by INSTRON (Norwood, MA, USA). In this case, a comparative assessment of the adhesion characteristics of two types of coatings was made by comparing the prints obtained after the intrusion of a cone indenter.

2.5. Friction Coefficient and Abrasion Resistance of DLC Coatings

The friction coefficient was evaluated in work on a Tetra Basalt N2 testing machine from TETRA GmbH to assess the DLC's effect on the transformation of frictional properties. During the tests, the coefficient of friction-sliding of rubbing pairs "hard alloy with DLC-coating—a counter body made of AlCuMg2 alloy" and "hard alloy with DLC-coating—a counter body made of 41Cr4 steel" were determined. The tests of all samples were carried out under conditions of dry friction at identical normal loads on the counter body (1 N), with the speed of relative displacement being 2 $mm{\cdot}s^{-1}$ and the friction path being 90 mm. We evaluated the coefficient of friction for samples with single-layer DLCs.

The abrasion resistance of the surface layer of DLC-coated carbide samples was carried out by testing on a Calotest unit by CSM Instruments. The test principle was that a rotating sphere was placed on a test sample and operated with a preselected load of 20 N, which was controlled by a particular sensor. The rotation of the sphere during the tests was carried out by the driveshaft, and the position of the sphere to the test samples and the load in the contact area was constant. A water-based abrasive suspension was fed to the test area, and its solid particles in the contact area of the sphere with the coated sample led to abrasion of the surface area and the formation of a wear hole (spherical recess). The suspension and sphere wear the coating and substrate in a controlled manner, which guarantees reproducible results. Optical analysis of the geometric dimensions of the worn hole makes it possible to judge the ability of hard-alloy samples with DLCs to resist abrasion [60,61].

3. Results

3.1. Structure and Chemical Composition of DLCs

Figure 5 shows SEM images of cross-sections of carbide samples with one-layer and double-layer DLCs. The presented photographs give a general idea of the structure of the investigated coatings and the thickness of their layers formed under the selected combination of technological modes. The thickness of the formed adhesive sublayer (CrAlSi)N was 1.4 ± 0.15 μm, and the thickness of the outer DLC layer was 2.6 ± 0.2 μm.

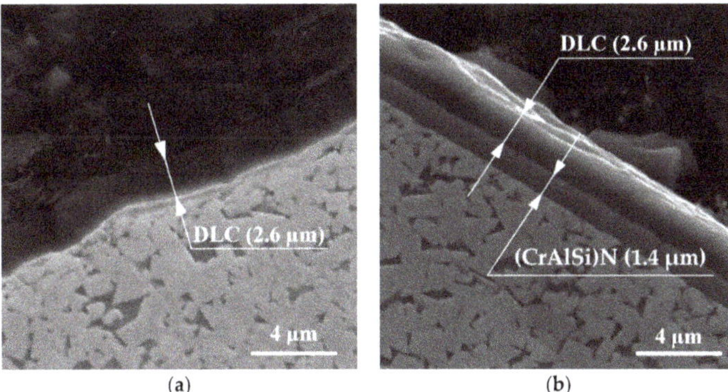

(a) (b)

Figure 5. SEM images in secondary electrons of cross-sections of carbide samples, high voltage of 10.0 kV, view field of 16.6 μm, 25.0 k×: (**a**) single-layer DLC; (**b**) double-layer DLC.

Tables 4 and 5 present data for processing and decoding the results of XPS analysis of two types of samples with DLCs without sublayer (Table 4) and with sublayer (CrAlSi)N (Table 5). These results were obtained using specialized software of K-alpha X-ray photoelectron spectrometer, an Avantage Data System (version 5.0) (Thermo Scientific, Bremen, Germany). The analysis was performed after the surface layer of the coating with a thickness of about 0.004 μm was etched with argon ions.

Table 4. Results of XPS-analysis software interpretation: composition and energy characteristics of DLC peaks (without sublayer).

XPS-Peak	Start BE	Peak BE	End BE	FWHM eV	Atomic %
C1s sp^2 (graphite)	297.98	284.5	279.18	1.1	16.03
C1s sp^3 (diamond)	297.98	285.3	279.18	1.3	47.92
C1s C=O	297.98	286.5	279.18	2.4	13.97
C1s C–O	297.98	286.1	279.18	2.5	10.58
O1s C–O	545.08	533.5	525.08	1.9	4.96
O1s C=O	545.08	531.2	525.08	1.6	4.03
Si $2p^3$ Si, Si–N, Si–C	110.08	101	95.08	2.7	2.51

Table 5. Results of f-analysis software interpretation: composition and energy characteristics of DLC peaks (with (CrAlSi)N sublayer).

XPS-Peak	Start BE	Peak BE	End BE	FWHM eV	Atomic %
C1s sp^2 (graphite)	297.98	284.3	279.18	1.0	15.24
C1s sp^3 (diamond)	297.98	284.9	279.18	1.2	51.22
C1s C=O	297.98	286.5	279.18	2.4	10.35
C1s C–O	297.98	286.1	279.18	2.5	7.58
O1s C–O	545.08	533.4	525.08	1.9	4.46
O1s C=O	545.08	531.8	525.08	1.6	4.18
O1s SiO_2	545.08	532.5	525.08	1.5	2.95
Si $2p^3$ Si, Si–N, Si–C	110.08	100.4	95.08	2.8	4.02

It should be noted that the binding energies of the coating components are very close for both coatings. The electron energy spectra of the primary coating peak (C1*s*) indicate that in the studied DLCs chemical bonds prevail between carbon atoms with a binding energy of about 284–285 eV, representing sp^2- and sp^3-hybridized states. The percentage of the diamond component (sp^3 hybridization) in coatings for a sample with a DLC without a sublayer is 47%, and 51% for a sample with a sublayer (we emphasize that here we are talking about a pure diamond component). The proportions of graphite-like carbon hybridization (sp^2) present in the coatings are very similar for the two samples studied, 16% and 15%, respectively. Besides, in the central peak of the C1*s* coating, the analysis revealed the presence of various impurities and contaminants, which is typical for DLC films. The percentage of this component for a sample with a DLC without a sublayer is 24%, and 17% for a sample with a sublayer. According to authoritative scientists' works, when choosing rational modes of deposition of the investigated type of DLCs, the formation of a diamond phase (sp^3 hybridization) in the range of 30–50% is characteristic [10,20,24,26]. That is, the results obtained are at the upper limit of the possible range of values.

XPS analysis of DLC films revealed various oxygen forms with increased binding energy values (O1*s* peak), 531.8 eV and 533.5 eV. Both samples contain oxygen in the form of surface adsorbed groups, and the proportion of the oxygen component is approximately the same and is about 9%. The only difference is that in a DLC film deposited on a sublayer (CrAlSi)N, in the primary oxygen peak of O1*s*, another component was found, identified as silicon dioxide SiO_2, its proportion is slightly less than 3%. A photoelectron peak of silicon Si $2p^3$ was also detected when examining the surface of the samples. The proportion of identified silicon compounds (such as Si, Si–N, SiC) for the studied samples was almost identical and amounted to 2.5–3.0%.

3.2. Microhardness, Modulus of Elasticity, Residual Stress and Adhesion Strength of DLCs

Table 6 shows the results of hardness measurements according to the Martens scale (HM), which characterizes both elastic and plastic properties of DLC-coated carbide samples. As follows from the experimental data obtained, with an increase in the applied load and, accordingly, with an increase in the penetration depth of the indenter, the hardness of the surface layer decreases. Within one load (1.0 mN), the hardness of the coated samples varies from 32 to 38 GPa, and from 23 to 29 GPa at a load of 4.0 mN, respectively. Simultaneously, the microhardness value is practically identical for samples with DLCs without a sublayer and with a (CrAlSi)N sublayer. Another type of change is observed when evaluating the modulus of elasticity of samples with DLCs. If at a load of 1.0 mN, the modulus of elasticity of the two types of samples had similar values, 284–300 GPa for samples with DLC without sublayer and 275–295 GPa for samples with (CrAlSi)N/DLC-coated, then with an increase in the load to 4.0 mN, significant differences were observed of 235–239 and 193–209 GPa, respectively. The data in Table 6 also gives an idea of the index (coefficient) of the plasticity of the studied samples. At an applied load of 1.0 mN, the index of the plasticity of the samples are very close and is 0.113 for samples without sublayer and 0.119 for samples with (CrAlSi)N/DLC, and with a four-fold increase in the load to 4.0 mN, the differences are quite noticeable, 0.10 and 0.129 respectively.

Table 6. Physical and mechanical characteristics of DLCs deposited on hard alloy samples.

No.	Type of Test Sample	Applied Load (mN)	Penetration Depth, (nm)	Hardness HM (GPa)	Modulus of Elasticity *E* (GPa)	Index of Plasticity (*H*/*E*)
1	Hard alloy/DLC	1.0	90	33 ± 3	292 ± 8	0.113
2	Hard alloy/(CrAlSiN/DLC			34 ± 2	285 ± 10	0.119
3	Hard alloy/DLC	4.0	380	25 ± 2	242 ± 7	0.10
4	Hard alloy/(CrAlSiN/DLC			26 ± 3	201 ± 8	0.129

It should be noted that the obtained value correlates to the known data of other research groups and show that the researched DLC is not fragile since the measured range of values for hardness is relatively low when sublayer increases its plasticity.

As follows from the results of the analysis of samples on an X-ray diffractometer (Figure 6), the character of the distribution of residual compressive stresses in the coating for samples with a single-layer DLC and a (CrAlSi)N/DLC do not change significantly. Attention is drawn to a noticeable decrease in residual stresses' average values when a DLC is applied to a sample with a (CrAlSi)N sublayer. Average residual stresses for single-layer DLCs are 1250–3650 MPa, while 650–2750 MPa for DLCs with a sublayer.

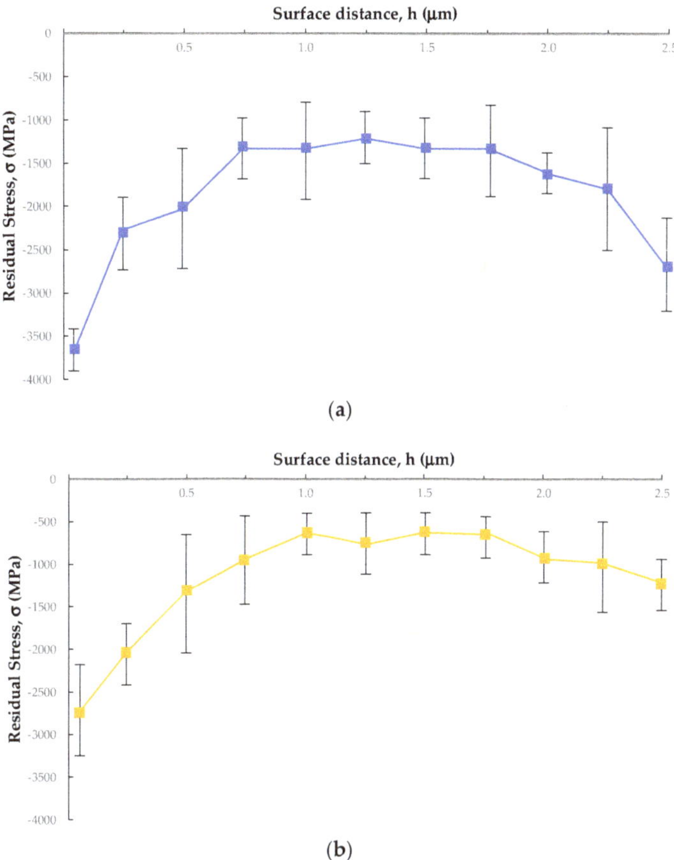

Figure 6. Values of residual stresses in DLCs deposited on different samples: (**a**) coating without sublayer and (**b**) coating with sublayer (CrAlSi)N.

Figure 7 shows optical images of indenter movement tracks on the surface of hard alloy samples with DLCs and their corresponding loads from 1 to 30 N obtained from scratch testing. The presented panoramas make it possible to quantify the value of critical loads at which the coatings break off from the carbide substrate and on which part of the indenter movement path this occurred. Acoustic emission sensors registered peak loads indicating the beginning of the indenter penetration into the coating (*P*1), the beginning of the first crack (*P*2), the peeling of local sections of the coating from the substrate (*P*3), and the critical (destructive) load (*P*4). It can be seen that, for samples with (CrAlSi)N/DLC, the peak loads are shifted to the right. In particular, the critical load *P*4 corresponds to significantly higher forces (about 23 N), while for a sample with a DLC without a sublayer the critical load corresponds to the force equal to 12 N. It should be noted that to exclude erroneous experimental results and subsequent erroneous conclusions, five samples of each type of DLCs were subjected to scratch testing. The data obtained on the quantitative values of critical loads for the samples under study were similar, and the spread of values was ±2 N.

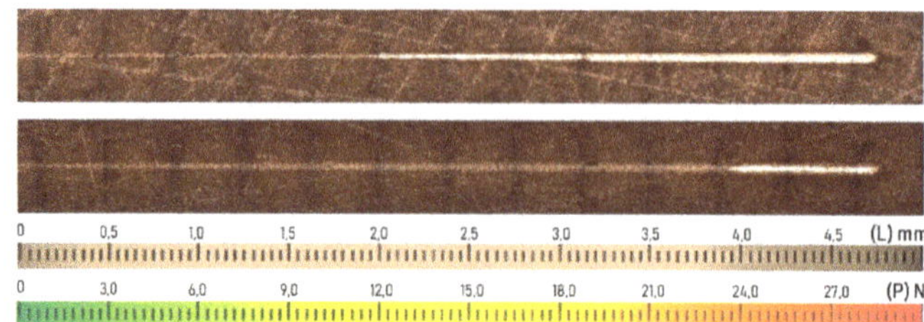

Figure 7. Tracks of indenter movements on the surface of 6WH10F hard alloy samples with DLCs and corresponding loads from 1 to 30 N; coating without sublayer (upper image), coating with sublayer (CrAlSi) N (lower image).

Observing such significant differences in the results of scratch testing of samples with DLCs, we additionally performed a qualitative rapid assessment of the strength of the adhesive bond by indentation of the Rockwell indenter. Figure 8 shows the micro images of the indenter insertion area in the samples. It is seen that the penetration of an indenter into a sample with a DLC without a sublayer is accompanied by the exfoliation of micro-sections of the coating in a circle at a certain distance from the indentation area (Figure 8a). In this case, the load applied to samples with (CrAlSi)N/DLC (Figure 8b) does not cause noticeable peeling and destruction of the outer coating.

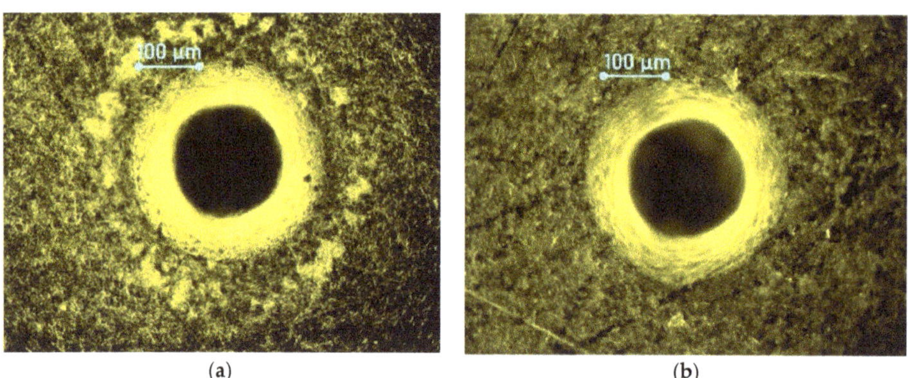

(**a**) (**b**)

Figure 8. Rockwell indenter indentation area in the surface layer of 6WH10F hard alloy samples with DLCs: (**a**) coating without sublayer and (**b**) coating with sublayer (CrAlSi)N.

3.3. Friction Coefficient and Abrasion Resistance of DLCs

As shown from Table 7, DLC significantly reduces the hard alloy's friction on the aluminum alloy from 0.29–0.32 to 0.15–0.16. The effect of coating deposition on friction against steel is also noticeable, the coefficient of friction coefficient was reduced from 0.41–0.44 to 0.25–0.27.

Table 7. Friction coefficient of DLCs deposited on hard alloy samples.

No.	Type of Sample	Type of Counter Body	Friction Coefficient Value
1	Hard alloy 6WH10F	AlCuMg2	0.29–0.32
2	Hard alloy 6WH10F/DLC	AlCuMg2	0.15–0.16
3	Hard alloy 6WH10F	41Cr4	0.41–0.44
4	Hard alloy 6WH10F/DLCs	41Cr4	0.25–0.27

Figure 8 shows two-dimensional optical images of wear holes on the surface of hard alloy samples after a rotating sphere is forcibly applied to them for 5 min in the presence of an abrasive suspension in the contact area. When measuring the linear dimensions (X and Y) of the section of material that has undergone abrasion, we judged the quantitative value of wear and subsequently concluded that the carbide samples are resistant to abrasion. The results shown in Figure 8 show marked differences for three types of samples: uncoated (Figure 9a), DLC-coated (Figure 9b), and (CrAlSi)N/DLC-coated (Figure 9c). There are differences in the shape of the formed wear hole and its linear dimensions. For an uncoated hard alloy sample, the worn surface area has an irregular shape with dimensions (X and Y) of 0.9 mm × 0.6 mm. For samples with DLC and (CrAlSi)N/DLCs, the worn areas have the shape of spherical recesses with dimensions of 0.6 mm × 0.55 mm and 0.4 mm × 0.4 mm, respectively.

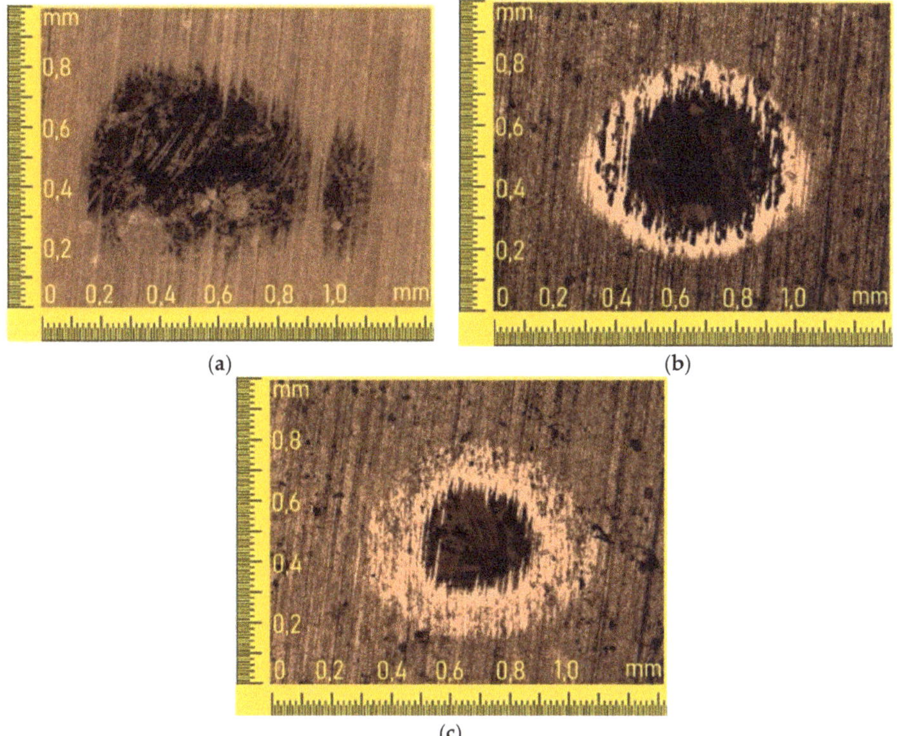

Figure 9. 2D images of the worn hole on 6WH10F carbide samples after surface layer abrasion resistance tests: (**a**) uncoated; (**b**) DLC-coated; and (**c**) (CrAlSi)N/DLC-coated.

3.4. DLC-Coated End Mill Performance Test

3.4.1. Processing of Aluminum Alloy AlCuMg2

The objects of study during the tests were the wear of three-flute end mills made of 6WH10F hard alloy (Figure 2) and the surface roughness of machined workpieces in the form of bars made of aluminum alloy AlCuMg2.

Figure 10 shows the experimentally obtained dependence of the flank wear chamfer (h) of end mills on the tool time (T) with various coatings, uncoated, with DLC and (CrAlSi)N/DLC.

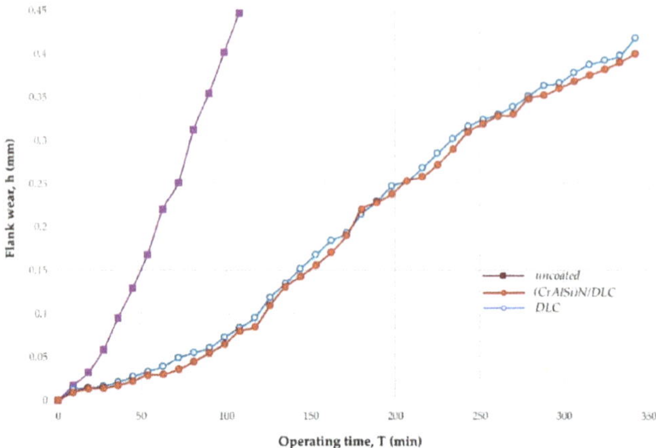

Figure 10. Dependence of the wear chamfer size on the flank surface of the end mills on the operating time when machining the aluminum alloy AlCuMg2 with end mills with different coatings.

These data demonstrate a pronounced positive effect of DLCs on the resistance of milling cutters (up to the maximum wear value of 0.4 mm) when processing the aluminum alloy. The development of the wear chamfer on the flank surface due to the coating slows down many times; if the tool life without coating was 99 min, the tool life with DLC is now 342 min. Optical images (Figure 11) showing the general appearance of the worn chamfer on the flank surface of the milling cutters after 72 min of operation, illustrating the above well—the coating significantly slows down the development of wear processes. Thus, we can conclude that the durability of a DLC-coated cutter is 3.45 times higher than that of an uncoated cutter when processing an aluminum alloy AlCuMg2.

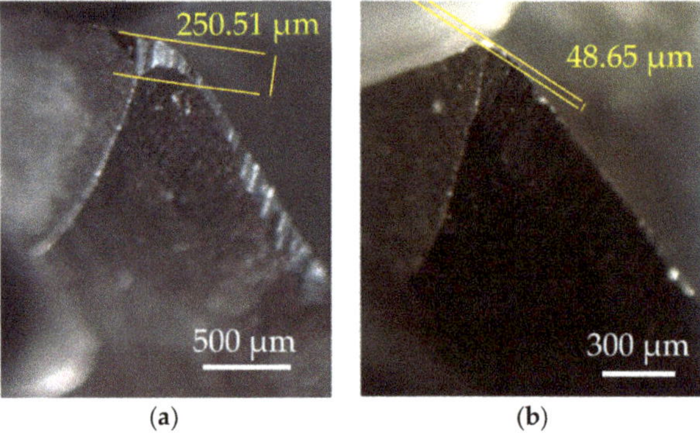

Figure 11. The size of flank wear chamfer after 72 min of operation when processing aluminum alloy AlCuMg2 with end mills: (**a**) uncoated, 250.51 μm; and (**b**) with a single-layer DLC, 48.65 μm.

An interesting fact is noteworthy, the application of a (CrAlSi)N sublayer before the formation of the DLC does not have any effect on the cutter's behavior during operation (Figure 10).

The experimental data obtained during the tests (Figure 12) show that when machining an aluminum alloy with uncoated end mills, the surface roughness (R) increases significantly after 27 min of operation and at the time of tool failure (99 min), is about 3 μm. When cutting an aluminum alloy with a diamond-like coated cutter, the roughness of the surface of the processed workpiece during 150 min of tool operation practically does not change and, on average, is 1.3 μm. At the moment of failure of the coated tool (342 min), the average roughness value was 2.5 μm. It is important to note that even in the first minutes of operation of the still unworn cutter, there is a noticeable difference in the surface quality of the workpiece, achieved by the tool without coating and with DLC of 1.75 μm and 1.25 μm, respectively. Experimental data presented in Figure 12 show that the deposition of a sublayer (CrAlSi)N before forming the DLC does not significantly affect the change in the roughness of the treated surface during the end mill operation.

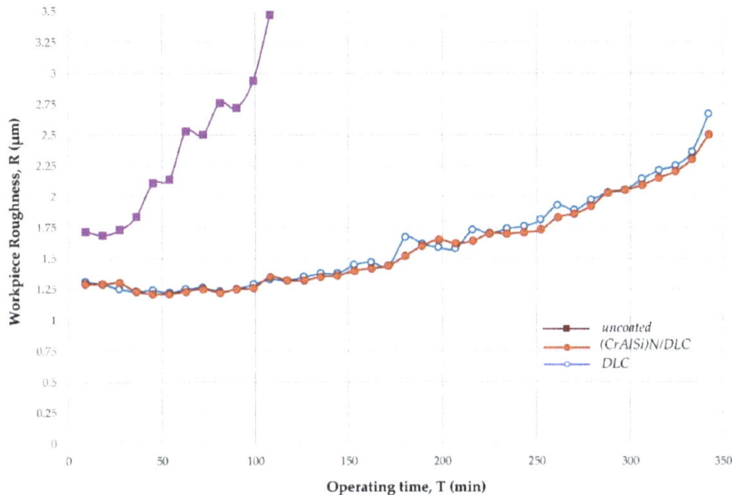

Figure 12. Dependences of the roughness of the treated surface of a workpiece made of aluminum alloy AlCuMg2 on the operating time of end mills with different coatings.

3.4.2. Processing of Structural Steel 41Cr4

The objects of study during the tests were the wear of five-flute end mills made of 6WH10F hard alloy (Figure 2) and the surface roughness of workpieces in the form of bars made of structural steel 41Cr4. As can be seen from the experimental data presented in Figure 13, in the process of machining structural steel, rather intense wear of the uncoated end mills occurs; the operating time to failure (until the wear limit value of 0.4 mm is reached) is slightly more than 30 min. Simultaneously, the application of a single-layer DLC to the cutter has little effect on the wear rate of the tool and only slightly increases its tool life (no more than 4 min). Optical images (Figure 14), showing the general view of the worn chamfer on the flank surface of the cutters after 17.5 min of operation, well illustrate the above well—the coating application practically does not slow down the development of wear processes.

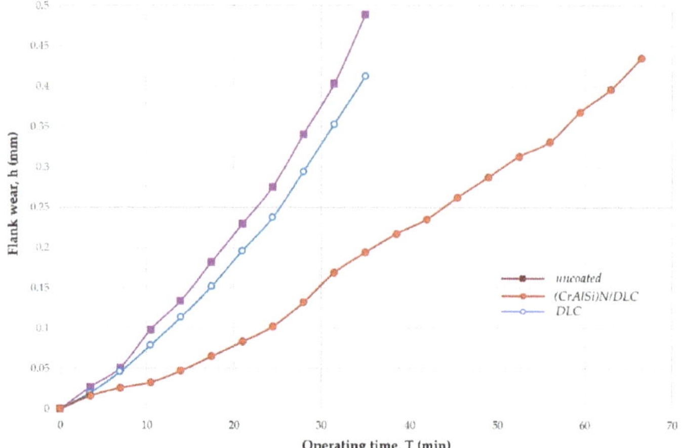

Figure 13. Dependences of the wear chamfer size on the flank surface of the end mills on the operating time when machining 41Cr4 structural steel with end mills with different coatings.

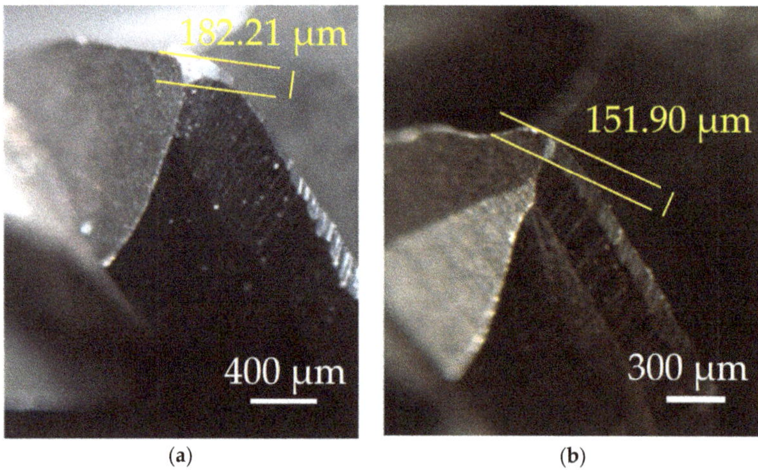

Figure 14. The size of the flank wear chamfer of end mills after 17.5 min of work when processing structural steel 41Cr4 with end mills; (**a**) uncoated, 182.21 μm; and (**b**) with a single-layer DLC, 151.90 μm.

Analysis of the development of wear in end mills after applying the sublayer (CrAlSi)N in combination with the formation of a DLC shows a noticeable effect of a two-layer coating on the tool wear rate when milling 41Cr4 structural steel. The tool life of end mills with (CrAlSi)N/DLC was 64 min, which is 2.0 times higher than the corresponding indicator of the uncoated tool (32 min) and 1.8 times higher than the tool with a single-layer DLC (35 min).

The dependences of the change in the surface roughness of the 41Cr4 steel workpiece on the operating time of the tool with different coatings shown in Figure 15 allow us to reveal the following regularity. From the very beginning of the cutting process, up to 25 min of work, the quantitative value of the roughness formed on the workpieces has a comparable value for all investigated mills and is in the range of 2.5–3.0 μm. After 25 min of operation, uncoated end mills show a noticeable deterioration in the quality of the machined surface and its sharp increase; at the time of tool failure (32 min), the roughness of the workpiece is more than 6 μm. For mills with a single-layer DLC, a significant increase in the roughness of the processed workpiece is also observed, and by the time of tool failure (35 min), it is about 5 μm. The (CrAlSi)N/DLC makes a slightly more significant contribution to the reduction of roughness. Thanks to this coating, the surface roughness gradually increases over time, reaching the value of 8.75 μm at the time of failure (64 min).

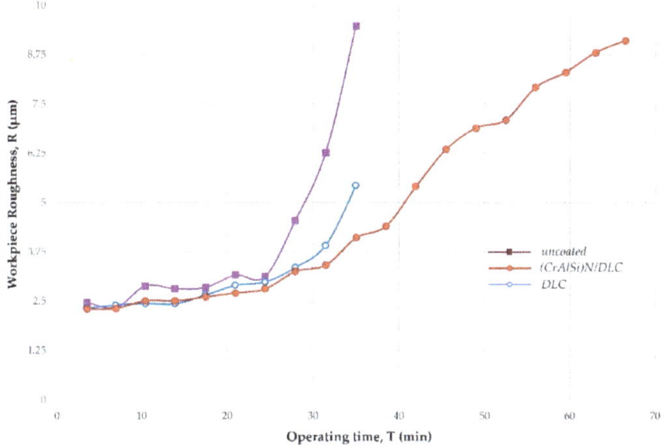

Figure 15. Dependence of the roughness of the treated workpiece surface of structural steel 41Cr4 on the operating time of end mills with different coatings.

4. Discussion of Research Results

The results of the XPS analysis allow us to judge the chemical composition and electronic state of atoms on the surface of the formed DLCs. From the experimental data obtained in Tables 4 and 5, it can be seen that DLCs of the two types studied (without sublayer and with sublayer (CrAlSi)N) have a very close ratio of components and approximately 64–66% consist of carbon atoms with different hybridization of valence electrons: sp^3-hybridization (diamond structure) and sp^2-hybridization (graphite structure). At the same time, the proportion of diamond hybridization (which is of particular interest to researchers) is almost identical for a single-layer DLC and a double-layer (CrAlSi)N/DLC. It is 47% and 51%, respectively, and the sp^3/sp^2 ratios of these coatings are similar at 3.0 and 3.4, respectively. First, these indicators show good reproducibility of PECVD technology used in work for the deposition of DLCs on a hard alloy. We add that this technology in a previous work was used by the authors for the deposition of DLC films on oxide-carbide ceramics [26,62,63], and the value of the sp^3/sp^2 ratio was close to that which was established when coating a hard alloy. Secondly, taking into account the fact that the volume fractions of the revealed silicon compounds on the surface of the DLC and (CrAlSi)N/DLCs were very close, which, taking into account the analysis error, allows us to conclude that the physicochemical processes occurring under the temperature conditions of PECVD technology do not lead to intensive chemical adsorption, mixing, and the formation of new phases in the formed DLC film. This is also indicated by the absence of compounds based on tungsten (the main component of the hard alloy), chromium, and aluminum (components of the CrAlSi sublayer) on the surface of DLCs, and the presence of silicon compounds in the two coatings are the result of discharge destruction of tetramethylsilane, which is a component of the gas mixture. It is necessary to clarify, speaking about the absence of chemisorption, we mean volumetric interaction processes inside the coating layer, while understanding that there is a chemical interaction at the substrate/coating interface, since it would be impossible to ensure good adhesive bond strength in its absence.

Thus, in reality, the structure (chemical composition and electronic state of atoms) of the formed DLC does not depend on the chemical composition and properties of the substrate on which it is deposited but is determined by the strategy and parameters of the implementation of PECVD technology. At the same time, studies have shown that the physical and mechanical characteristics and strength of the adhesive bond of the formed DLC, on the contrary, strongly depend on the properties of the substrate on which they are deposited. Therefore, DLC films that are very similar in chemical composition and electronic state of atoms can have very different values of physical and mechanical characteristics, particularly the modulus of elasticity (Table 6).

The Martens hardness values measured by nanoindentation were very similar (Table 6) for all samples with DLCs (without sublayer and with sublayer (CrAlSi)N). At a load of 1.0 mN, the samples had hardness 33 and 34 GPa, and with increasing loads up to 4.0 mN and, respectively, with the increase in the contact depth, the hardness of samples decreased to 25 and 26 GPa. Interesting regularities were established concerning the quantitative values of the modulus of elasticity of hard alloy samples with different coatings. The data on the modulus of elasticity (E) of samples calculated during nanoindentation and shown in Table 6 essentially characterize the degree of rigidity of the surface layer of the tool material and its ability to deform elastically under external loading. We know that the lower the modulus of elasticity, the less rigid the material is with low deformation resistance, but it is less brittle and more ductile. If at a low load (1.0 mN), the E values for all samples were similar, 292 GPa for the DLC and 285 GPa for (CrAlSi)N/DLC, then at a higher load (4.0 mN), significant differences were found, 242 and 201 GPa. Analysis of the obtained data suggests that it is possible to identify areas in which there is a different nature of changes in the properties of the coating. For example, when an indenter is inserted to a depth of 90–100 nm, the properties of DLCs are determined by the surface characteristics. If the indenter is embedded to a depth of 380 nm or more, the properties of DLCs largely depend on the state and properties of the substrate and, of course, on the stresses at the "substrate-coating" interface. Hardness and modulus of elasticity cannot be considered separately, since they are interrelated, and their H/E ratio carries information about an important characteristic—the index of plasticity. A high value of the index of plasticity provides an increased service life in conditions of cyclic loads typical for milling, and the closeness of the values of the modulus of elasticity of the coating and the substrate helps to reduce technological stresses on the interface and increase the adhesive strength. Authoritative studies show that the ratio $H/E = 0.15$ characterizes the "ideal elasticity" of the coating [64–66]. In the present study, this indicator at a load of 1.0 mN was similar for DLC and (CrAlSi)N/DLCs, 0.113 and 0.119, respectively. In tests with a fourfold increase in load, the influence of the properties of the substrate with a preformed sublayer (CrAlSi)N was immediately indicated, the index of plasticity was almost 0.13, while for a single-layer DLC, this indicator was at the level of 0.1.

It seems logical to explain that the higher plasticity of the surface layer (coating) of the samples and the less stressed state at the "substrate-coating" interface had a decisive effect on the subsequently established higher adhesive bond strength of the DLC deposited on a carbide substrate with a pre-formed sublayer (CrAlSi)N. Of course, the hardness of the formed coating is a primary characteristic, but it is the elastic characteristics of the substrate and coating that have the most significant influence on the level of occurring stresses [67–70]. As a confirmation of this, we can consider the empirical dependence of the stresses in the coating, which appear due to the loading of the substrate, when tangential stresses occur in the plane of the adhesive contact, which can cause adhesive destruction of the coating. The maximum tangential stresses (τ_{max}) in the plane of the adhesive contact are equal to [28]:

$$\tau_{max} = P\frac{L}{Ek} \quad (2)$$

where P is the load applied to the substrate (Pa); E is the modulus of elasticity of the substrate (GPa); L (Pa/m) and k (m^{-1}) are coefficients that depend on the modulus of elasticity of the coating.

It can be seen that by reducing the modulus of elasticity of the coating, it is possible to reduce the values of the maximum tangential stresses, the effect of which leads to the delamination of the coating, making it possible to increase its resistance to acting loads.

It is no coincidence that the experimental data shown in Figure 7 demonstrated that peeling of local areas of the coating from the substrate and especially the separation of the coating from a sample with a single-layer DLC, which has a higher modulus of elasticity, occurs at a significantly (almost two times) lower average force, at 12 N, while for the sample with (CrAlSi)N/DLC, which has a lower modulus of elasticity, at a load of 23 N.

The results of a comparative qualitative assessment of the adhesive bond strength of various coatings with hard alloy substrates once again confirm the above (Figure 8). Since local elastic-plastic deformation occurs around the indenter print during indentation, which causes swelling and destruction of the coating, it is the plasticity of the coating, which directly depends on the modulus of elasticity that is of paramount importance. A less plastic single-layer DLC at a uniform distance from the hole formed by the indenter has local areas of peeling of the coating and micro-chips, while a more plastic (CrAlSi)N/DLC after the load is removed, restores the surface layer, and there are no irreversible changes in it.

Since the resistance of the coating to abrasive wear is an important indicator that largely determines the performance of a coated product during operation, it is also necessary to discuss the results presented in Figure 9. The obtained images make it possible to approximate the surface area worn by external abrasive action, which can approximate the wear intensity. Unexpected results were not obtained here; the wear rate of hard alloy (CrAlSi)N/DLC-coated samples is 2.4 times lower than the uncoated samples and 1.5 times lower than the wear of single-layer DLC-coated samples.

Given that the hardness, which is the leading property that provides resistance to abrasion, two different coatings had very similar values (Table 6), such results can be explained as follows. First, it is the higher adhesive bond strength of the (CrAlSi)N/DLC with the substrate already discussed above. Secondly, it is necessary to consider the fact that after the external DLC layer is worn off; the nitride sublayer continues to function for some time and performs protective functions. Furthermore, the total thickness of (CrAlSi)N/DLCs is 4 μm, and the thickness of a single-layer DLC is 2.6 μm, which also affects the intensity of abrasive wear.

All of the above experimental results are very indicative and explain a lot, but they cannot consider the whole complex of actual operating loads. Therefore, it is possible to conclude the effectiveness of a particular coating only after discussing the results of operational tests of end mills with two types of DLCs when machining aluminum alloy and structural steel. In the operational tests carried out, the coatings showed significant differences in efficiency. The fact that when processing the aluminum alloy AlCuMg2, the application of a single-layer DLC significantly increased the durability of the end mill, and the application of a sublayer (CrAlSi)N did not contribute to an additional increase in the durability of the DLC-coated tool (Figure 10) can be explained as follows. The main task of the coating when processing aluminum-based alloys is to reduce the friction interaction on the contact pads of the cutter since the tool wear mechanism is associated with active cyclic adhesion of aluminum particles. The separation of particles from the contact surface of the tool and the workpiece is accompanied by the removal of micro-volumes of tool material from the working surfaces of the cutter, which are carried away from the cutting area by descending chips. However, a significant proportion sticks to the opposite surface of the part, reducing the quality of the treated surface. Therefore, due to the external DLC, which has excellent anti-friction characteristics, there is a decrease in friction interaction on the contact pads of the mill, and the development of the wear chamfer on the flank surface is significantly slowed down (Figures 10 and 11). The power loads when milling aluminum alloy are such that the coating on the cutting tool works in relatively favorable conditions and, according to experimental data (Figure 10), there is no need to form a sublayer under the external DLC (CrAlSi)N, thereby affecting the resistance of the coating to elastic and plastic deformations.

Analysis of the experimental data (Figure 13) obtained during the processing of 41Cr4 structural steel allows us to draw conclusions that differ significantly from those made above. This is not surprising, given that the maximum values of cutting forces that occur when processing structural steels can be four times higher than the corresponding values of force parameters when processing aluminum alloys. Judging by the experimentally obtained dependences of tool wear on operating time (Figures 14 and 15), the external DLC quickly ceases to perform protective functions, and the conditions of contact interaction on the front and flank surfaces of the tool are similar to those that occur when using end mills without coatings. The role of the (CrAlSi)N sublayer under the influence of heat and power loads when milling 41Cr4 steel is manifested in the fact that more favorable conditions are created for the functioning of the external DLC, its adhesive bond strength with the tool material increases, the stress level in the coating decreases and it is capable of resisting the current heat and power loads for a longer time. Even after the abrasion of the outer DLC layer, the nitride sublayer can act as a wear-resistant coating, slowing down the wear rate. A consequence of the described processes is an increase in the resistance of the (CrAlSi)N/DLC to the applied loads and providing a twofold increase in the wear resistance of end mills compared to an uncoated tool when milling 41Cr4 steel (Figure 13) and 1.8 times compared to milling cutters with a single layer DLC.

The experimentally obtained dependences of the roughness of the machined surfaces of workpieces made of aluminum alloy AlCuMg2 and structural steel 41Cr4 on the operating time of end mills with various DLCs, presented in Figures 12 and 15, are in good agreement with the characteristic curves of the development of mill wear in time discussed above (Figures 10 and 13) and obey the classical principles of the theory of metal cutting. The wear of the cutting tool leads to an increase in the roughness of the machined surface because there is an increase in the actual contact area of the cutting tool and the workpiece being processed, and the intensity of their frictional and adhesive interaction increases. The use of DLCs hinders the development of these processes (Figure 12), thereby significantly reducing the average height of microroughness of the surface layer of an aluminum alloy part, which, even at the moment of tool failure (i.e., with a heavily worn tool), did not exceed 2.5 μm. In contrast to the milling of aluminum alloys during the forming of structural steel 41Cr4, the contribution of DLCs is significantly lower in reducing the intensity of friction and adhesive interaction between the tool and the workpiece. Therefore the effect on the roughness of the treated surface is not so pronounced. Nevertheless, the coating (CrAlSi)N/DLC slightly reduces the average height of the micro-roughness of the surface layer of the structural steel part (Figure 15), which is 2.5 μm in the first minutes of tool operation, and 8.75 μm at the time of tool failure.

5. Conclusions

The complex of experimental studies allowed us to obtain some actual results that can be used for further development of research in the field of development of DLC for the needs of tool production.

- The volume fraction of diamond hybridization in DLCs formed by PECVD technology in a gas medium containing tetramethylsilane is about 50%. It does not depend on the properties and chemical composition of the substrate, since the physical and chemical processes do not lead to intensive chemical adsorption and the formation of new phases in the formed DLC film. The chemical composition and electronic state of the DLC atoms depend on the strategy and parameters of the PECVD process.
- The hardness and modulus of elasticity of DLCs, on the contrary, are very strongly dependent on the properties of the substrate on which they are deposited. DLC films that are very close in chemical composition and electronic state of atoms can have very different values of physical and mechanical characteristics, particularly, the modulus of elasticity. When nanoindenting with a load of 4.0 mN, the positive effect of the properties of the substrate with a preformed sublayer (CrAlSi)N is sharply manifested; the index of plasticity (ratio of hardness and modulus of elasticity) for the (CrAlSi)N/DLC was almost 0.13, while for a single-layer DLC, this indicator was at the level of 0.1. The formation of the (CrAlSi)N intermediate sublayer has a significant effect on the ability of the DLC to resist elastoplastic deformations.
- Higher plasticity of (CrAlSi)N/DLCs and a less stressed state at the substrate-coating interface had an essential effect on increasing the adhesive bond strength of the coating to the substrate compared to a single-layer DLC. The separation of the coating from the sample with a single-layer

DLC, which has lower plasticity (greater modulus of elasticity), occurs at a significantly (almost two times) lower load. Simultaneously, the wear rate under the influence of abrasive particles on (CrAlSi)N/DLC-coated hard alloy samples are 2.4 times lower than uncoated samples and 1.5 times lower than single-layer DLC-coated samples.

- When milling aluminum alloy AlCuMg2, applying a single layer DLC increases the tool life of end mills many times (3.45 times), while applying a sublayer (CrAlSi)N does not contribute to an additional increase in the durability of a DLC-coated tool. This is due to the low thermal and power loads when machining aluminum-based alloys and the need to reduce frictional interaction on the contact pads of the cutter, which is entirely handled by a single-layer DLC.
- When milling 41Cr4 structural steel, the single-layer DLC very quickly ceases to perform protective functions and the contact conditions on the front and flank surfaces of the tool approach those that occur when using end mills without coatings. The application of (CrAlSi)N/DLC alone provided an increase in the durability of end mills up to two times. The role of the (CrAlSi)N sublayer in milling steel 41Cr4 was manifested in providing more favorable conditions for the functioning of the external DLC, increasing its adhesive bond strength to the substrate, reducing the stress level in the coating, and, as a result, the (CrAlSi)N/DLC was able to resist the existing heat and power loads a longer time.
- The effect of DLCs on the roughness of the machined surface obeys the classical principles of the theory of metal cutting—an increase in the wear of the cutting tool leads to an increase in the roughness of the machined surface. The use of single-layer DLCs noticeably reduces the average height of microroughness of the surface layer of a workpiece made of aluminum alloy AlCuMg2, which, even at the moment of cutter failure, does not exceed 2.5 μm. In contrast to the milling of aluminum alloys, when forming structural steel 41Cr4, the role of DLCs in reducing the intensity of the frictional and adhesive interaction of the tool and the workpiece being processed is not so pronounced, and, as a consequence, in reducing the roughness of the processed surface. However, the (CrAlSi)N/DLC has a definite effect on reducing the average roughness of the surface layer of a structural steel part.

Author Contributions: Conceptualization, S.N.G.; methodology, M.A.V.; validation, S.V.F.; formal analysis, M.A.V. and S.V.F.; investigation, M.M. and S.V.F.; resources, S.N.G.; data curation, M.A.V.; writing—original draft preparation, S.V.F. and M.A.V.; writing—review and editing, S.N.G., M.A.V., S.V.F., and M.M.; visualization, M.M.; supervision, M.A.V.; project administration, S.N.G. All authors have read and agreed to the published version of the manuscript.

Funding: This work was supported by the Russian Science Foundation under grant 18-19-00599.

Acknowledgments: The work was carried out using the equipment of the Center of collective use of MSUT STANKIN.

Conflicts of Interest: The authors declare no conflict of interest.

References

1. Danek, M.; Fernandes, F.; Cavaleiro, A.; Polcar, T. Influence of Cr additions on the structure and oxidation resistance of multilayered TiAlCrN films. *Surf. Coat. Technol.* **2017**, *313*, 158–167. [CrossRef]
2. Vereschaka, A.A.; Volosova, M.A.; Grigoriev, S.N.; Vereschaka, A.S. Development of wear-resistant complex for high-speed steel tool when using process of combined cathodic vacuum arc deposition. *Procedia CIRP* **2013**, *9*, 8–12. [CrossRef]
3. Sousa, V.F.C.; Silva, F.J.G. Recent advances on coated milling tool technology—A Comprehensive review. *Coatings* **2020**, *10*, 235. [CrossRef]
4. Sobol, O.V.; Andreev, A.A.; Stolbovoj, V.A.; Grigor'ev, V.F.; Volosova, S.N.; Aleshin, S.V.; Gorban, V.F. Physical characteristics, structure and stress state of vacuum-arc TiN coating, deposition on the substrate when applying high-voltage pulse during the deposition. *Probl. Atom. Sci. Tech.* **2011**, *4*, 174–177.
5. Durmaz, Y.M.; Yildiz, F. The wear performance of carbide tools coated with TiAlSiN, AlCrN and TiAlN ceramic films in intelligent machining process. *Ceram. Int.* **2018**, *45*, 3839–3848. [CrossRef]

6. Grigoriev, S.; Metel, A. Plasma- and beam-assisted deposition methods. In *Nanostructured Thin Films and Nanodispersion Strengthened Coatings: NATO Science Series*; Voevodin, A.A., Shtansky, D.V., Levashov, E.A., Moore, J.J., Eds.; Springer: Dordrecht, The Netherlands, 2004; Volume 155, pp. 147–154.
7. Vereschaka, A.A.; Grigoriev, S.N.; Sitnikov, N.N.; Oganyan, G.V.; Batako, A. Working efficiency of cutting tools with multilayer nano-structured Ti-TiCN-(Ti,Al)CN and Ti-TiCN-(Ti,Al,Cr)CN coatings: Analysis of cutting properties, wear mechanism and diffusion processes. *Surf. Coat. Technol.* **2017**, *332*, 198–213. [CrossRef]
8. Li, Y.; Zheng, G.; Cheng, X.; Yang, X.; Xu, R.; Zhang, H. Cutting performance evaluation of the coated tools in high-speed milling of AISI 4340 steel. *Materials* **2019**, *12*, 3266. [CrossRef]
9. Grigoriev, S.N.; Sobol, O.V.; Beresnev, V.M.; Serdyuk, I.V.; Pogrebnyak, A.D.; Kolesnikov, D.A.; Nemchenko, U.S. Tribological characteristics of (TiZrHfVNbTa)N coatings applied using the vacuum arc deposition method. *J. Frict. Wear* **2014**, *35*, 359–364. [CrossRef]
10. Podgursky, V.; Bogatov, A.; Yashin, M.; Sobolev, S.; Gershman, I.S. Relation between self-organization and wear mechanisms of diamond films. *Entropy* **2018**, *20*, 279. [CrossRef]
11. Santecchia, E.; Hamouda, A.M.S.; Musharavati, F.; Zalnezhad, E.; Cabibbo, M.; Spigarelli, S. Wear resistance investigation of titanium nitride-based coatings. *Ceram. Int.* **2015**, *41*, 10349–10379. [CrossRef]
12. Vereschaka, A.A.; Vereschaka, A.S.; Grigoriev, S.N.; Kirillov, A.K.; Khaustova, O.Y. Development and research of environmentally friendly dry technological machining system with compensation of physical function of cutting fluids. *Procedia CIRP* **2013**, *7*, 311–316. [CrossRef]
13. Kohlscheen, J.; Bareiss, C. Effect of hexagonal phase content on wear behaviour of AlTiN Arc PVD coatings. *Coatings* **2018**, *8*, 72. [CrossRef]
14. Vereschaka, A.S.; Grigoriev, S.N.; Sotova, E.S.; Vereschaka, A.A. Improving the efficiency of the cutting tools made of mixed ceramics by applying modifying nano-scale multilayered coatings. *Adv. Mat. Res.* **2013**, *712*, 391–394. [CrossRef]
15. Siwawut, S.; Saikaew, C.; Wisitsoraat, A.; Surinphong, S. Cutting performances and wear characteristics of WC inserts coated with TiAlSiN and CrTiAlSiN by filtered cathodic arc in dry face milling of cast iron. *Int. J. Adv. Manuf. Technol.* **2018**, *97*, 3883–3892. [CrossRef]
16. Vereschaka, A.A.; Grigoriev, S.N.; Vereschaka, A.S.; Popov, A.Y.; Batako, A.D. Nano-scale multilayered composite coatings for cutting tools operating under heavy cutting conditions. *Procedia CIRP* **2014**, *14*, 239–244. [CrossRef]
17. Anders, S.; Callahan, D.L.; Pharr, G.M.; Tsui, T.Y.; Singh Bhatia, C. Multilayers of amorphous carbon prepared by cathodic arc deposition. *Surf. Coat. Technol.* **1997**, *94–95*, 189–194. [CrossRef]
18. Robertson, J. Diamond-like amorphous carbon. *Mater. Sci. Eng. R Rep.* **2002**, *37*, 129–281. [CrossRef]
19. Derakhshandeh, M.R.; Eshraghi, M.J.; Hadavi, M.M.; Javaheri, M.; Khamseh, S.; Sari, M.G.; Zarrintaj, P.; Saeb, M.R.; Mozafari, M. Diamond-like carbon thin films prepared by pulsed-DC PE-CVD for biomedical applications. *Surf. Innov.* **2018**, *6*, 167–175. [CrossRef]
20. Wei, C.; Yang, J.-F. A finite element analysis of the effects of residual stress, substrate roughness and non-uniform stress distribution on the mechanical properties of diamond-like carbon films. *Diam. Relat. Mater.* **2011**, *20*, 839–844. [CrossRef]
21. Volosova, M.; Grigoriev, S.; Metel, A.; Shein, A. The role of thin-film vacuum-plasma coatings and their influence on the efficiency of ceramic cutting inserts. *Coatings* **2018**, *8*, 287. [CrossRef]
22. Hainsworth, S.V.; Uhure, N.J. Diamond like carbon coatings for tribology: Production techniques, characterisation methods and applications. *Int. Mat. Rev.* **2007**, *52*, 153–174. [CrossRef]
23. Grigoriev, S.; Melnik, Y.; Metel, A. Broad fast neutral molecule beam sources for industrial scale beam-assisted deposition. *Surf. Coat. Technol.* **2002**, *156*, 44–49. [CrossRef]
24. Zou, C.W.; Wang, H.J.; Feng, L.; Xue, S.W. Effects of Cr concentrations on the microstructure, hardness, and temperature-dependent tribological properties of Cr-DLC coatings. *Appl. Surf. Sci.* **2013**, *286*, 137–141. [CrossRef]
25. Jeon, Y.; Park, Y.S.; Kim, H.J.; Hong, B.; Choi, W.S. Tribological properties of ultrathin DLC films with and without metal interlayers. *J. Korean Phys. Soc.* **2007**, *51*, 1124–1128. [CrossRef]
26. Grigoriev, S.; Volosova, M.; Fyodorov, S.; Lyakhovetskiy, M.; Seleznev, A. DLC-coating application to improve the durability of ceramic tools. *J. Mater. Eng. Perform.* **2019**, *28*, 4415–4426. [CrossRef]

27. Martinez-Martinez, D.; De Hosson, J.T.M. On the deposition and properties of DLC protective coatings on elastomers: A critical review. *Surf. Coat. Technol.* **2014**, *258*, 677–690. [CrossRef]
28. Dolgov, N.A. Analytical methods to determine the stress state in the substrate-coating system under mechanical loads. *Strength Mater.* **2016**, *48*, 658–667. [CrossRef]
29. Volosova, M.A.; Grigor'ev, S.N.; Kuzin, V.V. Effect of titanium nitride coating on stress structural inhomogeneity in oxide-carbide ceramic. Part 4. Action of heat flow. *Refract. Ind. Ceram.* **2015**, *56*, 91–96. [CrossRef]
30. Yi, J.; Chen, K.; Xu, Y. Microstructure, properties, and titanium cutting performance of AlTiN–Cu and AlTiN–Ni coatings. *Coatings* **2019**, *9*, 818. [CrossRef]
31. Vopát, T.; Sahul, M.; Haršáni, M.; Vortel, O.; Zlámal, T. The tool life and coating-substrate adhesion of AlCrSiN-coated carbide cutting tools prepared by LARC with respect to the edge preparation and surface finishing. *Micromachines* **2020**, *11*, 166. [CrossRef]
32. Vereschaka, A.S.; Grigoriev, S.N.; Tabakov, V.P.; Sotova, E.S.; Vereschaka, A.A.; Kulikov, M.Y. Improving the efficiency of the cutting tool made of ceramic when machining hardened steel by applying nano-dispersed multi-layered coatings. *Key Eng. Mater.* **2014**, *581*, 68–73. [CrossRef]
33. Georgiadis, A.; Fuentes, G.G.; Almandoz, E.; Medrano, A.; Palacio, J.F.; Miguel, A. Characterisation of cathodic arc evaporated CrTiAlN coatings: Tribological response at room temperature and at 400 °C. *Mater. Chem. Phys.* **2017**, *190*, 194–201. [CrossRef]
34. Kuzin, V.V.; Grigoriev, S. Method of investigation of the stress-strain state of surface layer of machine elements from a sintered nonuniform material. *Appl. Mech. Mater.* **2013**, *86*, 32–35. [CrossRef]
35. Aijaz, A.; Ferreira, F.; Oliveira, J.; Kubart, T. Mechanical properties of hydrogen free diamond-like carbon thin films deposited by high power impulse magnetron sputtering with Ne. *Coatings* **2018**, *8*, 385. [CrossRef]
36. Liu, X.Q.; Yang, J.; Hao, J.Y.; Zheng, J.Y.; Gong, Q.Y.; Liu, W.M. A near-frictionless and extremely elastic hydrogenated amorphous carbon film with self-assembled dual nanostructure. *Adv. Mater.* **2012**, *24*, 4614–4617. [CrossRef]
37. Lubwama, M.; Corcoran, B.; McDonnell, K.A.; Dowling, D. Flexibility and frictional behaviour of DLC and Si-DLC films deposited on nitrile rubber. *Surf. Coat. Technol.* **2014**, *239*, 84–94. [CrossRef]
38. Zhang, T.F.; Pu, J.J.; Xia, Q.X.; Son, M.J.; Kim, K.H. Microstructure and nano-wear property of Si-doped diamond like carbon films deposited by a hybrid sputtering system. *Mater. Today Proc.* **2016**, *3*, S190–S196. [CrossRef]
39. Wu, Y.; Li, H.; Ji, L.; Ye, Y.; Chen, J.; Zhou, H. Vacuum tribological properties of a-C:H film in relation to internal stress and applied load. *Tribol. Int.* **2014**, *71*, 82–87. [CrossRef]
40. Nakazawa, H.; Kamata, R.; Miura, S.; Okuno, S. Effects of frequency of pulsed substrate bias on structure and properties of silicon-doped diamond-like carbon films by plasma deposition. *Thin Solid Films* **2015**, *574*, 93–98. [CrossRef]
41. Vereschaka, A.; Tabakov, V.; Grigoriev, S.; Aksenenko, A.; Sitnikov, N.; Oganyan, G.; Seleznev, A.; Shevchenko, S. Effect of adhesion and the wear-resistant layer thickness ratio on mechanical and performance properties of ZrN—(Zr,Al,Si)N coatings. *Surf. Coat. Technol.* **2019**, *357*, 218–234. [CrossRef]
42. Hao, G.; Liu, Z. Experimental study on the formation of TCR and thermal behavior of hard machining using TiAlN coated tools. *Int. J. Heat Mass Transf.* **2019**, *140*, 1–11. [CrossRef]
43. Shuai, J.; Zuo, X.; Wang, Z.; Guo, P.; Xu, B.; Zhou, J.; Wang, A.; Ke, P. Comparative study on crack resistance of TiAlN monolithic and Ti/TiAlN multilayer coatings. *Ceram. Int.* **2020**, *46*, 6672–6681. [CrossRef]
44. Endrino, J.L.; Fox-Rabinovich, G.S.; Gey, C. Hard AlTiN, AlCrN PVD coatings for machining of austenitic stainless steel. *Surf. Coat. Technol.* **2006**, *200*, 6840–6845. [CrossRef]
45. Vereshchaka, A. Improvement of working efficiency of cutting tools by modifying its surface properties by application of wear-resistant complexes. *Adv. Mat. Res.* **2013**, *712–715*, 347–351. [CrossRef]
46. Varghese, V.; Chakradhar, D.; Ramesh, M.R. Micro-mechanical characterization and wear performance of TiAlN/NbN PVD coated carbide inserts during end milling of AISI 304 austenitic stainless steel. *Mater. Today Proc.* **2018**, *5*, 12855–12862. [CrossRef]
47. Bag, R.; Panda, A.; Sahoo, A.K.; Kumar, R. Cutting tools characteristics and coating depositions for hard part turning of AISI 4340 martensitic steel: A review study. *Mater. Today Proc.* **2020**, *26*, 2073–2078. [CrossRef]

48. Chang, Y.-Y.; Chuang, C.-C. Deposition of multicomponent AlTiCrMoN protective coatings for metal cutting applications. *Coatings* **2020**, *10*, 605. [CrossRef]
49. Kuzin, V.V.; Grigor'ev, S.N.; Volosova, M.A. Effect of a TiC coating on the stress-strain state of a plate of a high-density nitride ceramic under nonsteady thermoelastic conditions. *Refract. Ind. Ceram.* **2014**, *54*, 376–380. [CrossRef]
50. Kuzin, V.V.; Grigoriev, S.N.; Fedorov, M.Y. Role of the thermal factor in the wear mechanism of ceramic tools. Part 2: Microlevel. *J. Frict. Wear* **2015**, *36*, 40–44. [CrossRef]
51. Grigoriev, S.N.; Volosova, M.A.; Vereschaka, A.A.; Sitnikov, N.N.; Milovich, F.; Bublikov, J.I.; Fyodorov, S.V.; Seleznev, A.E. Properties of (Cr,Al,Si)N-(DLC-Si) composite coatings deposited on a cutting ceramic substrate. *Ceram. Int.* **2020**, *46*, 18241–18255. [CrossRef]
52. Li, G.; Li, L.; Han, M.; Luo, S.; Jin, J.; Wang, L.; Gu, J.; Miao, H.J.M. The performance of TiAlSiN coated cemented carbide tools enhanced by inserting Ti interlayers. *Metals* **2019**, *9*, 918. [CrossRef]
53. Alamgir, A.; Yashin, M.; Bogatov, A.; Viljus, M.; Traksmaa, R.; Sondor, J.; Lümkemann, A.; Sergejev, F.; Podgursky, V. High-temperature tribological performance of hard multilayer TiN-AlTiN/nACo-CrN/AlCrN-AlCrO-AlTiCrN coating deposited on WC-Co substrate. *Coatings* **2020**, *10*, 909. [CrossRef]
54. Bobzin, K.; Brögelmann, T.; Kruppe, N.C.; Carlet, M. Nanocomposite (Ti,Al,Cr,Si)N HPPMS coatings for high performance cutting tools. *Surf. Coat. Technol.* **2019**, *378*, 124857. [CrossRef]
55. Wang, Y.; Su, D.; Ji, H.; Li, X.; Zhao, Z.; Tang, H. Gradient structure high emissivity $MoSi_2$-SiO_2-SiOC coating for thermal protective application. *J. Alloy. Compd.* **2017**, *703*, 437–447. [CrossRef]
56. Oliver, W.C.; Pharr, G.M. An improved technique for determining hardness and elastic modulus using load and displacement sensing indentation experiments. *J. Mater. Res.* **1992**, *7*, 1564–1583. [CrossRef]
57. Fox-Rabinovich, G.S.; Yamomoto, K.; Veldhuis, S.C.; Kovalev, A.I.; Dosbaeva, G.K. Tribological adaptability of TiAlCrN PVD coatings under high performance dry machining conditions. *Surf. Coat. Technol.* **2005**, *200*, 1804–1813. [CrossRef]
58. Metel, A.S.; Grigoriev, S.N.; Melnik, Y.A.; Prudnikov, V.V. Glow discharge with electrostatic confinement of electrons in a chamber bombarded by fast electrons. *Plasma Phys. Rep.* **2011**, *37*, 628–637. [CrossRef]
59. Kim, Y.S.; Park, H.J.; Lim, K.S.; Hong, S.H.; Kim, K.B. Structural and mechanical properties of AlCoCrNi high entropy nitride films: Influence of process pressure. *Coatings* **2020**, *10*, 10. [CrossRef]
60. Kuzin, V.; Gurin, V.; Shein, A.; Kochetkova, A.; Mikhailova, M. The influence of duplex vacuum-plasma treatment on the mechanics of complex-profile cutting tool wearing in the production of aircraft engine parts. *Mech. Ind.* **2018**, *19*, 704. [CrossRef]
61. Metel, A.S.; Grigoriev, S.N.; Melnik, Y.A.; Bolbukov, V.P. Broad beam sources of fast molecules with segmented cold cathodes and emissive grids. *Instrum. Exp. Tech.* **2012**, *55*, 122–130. [CrossRef]
62. Metel, A.; Bolbukov, V.; Volosova, M.; Grigoriev, S.; Melnik, Y. Source of metal atoms and fast gas molecules for coating deposition on complex shaped dielectric products. *Surf. Coat. Technol.* **2013**, *225*, 34–39. [CrossRef]
63. Grigoriev, S.N.; Melnik, Y.A.; Metel, A.S.; Panin, V.V.; Prudnikov, V.V. Broad beam source of fast atoms produced as a result of charge exchange collisions of ions accelerated between two plasmas. *Instrum. Exp. Tech.* **2009**, *52*, 602–608. [CrossRef]
64. Metel, A.S. Ionization effect in cathode layers on glow-discharge characteristics with oscillating electrons. 1. Discharge with the hollow cathode. *Zhurnal Tekhnicheskoi Fiziki* **1985**, *55*, 1928–1934.
65. Vereshchaka, A. Development of assisted filtered cathodic vacuum arc deposition of nano-dispersed multi-layered composite coatings on cutting tools. *Key Eng. Mater.* **2014**, *581*, 62–67. [CrossRef]
66. Metel, A.; Bolbukov, V.; Volosova, M.; Grigoriev, S.; Melnik, Y. Equipment for deposition of thin metallic films bombarded by fast argon atoms. *Instrum. Exp. Tech.* **2014**, *57*, 345–351. [CrossRef]
67. Beake, B.D.; Endrino, J.L.; Kimpton, C.; Fox-Rabinovich, G.S.; Veldhuis, S.C. Elevated temperature repetitive micro-scratch testing of AlCrN, TiAlN and AlTiN PVD coatings. *Int. J. Refract. Met. Hard Mater.* **2017**, *69*, 215–226. [CrossRef]
68. Boing, D.; Oliveira, A.J.; Schroeter, R.B. Limiting conditions for application of PVD (TiAlN) and CVD (TiCN/Al_2O_3/TiN) coated cemented carbide grades in the turning of hardened steels. *Wear* **2018**, *416–417*, 54–61. [CrossRef]

69. Xu, Q.; Zhao, J.; Ai, X. Fabrication and cutting performance of Ti(C,N)-based cermet tools used for machining of high-strength steels. *Ceram. Int.* **2017**, *43*, 6286–6294. [CrossRef]
70. Paskvale, S.; Remskar, M.; Cekada, M. Tribological performance of TiN, TiAlN and CrN hard coatings lubricated by MoS_2 nanotubes in Polyalphaolefin oil. *Wear* **2016**, *352–353*, 72–78. [CrossRef]

Publisher's Note: MDPI stays neutral with regard to jurisdictional claims in published maps and institutional affiliations.

Article

Structure, Oxidation Resistance, Mechanical, and Tribological Properties of N- and C-Doped Ta-Zr-Si-B Hard Protective Coatings Obtained by Reactive D.C. Magnetron Sputtering of TaZrSiB Ceramic Cathode

Ph. V. Kiryukhantsev-Korneev [1,*], A. D. Sytchenko [1], S. A. Vorotilo [1], V. V. Klechkovskaya [2], V. Yu. Lopatin [1] and E. A. Levashov [1]

1 National University of Science and Technology «MISiS», Leninsky pr. 4, 119049 Moscow, Russia; alina-sytchenko@yandex.ru (A.D.S.); stepan.vorotylo@gmail.com (S.A.V.); lopatin63@mail.ru (V.Y.L.); levashov@shs.misis.ru (E.A.L.)

2 Shubnikov Institute of Crystallography, Federal Scientific Research Centre "Crystallography and Photonics", Russian Academy of Sciences, Leninsky pr. 59, 119333 Moscow, Russia; klechvv@crys.ras.ru

* Correspondence: kiruhancev-korneev@yandex.ru; Tel.: +7-(495)-638-46-59

Received: 27 August 2020; Accepted: 29 September 2020; Published: 30 September 2020

Abstract: Coatings in the Ta-Zr-Si-B-C-N system were produced by magnetron sputtering of a $TaSi_2$-Ta_3B_4-$(Ta,Zr)B_2$ ceramic target in the Ar medium and Ar-N_2 and Ar-C_2H_4 gas mixtures. The structure and composition of coatings were studied using scanning electron microscopy, glow discharge optical emission spectroscopy, energy-dispersion spectroscopy, and X-ray diffraction. Mechanical and tribological properties of coatings were determined using nanoindentation and pin-on-disk tests using 100Cr6 and Al_2O_3 balls. The oxidation resistance of coatings was evaluated by microscopy and X-ray diffraction after annealing in air at temperatures up to 1200 °C. The reactively-deposited coatings containing from 30% to 40% nitrogen or carbon have the highest hardness up to 29 GPa and elastic recovery up to 78%. Additionally, coatings with a high carbon content demonstrated a low coefficient of friction of 0.2 and no visible signs of wear when tested against 100Cr6 ball. All coatings except for the non-reactive ones can resist oxidation up to a temperature of 1200 °C thanks to the formation of a protective film based on Ta_2O_5 and SiO_2 on their surface. Coatings deposited in Ar-N_2 and Ar-C_2H_4 demonstrated superior resistance to thermal cycling in conditions 20-T−20 °C (where T = 200–1000 °C). The present article compares the structure and properties of reactive and "standard-inert atmosphere" deposited coatings to develop recommendations for optimizing the composition.

Keywords: reactive magnetron sputtering; argon; nitrogen and ethylene; $TaSi_2$; Ta_3B_4 and ZrB_2; SHS and hot pressing; composition and structure; hardness and elastic modulus; friction coefficient and wear resistance; oxidation resistance

1. Introduction

One of the prospective avenues of surface engineering is the deposition of hard, wear-, and oxidation-resistant transitional-metal based coatings for high-performance metal machining instruments by arc-evaporation and magnetron sputtering of composite targets [1–5]. Additionally, such coatings can be employed for heavy-duty friction pairs, high-temperature sensors, and resistive elements, critical parts for the aerospace industry. Tantalum disilicide ($TaSi_2$) is often the material of choice for the engineering of coatings thanks to its high hardness (10–13 GPa) and refractoriness (2200 °C), low thermal expansion coefficient (8.8×10^{-6} °C^{-1}), high strength at temperatures above 1000 °C, and oxidation resistance (up to 1700 °C) [6–8].

$TaSi_2$-based coatings can be produced by chemical vapor deposition and subsequent diffusion saturation [9,10], plasma spraying [11], sintering [12], slurry impregnation [13], electron beam evaporation [14], low-pressure chemical vapor deposition [15], diode or high-frequency sputtering [16], magnetron sputtering [17,18], and double-cathode sputtering [19]. In the case of surface engineering of metal-machining tools, chemical (CVD) and physical vapor deposition (PVD) techniques outperform other methods owing to the unrivaled capability to retain blade geometry, high quality, and homogeneity of the coating, as well as the general flexibility and possibility to fine-tune the properties of coatings.

The functional properties of $TaSi_2$-based coatings can be enhanced by alloying. The addition of Al increases the adhesion strength of $TaSi_2$ coatings by ~28%, despite an 8% decrease in hardness and a ~12% decrease in elastic modulus in [19,20]. Additionally, Al decreases the corrosion current density in $Ta(Si_{1-x}Al_x)_2$ coatings from 7.08×10^{-6} to 3.05×10^{-6} A·cm^{-2} [20]. Alloying of Ta-Si coatings by Zr and Ti improved the coatings' hardness (12–16 GPa) and corrosion resistance [21], whereas alloying by Ni produced amorphous coatings stable up to 900 °C for microelectronic applications [22]. Alloying by nitrogen and carbon is another promising avenue for enhancing the properties of $TaSi_2$-based coatings. Such coatings were deposited by magnetron sputtering of composite $TaSi_2$-SiC targets in the atmosphere with varied nitrogen content, resulting in hardness up to 26 GPa, friction coefficient below 0.3 at temperatures above 600 °C, and heat resistance up to 800 °C [23]. The introduction of Si_3N_4 into the $TaSi_2$ coatings allows for a substantial increase in oxidation resistance [24]. The Ta-Si-N coatings have high thermal stability up to 900 °C [25]. Moreover, Ta-Si-N coatings possess good diffusion barrier properties [26] and are used to prolongate the lifespan of high-speed steel cutting tools [27] and glass forms [28]. Increased Si content endows the Ta-Si-N coatings with oxidation resistance up to 1300 °C [29]. A similar effect is produced by boron alloying, as it promotes the formation of borosilicate glass in the oxide layer, thereby reducing the viscosity and increasing the protective properties of the oxide [30].

The aim of this work is the comparative study of the structures and properties of Ta-Zr-Si-B-(C,N) coatings deposited by magnetron sputtering of composite target $TaSi_2$-Ta_3B_4-$(Ta,Zr)B_2$ in inert (Ar) and reactive atmospheres (Ar-N_2 and Ar-C_2H_4).

2. Materials and Methods

The coatings were deposited by magnetron sputtering in the direct current mode. The sputtered composite target TaZrSiB—with element composition of 70.8 wt.% Ta, 18.6 wt.% Si, 7.4 wt.% Zr, and 2.9 wt.% B, and phase composition of $TaSi_2$-Ta_3B_4-$(Ta,Zr)B_2$, 120 mm diameter, and 6 mm thickness—was produced by hot pressing on DSP-515 SA installation ("Dr. Fritsch", Germany). The feedstock powders for hot pressing were fabricated by the ball-milling of the combustion products of a reactive mixture of Ta, Zr, Si, and B. The details of the combustion synthesis protocol can be found elsewhere [31].

VK-100-1 grade alumina plates (99.7% Al_2O_3, JSC Policor, Kineshma, Ivanovo region, Russia), and alumina discs VOK (VNIITS, Russia) and VT-1-0-grade (OAO Conmet, Moscow, Russia) were used as the substrates. Before deposition, the substrates were ultrasonically cleaned in isopropanol at a frequency of 22 kHz for 5 min using the UZDN-2T installation (JSC «UKRROSPRIBOR», Ukraine). Additional cleaning via ion etching was performed right before the coating deposition using the slit-type ion source (Ar+ ions, 2 keV) for 20 min. The coatings were deposited under the following conditions: 80 mm distance between the substrate and the target, 10^{-3} Pa residual pressure, and 0.1–0.2 Pa working pressure in the vacuum chamber. Ar (99.9995%) was used as the working gas, as well as its mixtures with N_2 (99,999%) and C_2H_4 (99.95%). The gas flow rate (Table 1) was controlled by the gas supply system ELTOCHPRIBOR, LTD (Moscow, Russia). The constant current (1 kW power) was supplied to the magnetron using Pinnacle+ source (Advanced Energy, Fort Collins, CO, USA) during the whole deposition duration (40 min). Experiments were conducted without applying a bias voltage to the substrate. The schematics of the installation are given in the work [32].

Table 1. Composition, thickness, and deposition rate of coatings.

Sample	Gas Flow Rate, sccm			Chemical Composition, at.%						Thickness,	Deposition Rate,
	Ar	N_2	C_2H_4	Ta	Zr	Si	B	N	C	μm	nm/min
1	25	0	0	40.0	7.6	28.0	24.6	0	0	7.3	182
2	20	5	0	27.2	7.7	22.3	22.2	20.4	0	7.2	180
3	15	10	0	19.3	5.4	17.1	15.6	42.4	0	8.0	200
4	20	0	5	28.1	10.1	25.6	23.5	0	12.4	6.3	157
5	15	0	10	22.1	8.0	21.1	18.3	0	30.3	7.0	175

The elemental composition and structure of coatings were studied using scanning electron microscopy (SEM) using the S-3400 microscope (Hitachi, Tokyo, Japan) equipped with the energy dispersive spectroscopy add-on (EDS) Noran-7 Thermo. X-ray diffraction analysis (XRD) was performed on a D2 Phaser Bruker diffractometer using CuKα radiation. Elemental distribution profiles across the thickness of the coating were studied by glow discharge optical emission spectroscopy (GDOES) using Profiler 2 installation ("Horiba Jobin Yvon", Longjumeau, France). Studies of the fine structure were performed using a high-resolution transmission electron microscope (HR TEM) JEM-2100 Jeol. The foils for HR TEM were prepared using the ion-beam etching unit PIPS II System (Gatan, Pleasanton, CA, USA) and FIB (FEI Quanta 200 3D FIB instrument, Hillsboro, OR, USA). The coatings' mechanical properties were tested at a load of 2 mN on a nano-hardness tester (CSM Instruments, Peuseux, Switzerland) equipped with a Berkovich indenter. The tribological performance of the coatings was tested on an automated friction machine Tribometer (CSM Instruments) using a «pin-on-disc» scheme and a normal load of 1 N. Balls (d = 6 mm) made of alumina (hardness 19 GPa) and 100Cr6 grade steel (hardness of 8 GPa) were used as the counterparts for tribological experiments. Wear tracks were studied using an optical profiler Wyko-1100NT (Veeco, Plainview, NY, USA). To assess the coatings oxidation resistance, step-by-step annealing was performed in the air in a muffle furnace SNOL 7.2/1200 (Umega, Ukmerge, Lithuania) with a step of 200 °C, a maximum temperature of 1200 °C, and 1 h exposure. The specimens were inserted into a furnace heated to a given temperature T, dwelled during 1 h in the air, and afterward ejected from the furnace and cooled to 20 °C. The temperature was increased using 200 °C steps, and 1000 °C was the highest temperature the coatings could endure for 1 h without complete oxidation. At 1200 °C, all coatings were completely oxidized after 1 h long experiment; therefore, the oxidation duration for the test at this temperature was decreased to 10 min. Annealed coatings were studied by SEM and EDS. A more detailed description of the employed methods and equipment can be found elsewhere [33].

3. Results

The GDOES data (Table 1) were used to quantify the thickness of the coatings and assess the homogeneity of the elemental distribution within the coatings. The elemental profiles show all the main elements that make up the target, such as tantalum, zirconium, silicon, and boron. In coatings applied in environments containing nitrogen and ethylene, there is also a signal from nitrogen and carbon, respectively (Figure 1). The emergence of Ti lines signifies the boundaries between the coatings and Ti-alloys substrate and allows for an assessment of the coatings' thickness (6.6 μm in the case of coating 3 and 5.8 μm in the case of coating 5).

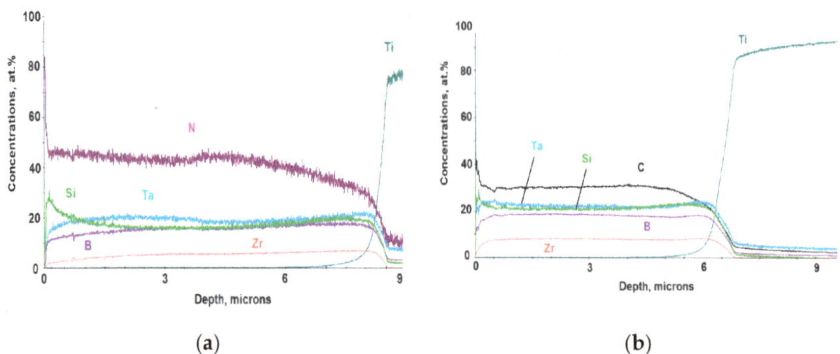

Figure 1. Typical glow discharge optical emission spectroscopy (GDOES) profiles of coatings 3 (10 sccm N_2) (**a**) and 5 (10 sccm C_2H_4) (**b**).

According to the GDOES data, all elements were evenly distributed across the thickness of coatings. The concentration of oxygen impurities was below 1 at.%. The minor oxygen contamination resulted from the presence of corresponding impurities both in the ceramic target and in the working gas. The depth-averaged coating composition is shown in Table 1.

The coatings deposited in argon contained 47.6 at.% of metals and 52.4 at.% of non-metals, so the metal-to-nonmetal species ratio was close to 1. The increased flow of reactive gases resulted in higher nitrogen or carbon content in coatings and a decreased concentration of elements inherited from the target (except for Zr). The maximum N and C concentrations (42.4 and 30.3 at.%, respectively) were achieved at 10 sccm of N_2 and C_2H_4, correspondingly.

Non-reactive deposited coating 1 had a dense structure with pronounced columnar grains, which is typical for ion-plasma deposited coatings [34,35] (Figure 2a).

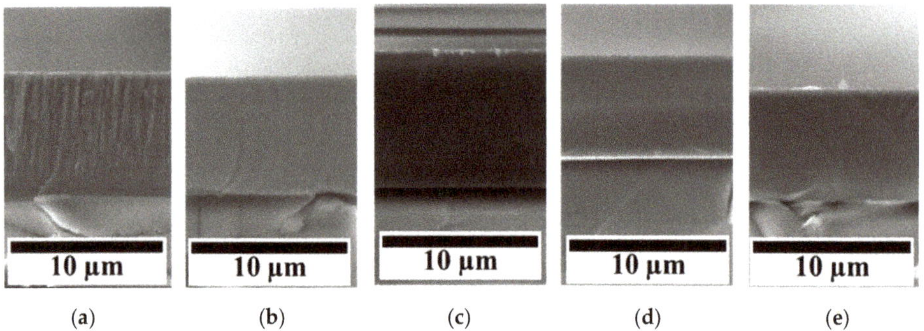

Figure 2. Cross-section scanning electron microscopy (SEM) images of as-deposited coatings 1 (Ar) (**a**), 2 (5 sccm N_2) (**b**), 3 (10 sccm N_2) (**c**), 4 (5 sccm C_2H_4) (**d**), and 5 (10 sccm C_2H_4) (**e**).

A similar columnar structure was reported for $TaSi_2$-based coatings in [18,28]. When N_2 (coatings 2 and 3) or C_2H_4 (coatings 4 and 5) were added to the working gas, the columnar structure became less pronounced or completely disappeared. The coatings produced by reactive sputtering featured a nearly perfect, defect-free structure.

The coatings' thickness, depending on the deposition modes, ranged from 6.3 to 8.0 microns (Table 1). The introduction of a minor amount of nitrogen (coating 2) had no significant effects on the growth rate (180 nm/min). With an increase in nitrogen concentration (coating 3), a boost in the growth rate of up to 200 nm/min was observed. When carbon-containing gas was added to the sputtering atmosphere, the growth rate decreased to 157 nm/min (coating 4) or 175 nm/min (coating 5).

The XRD patterns and HRTEM-imaged structure of the coatings 1–5 are shown in Figure 3.

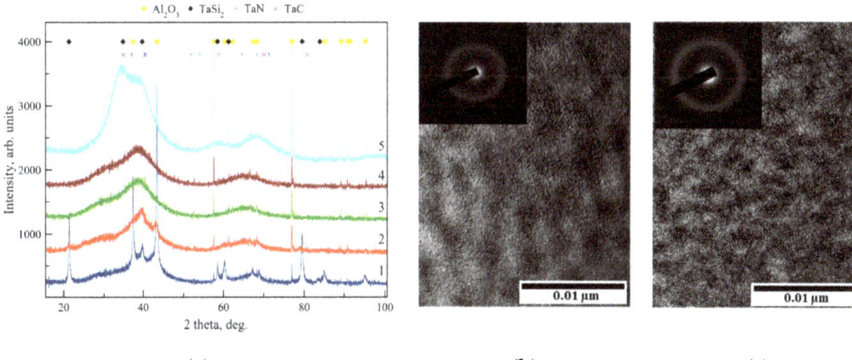

(a) (b) (c)

Figure 3. X-ray diffraction analysis (XRD) patterns of coatings 1(Ar), 2 (5 sccm N_2), 3 (10 sccm N_2), 4 (5 sccm C_2H_4), and 5 (10 sccm C_2H_4) deposited onto alumina substrate (**a**); selected area electron diffraction (SAED) and high-resolution transmission electron microscope (HR TEM) image of coatings 3 (10 sccm N_2) (**b**) and 5 (10 sccm C_2H_4) (**c**).

A signal from the substrate of Al_2O_3 (JCPDS 88-0107 card) was detected for all samples. The XRD pattern for coating 1 contained peaks associated with reflections from planes (100), (102), (111), (210), (203), (220), and (115) of the h-$TaSi_2$ hexagonal phase (JCPDS 89-2941), similar to [36]. The size of the h-$TaSi_2$ crystallites determined by the Scherrer formula for the coating obtained in Ar was approximately 11 nm. The transition to reactive sputtering (in Ar-N_2 and Ar-C_2H_4) resulted in a considerable structural refinement of the coatings. The most intense peak in the 2Θ = 25–45° range for coatings obtained in Ar-N_2 can be attributed to the Ta-N bonds in the coating (FCC-TaN, JCPDS 89-5198), which agrees with the previously obtained XRD results for Ta-Si-N coatings deposited at a different N_2/(N_2 + Ar) ratio [37]. The crystallite size of the phase h-$TaSi_2$ for coatings 2 and 3 produced by reactive sputtering was evaluated according to the most distinct lines. Coating 2, deposited at a flow rate of 5 sccm N_2, had a grain size of the h-$TaSi_2$ phase equal to ~6 nm. As the N_2 flow rate increased, the crystallite size decreased to 4.5 nm. Coatings 4 and 5 deposited in the Ar + C_2H_4 atmosphere show broad peaks on XRD patterns at 2Θ from 30° to 45°. Their position can be explained by the presence of Ta-Si and Ta-C bonds (TaC, JCPDS 89-3831). The introduction of nitrogen and carbon into the coatings resulted in grain refinement of h-$TaSi_2$ phase to 3–6 nm according to HR TEM иselected area electron diffraction (SAED) data (Figure 3b,c) The sizes of the h-$TaSi_2$ phase crystallites for coatings deposited at 5 and 10 sccm C_2H_4 flow rates were equal to 3.5 and 3.0 nm, respectively. The TEM data correlated well with the results of XRD.

The mechanical properties of coatings, such as hardness (H), elastic modulus (E), elastic recovery (W), plasticity index (H/E), and the resistance to plastic deformation (H^3/E^2) and compressive stress (σ), are shown in Table 2.

Table 2. Mechanical and chemical properties of coatings.

Sample	H, GPa	E, GPa	H/E	H^3/E^2, GPa	W, %	σ, GPa	Oxide Layer Thickness, μm	
							T = 1000 °C	T = 1200 °C
1	12.5	208	0.060	0.045	43.4	N/A	3.5	12.2 (complete oxidation)
2	18.2	233	0.078	0.111	57.3	N/A	4.4	8.2
3	29.2	279	0.105	0.320	77.9	−0.53	3.4	8.0
4	21.3	267	0.080	0.136	62.4	−0.77	3.6	3.6
5	28.3	288	0.098	0.273	76.4	−0.72	3.8	6.9

Coating 1 deposited in Ar demonstrated H = 12.5 GPa, E = 208 GPa, and W = 43.4%. The transition to reactive deposition at a flow rate of 5 sccm N_2 resulted in an increase in H by 46% as well as an increase in E and W by 12% and 32%, respectively. When the N_2 flow rate was increased to 10 sccm, H rose further to 29.2 GPa, while E and W increased to 279 GPa and 77.9%, respectively. This effect might be related to the formation of tantalum nitride phase [38,39] or structure refinement [40]. The introduction of carbon atoms into the coatings also resulted in an increase of mechanical properties. Coating 4 deposited at a flow rate of 5 sccm C_2H_4 had H = 21.3 GPa, E = 267 GPa, and W = 62.4%, which is 70%, 28%, and 44% higher, respectively, than the values obtained for the non-reactive sample. Coating 5 demonstrated hardness H = 28.3 GPa, Young's modulus E = 288 GPa, and elastic recovery W = 76.4%. The mechanical properties of coatings increased at higher carbon concentrations, similar to the results obtained in [41,42]. The increase in mechanical properties can be partially attributed to the increase of compressive stresses in the carbon-rich coatings (Table 2). Table 1 shows the values of the parameters H/E and H^3/E^2, which are important in terms of assessing the level of wear resistance and determining the mechanism of localized deformation [43,44]. In indentation experiments, the coatings with H^3/E^2 below 0.5 GPa are deformed in a heterogeneous fashion with the associated formation of shear bands, whereas the coatings with H^3/E^2 above 0.6 GPa the deformation are homogeneous, with no structural transformations of the surface [44].

The tribological testing of the coatings was carried using counterparts made of 100Cr6 and Al_2O_3, which differ significantly from each other in terms of physical and mechanical characteristics. The results of tests determining the coatings' friction coefficient are shown in Figure 4.

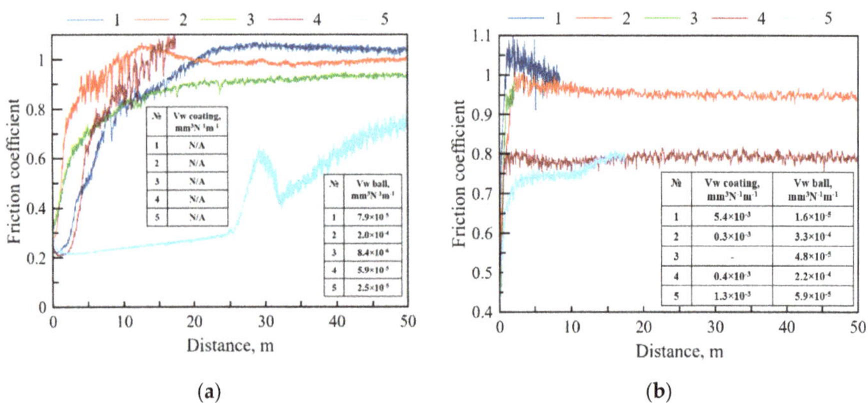

Figure 4. Friction coefficient vs. distance dependence during testing of coatings against 100Cr6 (**a**) and Al_2O_3 (**b**) counterparts. Inserts show the values of wear rate for coatings and counter-bodies.

When the 100Cr6 ball was used as a counterpart, the behavior of coatings 1–3 was similar. The friction coefficient increased from the initial values of 0.2–0.3 to 0.9–1.1 at a distance of 10–20 m and then stabilized at 1.04 (coating 1), 0.98 (coating 2), and 0.92 (coating 3) and remained relatively constant until the end of the measurements at a distance of 50 m. The behavior of coating 4 obtained at 5 sccm C_2H_4 at the initial moment of the test was similar to that of coatings 1–3. However, when reaching a distance of 17.5 m, the friction machine automatically stopped the measurement as a result of exceeding the permissible friction coefficient (>1.1). Coating 5 with the highest carbon content showed the lowest and the most stable f ~0.3 at a distance up to 25.7 m, followed by a jump to 0.65 (25.8–32.4 m), and further monotonous increase from 0.39 up to 0.76 at the distance from 32 to 50 m. This behavior can be explained by the positive role of carbon [45,46], as well as by specific tribochemical reactions [47].

In the tests where Al_2O_3 ball was used as a counterpart, for coatings 1–3, the friction coefficient f increased sharply to values of 0.95–1.05 at a 0–2 m distance. This resulted in the premature termination of tests for samples 1 and 3 at a distance below 10 m as a result of exceeding the permissible values of f.

The friction coefficient of sample 2 obtained at 5 sccm N_2 was stable after a short run-in period and remained within a 0.95–1 range until the end of the test at 50 m. Coating 4 deposited in a mixture of Ar and C_2H_4 showed a rapid rise of the f value to 0.8 and then remained stable until the end of the test. Coating 5 with a high carbon content had similar behavior with slightly lower values at distances up to 20 m.

Figure 5 provides three-dimensional profiles of wear tracks after testing with 100Cr6 and Al_2O_3 balls for coatings 1, 2, and 4.

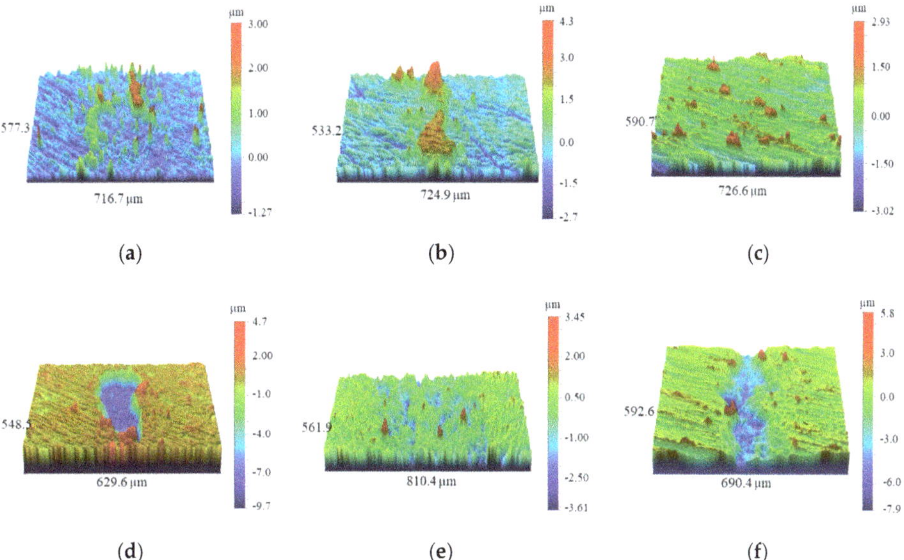

(**a**) (**b**) (**c**)

(**d**) (**e**) (**f**)

Figure 5. 3D images of wear tracks on the surface of coatings 1(Ar), 2 (5 sccm N_2), and 5 (10 sccm C_2H_4) after testing against 100Cr6 (**a–c**) and Al_2O_3 (**d–f**).

When tested against the steel ball counterpart, none of the coatings 1–5 showed any signs of surface wear; however, noticeable pile-ups of wear debris were present, suggesting that wear of steel ball and adhesion of steel debris to the coatings were the main wear mechanisms. Coatings 1, 2, and 4 had pronounced areas of wear debris accumulation, concentrated in local zones and having a height of 3–5 microns. In the case of coatings 3 and 5 with the maximum nitrogen and carbon concentration, the wear debris was distributed more evenly in the tribocontact zone and had a height of 2–3 microns. The wear rate (Vw) of the steel ball used to test coating 1 was 7.9×10^{-5} $mm^3 \cdot N^{-1} \cdot m^{-1}$ (Figure 4a, insert). The highest (2.0×10^{-4} $mm^3 \cdot N^{-1} \cdot m^{-1}$) and lowest ($8.4 \times 10^{-6}$ $mm^3 \cdot N^{-1} \cdot m^{-1}$) ball wear rates were achieved for nitrogen-rich coatings 2 and 3. The increase of carbon content was associated with the decrease of Vw from 5.9×10^{-5} (coating 4) to 2.5×10^{-5} $mm^3 \cdot N^{-1} \cdot m^{-1}$ (coating 5).

During the testing with alumina ball counterpart, coating 1 deposited in Ar experienced uneven wear; that is, both segments with no visible wear and areas of complete wear with a depth of ~8 microns were present (Figure 5d). The low wear resistance of coating 1 can be associated with its pronounced columnar structure, characterized by low fracture toughness [35]. The wear rate of the coating was 5.4 $\times 10^{-3}$ $mm^3 \cdot N^{-1} \cdot m^{-1}$ (Figure 4b, insert). In coatings 2–5, deposited by reactive sputtering, the wear depth did not exceed the coating thickness. In the case of nitrogen-containing coating 2, wear track with a width of 450 microns and depth below 2 microns was observed. The wear rate of the coating was 0.3×10^{-3} $mm^3 \cdot N^{-1} \cdot m^{-1}$, which is ~20 times lower than the values obtained for the coating 1 deposited in Ar. Coating 3 with the maximum nitrogen content showed no signs of wear (test run ~2 m). Sample 4 obtained at 5 sccm C_2H_4 demonstrated a wear rate of 0.4×10^{-3} mm^3 N^{-1} m^{-1}, which is equivalent to the value for a coating deposited in nitrogen at the same gas flow rate. When the C_2H_4 flow rate increased to 10 sccm, an increase in the wear rate was observed (1.3×10^{-3} $mm^3 \cdot N^{-1} \cdot m^{-1}$). Our previous work [23] addressed the case of Ta-Si-C and Ta-Si-C-N coatings produced by sputtering of composite SHS-cathodes TaSiC. The nitrogen-rich coatings (Ta-Si-C-N) showed a low wear rate against Al_2O_3 ball (1.7–2.8 $\times 10^{-5}$ $mm^3 \cdot N^{-1} \cdot m^{-1}$, whereas their nitrogen-free counterparts suffered complete wear.

The wear areas on alumina balls show that the lowest achieved wear rate was 1.6×10^{-5} $mm^3 \cdot N^{-1} \cdot m^{-1}$ and was achieved for coating 1 (Figure 4b, insert). Nitrogen-rich coatings 2 and 3 were worn at rates 3.3×10^{-4} and 4.8×10^{-5} $mm^3 \cdot N^{-1} \cdot m^{-1}$, respectively. The wear of counter-bodies for coatings 4 and 5 was 2.2×10^{-4} and 5.9×10^{-5} $mm^3 \cdot N^{-1} \cdot m^{-1}$, respectively. Therefore, the increase in both carbon and nitrogen content in reactively-deposited coatings is beneficial for the decreased wear of alumina counter-bodies.

The SEM images of the surface of coatings 1–5 annealed at the maximal temperature of 1000 °C are shown in Figure 6.

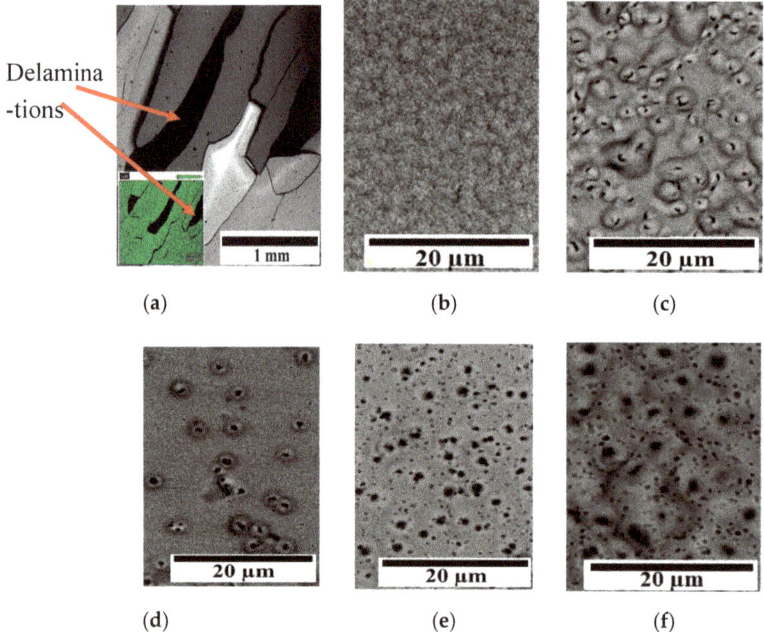

Figure 6. Top-view SEM images of coatings 1(Ar) (**a**,**b**), 2 (5 sccm N_2) (**c**), 3 (10 sccm N_2) (**d**), 4 (5 sccm C_2H_4) (**e**), and 5 (10 sccm C_2H_4) (**f**) after air annealing with maximal temperature of 1000 °C. Insert on 1a shows the energy dispersive spectroscopy (EDS) map for Ta signal.

The surface of coating 1 deposited in Ar contained visible cracks and delaminations. However, a local uniform microstructure with no visible defects can be seen at ×2000 magnification. The cracks formation likely resulted from the difference in thermal expansion coefficients between the coating and the substrate. As a result, tensile stresses develop along the perimeter of the defect, which leads to the coating's destruction [48]. For coating 2 obtained at 5 sccm N_2, a specific structure (Figure 6b), resembling blisters or bubbles was observed after the annealing. The formation of cavities or pores can be associated with the oxidation of nitrogen-containing phases in the coating and subsequent release of gaseous nitrogen oxides through the oxide film at high temperatures [37]. The bubbles occupied about 85% of the surface area of coating 2. When the N_2 flow rate increased to 10 sccm (coating 3), a reduction in the number of bubbles was observed and the total area of defects was ~10% of the surface area. Samples 4 and 5 showed the same defective structure, with open pores (~0.1–2 microns) occupying up to ~75% of the surface. The formation of pores is caused by the accumulation of CO_x in the upper layer of the coating and the formation of blisters, which, under a certain gas pressure, burst and form pores on the surface [49].

SEM cross-section images of annealed coatings 1–5 are shown in Figure 7.

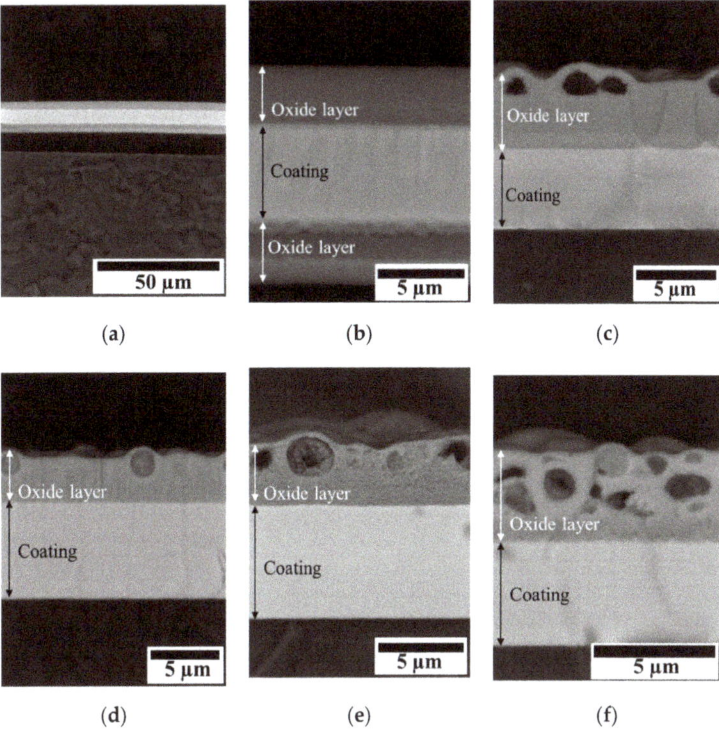

Figure 7. Cross-section SEM images of coatings 1 (Ar) (**a**,**b**), 2 (5 sccm N_2) (**c**), 3 (10 sccm N_2) (**d**), 4 (5 sccm C_2H_4) (**e**), and 5 (10 sccm C_2H_4) (**f**) after air annealing with maximal temperature of 1000 °C.

Coating 1 has experienced delamination from the substrate during the annealing (Figure 7a) and the formation of oxide layers on both sides of the coating. The oxide layer was pore-free, with a maximum thickness of 3.5 microns, and was comprised mainly of SiO_x and TaO_x. Cross-section SEM images of sample 2 show a porous oxide layer with a thickness of 4.4 microns. Pores (d = 0.2–2.0 microns) were concentrated in the upper part of the oxide layer. With an increase in the nitrogen content, the thickness of the oxide layer decreased by 23% and amounted to 3.4 microns. Coating 4 obtained at 5 sccm C_2H_4 showed an oxide layer thickness of 3.6 microns. The formed SiO_x-TaO_x oxide film had a loose structure with a pore size up to ~2.6 microns. The oxide layer of sample 5 with a thickness of 3.8 microns also had a porous structure. Pores with 0.2–1.5 microns diameter were randomly distributed over the thickness of the oxide layer. Coating 2 obtained at 5 sccm N_2 had the maximum thickness of the oxide layer (Table 2), and the minimum thickness of the oxide layer was characteristic of sample 3, deposited at a flow rate of N_2 10 sccm.

4. Discussion

4.1. Dependence of Growth Kinetics and Microstructure on the Gas Environment

The coating deposition kinetics was dependent on the composition of the sputtering atmosphere. The introduction of N_2 (10 sccm) into Ar flow resulted in an 11% increase in growth rate (coating 3, Table 2), whereas the introduction of a similar amount of C_2H_4 slightly decreased the growth rate. Both nitrogen and carbon formed related phases in the coatings, suggesting that the difference in growth rates arises from different ionization behavior of N_2 and C_2H_4 and higher scattering of atoms of sputtered material by C_2H_4 molecules (Table 1). The decrease in the size of h-$TaSi_2$ crystallites and the amorphization of coatings deposited in a reactive atmosphere is associated with the formation of new TaN and TaC phases, which apparently interrupt the growth of h-$TaSi_2$ crystallites. Reactive sputtering was instrumental in eliminating the unwanted columnar structures in the coatings (Figure 2). Columnar structures are known to adversely affect the mechanical and tribological properties of coatings, as well as their oxidation resistance [34,35].

4.2. Mechanical and Tribological Properties

Reactive deposition in both Ar-N_2 and Ar-C_2H_4 resulted in the enhancement of mechanical properties and tribological performance of the coatings. The increase of mechanical properties can be partially attributed to the growth of internal stresses (Table 2). In the case of N-containing coatings, this effect might be associated with the formation of a crystalline/amorphous phase based on TaN [38,39], as well as a decrease in the size of crystallites [40]. In C-doped coatings, the rise in hardness is due to the formation of a hard DLC-phase, as well as the positive effect of microstructural modification via the introduction of carbon into the composition of the coating. The complete understanding of structural features and phase composition of reactive coatings requires additional studies, which we plan to do next.

All the coatings investigated in this work were characterized by a high friction coefficient when relatively soft material (steel 100Cr6, H = 8 GPa) was used as a counterpart. The surface of the coating is much harder (H = 12–30 GPa), and has high adhesion to steel, similar to [5]. The distinct behavior of coating 5 at the initial part of testing distance can be associated with the positive role of free carbon, which could be released owing to the oversaturation of the crystalline carbide phase and plays the role of a solid lubricant during friction [45,46]. Thus, coating 5 obtained at a flow rate of 10 sccm C_2H_4 had the lowest friction coefficient f in contact with the 100Cr6 steel ball. Under more stringent conditions (Al_2O_3 counterpart), the lowest f was demonstrated by sample 4 deposited at 5 sccm C_2H_4. The coating obtained at 5 sccm N_2 had a stable friction coefficient against both counterparts (100Cr6 steel and Al_2O_3). The transition to reactive sputtering leads to a decrease in the coatings' wear rate, which coincides with the results of other works [41,47]. The best wear resistance was recorded for coatings 2 and 4, obtained at 5 sccm in N_2 and C_2H_4. This effect can be attributed to the presence of amorphous solid lubricant phases.

4.3. Oxidation Resistance

The oxidation testing confirmed that the non-reactive coating was characterized by the lowest adhesion to the substrates, as this was the only coating that experienced complete delamination (Figure 7). However, its oxidation behavior was superior to other coatings, because the as-formed oxide layer was dense and contained neither pores nor blisters. The sublimation of boron oxide is a known problem for refractory borides [50–52], so the lack of gas porosity in the Zr-Ta-Si-B coatings is remarkable. Despite increasing the adhesion, preventing the delamination, and enhancing the mechanical and tribological performance, the reactive sputtering clearly resulted in pronounced pore formation in the as-formed oxide layer upon annealing. The pores were formed by the oxidation of nitrogen- and carbon-containing phases to gaseous oxides. The blistering of the oxide layer is to be avoided if one strives to produce coatings with the highest possible oxidation resistance. The best trade-off between tribological performance and oxide layer porosity was achieved for coating 4. This coating was characterized by the lowest wear rate and propinquity to pore-formation upon oxidation among the investigated reactively deposited coatings. A further improvement might be achieved by the deposition of functionally graded coatings with the lower layer produced by reactive sputtering and upper layer deposited in Ar.

In addition, short-term annealing in air at 1200 °C for 10 min was also performed to assess the maximum operating temperature. It was found that the non-reactive coating completely oxidizes under these conditions. The thickness of the oxide layer for nitrogen-containing coatings 2 and 3 was about 8 microns (Table 2). Coating 5 with the maximum carbon content formed a 6.9 micron thick oxide layer. The lowest oxide thickness of 3.6 microns and the highest non-oxidized layer thickness of 6.2 microns were observed for coating 4 (Figure 8), in line with the results at 1000 °C.

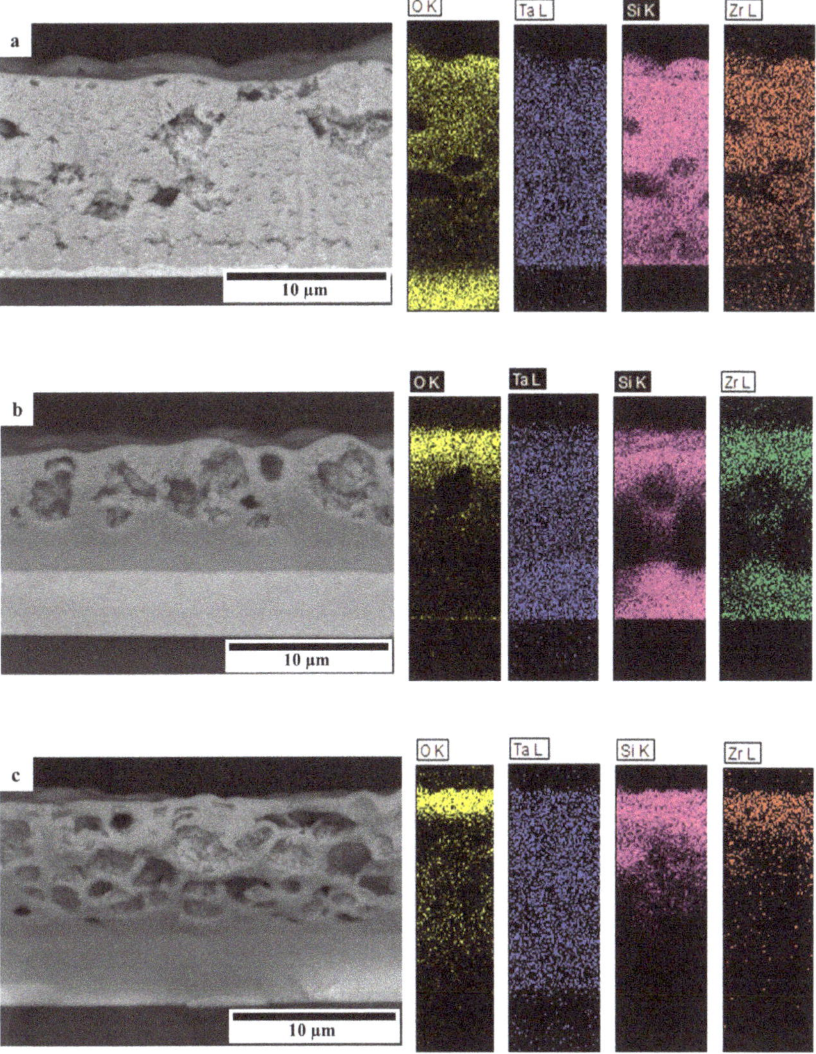

Figure 8. *Cont.*

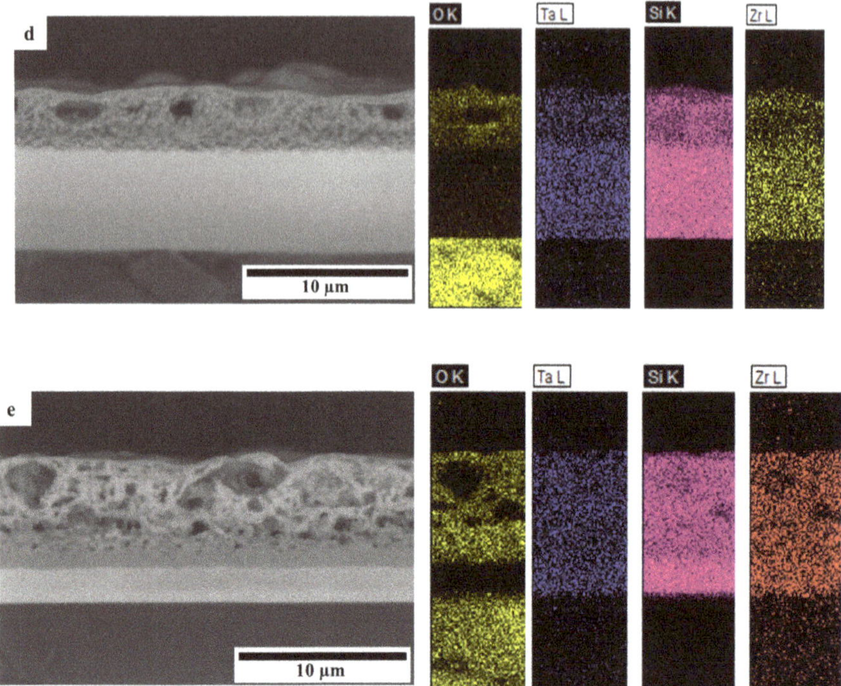

Figure 8. Cross-section SEM images and EDS maps of coatings 1 (Ar) (**a**), 2 (5 sccm N_2) (**b**), 3 (10 sccm N_2) (**c**), 4 (5 sccm C_2H_4) (**d**), and 5 (10 sccm C_2H_4) (**e**) annealed at 1200 °C for 10 min.

From the SEM-EDS data, it can be seen that a top-layer is formed on the surface of the coating, consisting mainly of silicon and tantalum oxides. Under the oxide film is an oxygen-free layer containing all the main elements of the coating, which has not undergone noticeable recrystallization.

Further analysis of oxidation mechanisms was based on XRD screening of coatings 2–5 annealed at 1200 °C for 10 min (Figure 9).

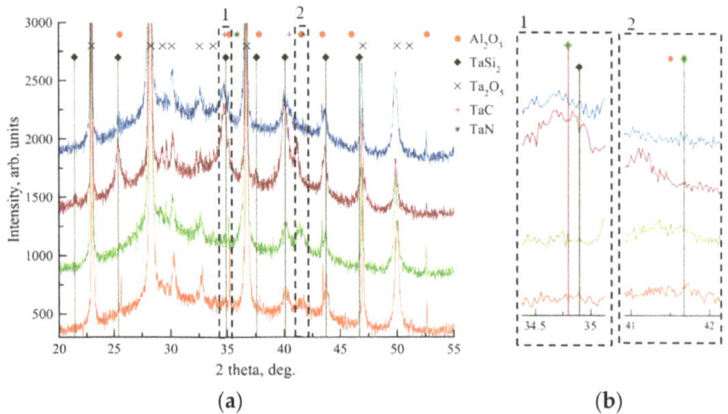

Figure 9. XRD patterns of coatings 2 (5 sccm N_2), 3 (10 sccm N_2), 4 (5 sccm C_2H_4), and 5 (10 sccm C_2H_4) after annealing at 1200 °C for 10 min: overview diffractograms (**a**) and enlarged areas (**b**).

A signal from alumina substrate can be recognized in all tested specimens after exposure to 1200 °C. A set of peaks that correlate well with (100), (101), (102), (110), (111), (200), (201), and (103) planes of the hexagonal h-$TaSi_2$. The highest characteristic peaks h-$TaSi_2$ (111) and (200) are located at 2Θ = 40.1° and 43.7°, respectively. Intense peaks at 2Θ = 23.0, 28.3, 36.7, and 47.0 are related to oxidation product Ta_2O_5. In the case of nitrogen-rich coatings 2 and 3, the peaks for cubic TaN were found, with the highest peak at 2Θ= 41.7° (Figure 9b). The coatings deposited in C_2H_4 show the presence of (111) and (200) planes of FCC-TaC, with a corresponding highest peak at 2Θ = 34.8° (Figure 9b). The crystallite structure of h-$TaSi_2$ calculated based on the broadening of (111) peak was 6–10 nm for all coatings. XRD patters corroborate that the reactive-sputtered coatings 2–5 are relatively stable and resistant to corrosion at temperatures up to 1200 °C.

5. Conclusions

Coatings with a thickness of 6–8 microns were obtained by magnetron sputtering of the $TaSi_2$-Ta_3B_4-(Ta,Zr)B_2 target in the Ar environment and gas mixtures Ar-N_2 and Ar-C_2H_4. The coating deposited in argon had a pronounced columnar structure, while the transition to reactive sputtering suppressed the column growth. The main constituent of Ar-sputtered coatings was the h-$TaSi_2$ phase with a crystallite size of ~10 nm. During the deposition in Ar-N_2 and Ar-C_2H_4, a decrease in grain size, partial amorphization, and the formation of new phases based on TaN and TaC were observed. The coatings obtained at the maximum flow rate of N_2 and C_2H_4 had high hardness at the level of 28–29 GPa and elastic recovery of 76%–78%. The transition to reactive deposition reduced the friction coefficient and the coatings' wear rate. All studied coatings resisted oxidation in the air up to 1000 °C owing to the formation of protective Ta-Si-O oxide films. Sputtering at optimal nitrogen and carbon concentrations modified the structure and increased the heat resistance and thermal cycling resistance of the coatings. The reactively-deposited coatings can briefly resist oxidation at 1200 °C.

The combination of high mechanical and tribological characteristics with high oxidation resistance makes the coatings in the Ta-Zr-Si-B-C-N system promising as protective coatings for various high-temperature applications, including high-performance dry machining tools and parts of heavily loaded friction units operating in aggressive environments, among others.

Author Contributions: Supervision, Deposition, Annealing, Investigation, Conceptualization, Writing & Editing: P.V.K.-K.; Tribo-Investigation, Profilometry, Analysis, Calculations, Writing: A.D.S.; Translation and Editing: S.A.V.; Investigation: V.V.K.; Formal Analysis: V.Y.L.; Project Administration, Resources: E.A.L. All authors have read and agreed to the published version of the manuscript.

Funding: The research was funded by the Ministry of Science and Higher Education of the Russian Federation (Project No. 0718-2020-0034 of State assignment).

Acknowledgments: The authors are grateful to M.I. Petrzhik for the nanoindentation tests, N.V. Shvyndina for the SEM-EDS study, and P.A. Loginov for the XRD and HR TEM studies.

Conflicts of Interest: The authors declare no conflict of interest.

References

1. He, Q.; Paiva, J.M.; Kohlscheen, J.; Beake, B.D.; Veldhuis, S.C. An integrative approach to coating/carbide substrate design of CVD and PVD coated cutting tools during the machining of austenitic stainless steel. *Ceram. Int.* **2020**, *46*, 5149–5158. [CrossRef]
2. Vereschaka, A.; Tabakov, V.; Grigoriev, S.; Sitnikov, N.; Oganyan, G.; Andreev, N.; Milovich, F. Investigation of wear dynamics for cutting tools with multilayer composite nanostructured coatings in turning constructional steel. *Wear* **2019**, *420–421*, 17–37. [CrossRef]
3. Kiryukhantsev-Korneev, F.V.; Sheveiko, A.N.; Komarov, V.A.; Blanter, M.S.; Skryleva, E.A.; Shirmanov, N.A.; Levashov, E.A.; Shtansky, D.V. Nanostructured Ti-Cr-B-N and Ti-Cr-Si-C-N coatings for hard-alloy cutting tools. *Russ. J. Non-Ferrous Met.* **2011**, *52*, 311–318. [CrossRef]

4. Vereschaka, A.; Tabakov, V.; Grigoriev, S.; Sitnikov, N.; Milovich, F.; Andreev, N.; Bublikov, J. Investigation of wear mechanisms for the rake face of a cutting tool with a multilayer composite nanostructured Cr–CrN-(Ti,Cr,Al,Si)N coating in high-speed steel turning. *Wear* **2019**, *438–439*, 203069. [CrossRef]
5. Kiryukhantsev-Korneev, P.V.; Pierson, J.F.; Bychkova, M.Y.; Manakova, O.S.; Levashov, E.A.; Shtansky, D.V. Comparative Study of Sliding, Scratching, and Impact-Loading Behavior of Hard CrB_2 and Cr–B–N Films. *Tribol. Lett.* **2016**, *63*, 44–55. [CrossRef]
6. Shon, I.-J.; Ko, I.-Y.; Chae, S.-M.; Na, K. Rapid consolidation of nanostructured $TaSi_2$ from mechanochemically synthesized powder by high frequency induction heated sintering. *Ceram. Int.* **2011**, *37*, 679–682. [CrossRef]
7. Samsonov, G.V.; Vinitsky, I.V. *Refractory Compounds: Handbook*; Metallurgy: Moscow, Russia, 1976; p. 558.
8. Buecher, W.; Schliebs, R.; Winter, G.; Buechel, K.H. *Industrielle Angewandte Chemie*; Verlag Chemie: Weinheim, Germany, 1984.
9. Zhang, Y.-L.; Ren, J.; Tian, S.; Li, H.; Ren, X.; Hu, Z. HfC nanowire-toughened $TaSi_2$–TaC–SiC–Si multiphase coating for C/C composites against oxidation. *Corro. Sci.* **2015**, *90*, 554–561. [CrossRef]
10. Shi, X.H.; Zeng, X.R.; Li, H.J.; Fu, Q.G.; Zou, J.Z. TaSi 2 Oxidation Protective Coating for SiC Coated Carbon/Carbon Composites. *Rare Met. Mater. Eng.* **2011**, *40*, 403–406. [CrossRef]
11. Liu, F.; Li, H.; Gu, S.; Yao, X.; Fu, Q. Ablation behavior and thermal protection performance of $TaSi_2$ coating for SiC coated carbon/carbon composites. *Ceram. Int.* **2019**, *45*, 3256–3262. [CrossRef]
12. Du, B.; Zhou, S.; Zhang, X.; Hong, C.; Qu, Q. Preparation of a high spectral emissivity $TaSi_2$-based hybrid coating on SiOC-modified carbon-bonded carbon fiber composite by a flash sintering method. *Surf. Coat. Technol.* **2018**, *350*, 146–153. [CrossRef]
13. Tao, X.; Li, X.; Guo, L.; Xu, X.; Guo, A.; Hou, F.; Liu, J. Effect of $TaSi_2$ content on the structure and properties of $TaSi_2$-$MoSi_2$-borosilicate glass coating on fibrous insulations for enhanced surficial thermal radiation. *Surf. Coat. Technol.* **2017**, *316*, 122–130. [CrossRef]
14. Mansour, A.N. Effect of temperature on microstructure and electrical properties of $TaSi_2$ thin films grown on Si substrates. *Vacuum* **2011**, *85*, 667–671. [CrossRef]
15. Ringeisen, F.; Alaoui, M.; Bolmont, D.; Koulmann, J.J. Oxidation of $TaSi_2$ thin films on polycrystalline tantalum. *Appl. Surf. Sci.* **1992**, *62*, 167–173. [CrossRef]
16. Schultes, G.; Schmitt, M.; Goettel, D.; Freitag-Weber, O. Strain sensitivity of TiB_2, $TiSi_2$, $TaSi_2$ and WSi_2 thin films as possible candidates for high temperature strain gauges. *Sens. Actuators A Phys.* **2006**, *126*, 287–291. [CrossRef]
17. Sidorenko, S.I. Formation of Nanocrystalline Structure of $TaSi_2$ Films on Silicon. *Powder Metall. Met. Ceram.* **2003**, *42*, 14–18. [CrossRef]
18. Inui, H.; Fujii, A.; Hashimoto, T.; Tanaka, K.; Yamaguchi, M.; Ishizuk, K. Defect structures in $TaSi_2$ thin films produced by co-sputtering. *Acta Mater.* **2003**, *51*, 2285–2296. [CrossRef]
19. Xu, J.; Xie, Q.; Peng, S.; Li, Z.; Jiang, S. Investigation of slurry erosion-corrosion behavior of $Ta(Si_{1-x}Al_x)_2$ nanocrystalline coatings. *Mater. Res. Express* **2020**, *7*, 026408. [CrossRef]
20. Xu, J.; Zhang, S.K.; Lu, X.L.; Jiang, S.; Munro, P.; Xie, Z.-H. Effect of Al alloying on cavitation erosion behavior of $TaSi_2$ nanocrystalline coatings. *Ultrason. Sonochem.* **2019**, *59*, 104742. [CrossRef]
21. Lai, J.J.; Lin, Y.S.; Chang, C.H.; Wei, T.Y.; Huang, J.C.; Liao, Z.X.; Lin, C.H.; Chen, C.H. Promising Ta-Ti-Zr-Si metallic glass coating without cytotoxic elements for bio-implant applications. *Appl. Surf. Sci.* **2018**, *427*, 485–495. [CrossRef]
22. Oleksak, R.P.; Devaraj, A.; Herman, G.S. Atomic-scale structural evolution of Ta–Ni–Si amorphous metal thin films. *Mater. Lett.* **2016**, *164*, 9–14. [CrossRef]
23. Bondarev, A.V.; Vorotilo, S.; Shchetinin, I.V.; Levashov, E.A.; Shtansky, D.V. Fabrication of Ta-Si-C targets and their utilization for deposition of low friction wear resistant nanocomposite Si-Ta-C-(N) coatings intended for wide temperature range tribological applications. *Surf. Coat. Technol.* **2019**, *359*, 342–353. [CrossRef]
24. Yoon, J.-K.; Kim, G.-H.; Kim, H.-S.; Shon, I.-J.; Kim, J.-S.; Doh, J.-M. Microstructure and oxidation behavior of in situ formed $TaSi_2$–Si_3N_4 nanocomposite coating grown on Ta substrate. *Intermetallics* **2008**, *16*, 1263–1272. [CrossRef]
25. Chung, C.K.; Chen, T.S. Effect of Si/Ta and nitrogen ratios on the thermal stability of Ta–Si–N thin films. *Microelectron. Eng.* **2020**, *87*, 129–134. [CrossRef]
26. Olowolafe, J.O.; Rau, I.; Unruh, K.M.; Swann, C.P.; Jawad, Z.S.; Alford, T. Effect of composition on thermal stability and electrical resistivity of Ta–Si–N films. *Thin Solid Films* **2000**, *365*, 19–21. [CrossRef]

27. Nah, J.W.; Choi, W.S.; Hwang, S.K.; Lee, C.M. Chemical state of (Ta, Si)N reactively sputtered coating on a high-speed steel substrate. *Surf. Coat. Technol.* **2000**, *123*, 1–7. [CrossRef]
28. Chen, Y.-I.; Cheng, Y.-R.; Chang, L.-C.; Lu, T.-S. Chemical inertness of Ta–Si–N coatings in glass molding. *Thin Solid Films.* **2015**, *584*, 66–71. [CrossRef]
29. Zeman, P.; Musil, J.; Daniel, R. High-temperature oxidation resistance of Ta–Si–N films with a high Si content. *Surf. Coat. Technol.* **2006**, *200*, 4091–4096. [CrossRef]
30. Kiryukhantsev-Korneev, P.V.; Iatsyuk, I.V.; Shvindina, N.V.; Levashov, E.A.; Shtansky, D.V. Comparative investigation of structure, mechanical properties, and oxidation resistance of Mo-Si-B and Mo-Al-Si-B coatings. *Corr. Sci.* **2017**, *123*, 319–327. [CrossRef]
31. Levashov, E.A.; Mukasyan, A.S.; Rogachev, A.S.; Shtansky, D.V. Self-propagating high-temperature synthesis of advanced materials and coatings. *Int. Mater. Rev.* **2017**, *62*, 203–239. [CrossRef]
32. Shtansky, D.V.; Sheveyko, A.N.; Sorokin, D.I.; Lev, L.C.; Mavrin, B.N.; Kiryukhantsev-Korneev, P.V. Structure and properties of multi-component and multilayer TiCrBN/WSe_x coatings deposited by sputtering of TiCrB and WSe_2 targets. *Surf. Coat. Technol.* **2008**, *202*, 5953–5961. [CrossRef]
33. Levashov, E.A.; Shtansky, D.V.; KiryukhantsevKorneev, P.V.; Petrzhik, M.I.; Tyurina, M.Y.; Sheveiko, A.N. Multifunctional nanostructured coatings: Formation, structure, and the uniformity of measuring their mechanical and tribological properties. *Russ. Metall.* **2010**, *10*, 917–935. [CrossRef]
34. Avila, P.R.T.; da Silva, E.P.; Rodrigues, A.M.; Aristizabal, K.; Pineda, F.; Coelho, R.S.; Garcia, G.L.; Soldera, F.; Walczak, M.; Pinto, H.C. On manufacturing multilayer-like nanostructures using misorientation gradients in PVD films. *Sci. Rep.* **2019**, *9*, 15898. [CrossRef] [PubMed]
35. Szala, M.; Walczak, M.; Pasierbiewicz, K.; Kamiński, M. Cavitation Erosion and Sliding Wear Mechanisms of AlTiN and TiAlN Films Deposited on Stainless Steel Substrate. *Coatings* **2019**, *9*, 340. [CrossRef]
36. Niu, Y.; Huang, L.; Zhai, C.; Zeng, Y.; Zheng, X.; Ding, C. Microstructure and thermal stability of $TaSi_2$ coating fabricated by vacuum plasma spray. *Surf. Coat. Technol.* **2015**, *279*, 1–8. [CrossRef]
37. Chen, Y.-I.; Lin, K.-Y.; Wang, H.-H.; Cheng, Y.-R. Characterization of Ta–Si–N coatings prepared using direct current magnetron co-sputtering. *Appl. Surf. Sci.* **2014**, *305*, 805–816. [CrossRef]
38. Nah, J.W.; Hwang, S.K.; Lee, C.M. Development of a complex heat resistant hard coating based on (Ta, Si)N by reactive sputtering. *Mater. Chem. Phys.* **2000**, *62*, 115–121. [CrossRef]
39. Koller, C.M.; Marihart, H.; Bolvardi, H.; Kolozsvari, S.; Mayrhofer, P.H. Structure, phase evolution, and mechanical properties of DC, pulsed DC, and high power impulse magnetron sputtered Ta–N films. *Surf. Coat. Technol.* **2018**, *347*, 304–312. [CrossRef]
40. Liu, X.; Yuan, F.; Wei, Y. Grain size effect on the hardness of nanocrystal measured by the nanosize indenter. *Appl. Surf. Sci.* **2013**, *279*, 159–166. [CrossRef]
41. Poladi, A.; Mohammadian Semnani, H.R.; Emadoddin, E.; Mahboubi, F.; Ghomi, H.R. Nanostructured TaC film deposited by reactive magnetron sputtering: Influence of gas concentration on structural, mechanical, wear and corrosion properties. *Ceram. Int.* **2019**, *45*, 8095–8107. [CrossRef]
42. Hu, J.; Li, H.; Li, J.; Huang, J.; Kong, J.; Zhu, H.; Xiong, D. Structure, mechanical and tribological properties of TaC_x composite films with different graphite powers. *J. Alloys Compd.* **2020**, *832*, 153769. [CrossRef]
43. Leyland, A.; Matthews, A. On the significance of the H/E ratio in wear control: A nanocomposite coating approach to optimised tribological behavior. *Wear* **2000**, *246*, 1–11. [CrossRef]
44. Levashov, E.A.; Petrzhik, M.I.; Shtansky, D.V.; Kiryukhantsev-Korneev, P.V.; Sheveyko, A.N.; Valiev, R.Z.; Gunderov, D.V.; Prokoshkin, S.D.; Korotitskiy, A.V.; Smolin, A.Y. Nanostructured titanium alloys and multicomponent bioactive films: Mechanical behavior at indentation. *Mater. Sci. Eng. A* **2013**, *570*, 51–62. [CrossRef]
45. Shtansky, D.V.; Bondarev, A.V.; Kiryukhantsev-Korneev, P.V.; Sheveyko, A.N.; Pogozhev, Y.S. Influence of Zr and O on the structure and properties of TiC(N) coatings deposited by magnetron sputtering of composite $TiC_{0.5}+ZrO_2$ and (Ti, Zr)$C_{0.5}+ZrO_2$ targets. *Surf. Coat. Technol.* **2012**, *206*, 2506–2514. [CrossRef]
46. Martinez-Martinez, D.; Lopez-Cartes, C.; Justo, A.; Fernandez, A.; Sanchez-Lopez, J.C. Self-lubricating Ti–C–N nanocomposite coatings prepared by double magnetron sputtering. *Solid State Sci.* **2009**, *11*, 660–670. [CrossRef]
47. Du, S.; Zhang, K.; Meng, Q.; Ren, P.; Hu, C.; Wen, M.; Zheng, W. N dependent tribochemistry: Achieving superhard wear-resistant low-friction TaC_xN_y films. *Surf. Coat. Technol.* **2017**, *328*, 378–389. [CrossRef]

48. Khanna, A.S. *Introduction to High Temperature Oxidation and Corrosion*; ASM International: Materials Park, OH, USA, 2002; p. 101.
49. Evans, H.E. Cracking and spalling of protective oxide layers. *Mater. Sci. Eng. A* **1989**, *120–121*, 139–146. [CrossRef]
50. Pan, X.; Li, C.; Niu, Y.; Zhong, X.; Huang, L.; Zeng, Y.; Zheng, X. Effect of Yb_2O_3 addition on oxidation/ablation behaviors of ZrB_2-$MoSi_2$ composite coating under different environment. *Corr. Sci.* **2020**, *175*, 108882. [CrossRef]
51. Wang, J.; Li, B.; Li, R.; Chen, X.; Wang, T.; Zhang, G. Unprecedented oxidation resistance at 900 °C of Mo–Si–B composite with addition of ZrB_2. *Ceram. Int.* **2020**, *46*, 14632–14639. [CrossRef]
52. Hassan, R.; Kundu, R.; Balani, K. Oxidation behavior of coarse and fine SiC reinforced ZrB2 at re-entry and atmospheric oxygen pressures. *Ceram. Int.* **2020**, *46*, 11056–11065. [CrossRef]

Article

Convection–Diffusion Model for the Synthesis of PVD Coatings and the Influence of Nanolayer Parameters on the Formation of Fractal and Hierarchical Structures

Alexey Vereschaka [1,*], Sergey Grigoriev [2], Anatoli Chigarev [3], Filipp Milovich [4], Nikolay Sitnikov [5], Nikolay Andreev [4], Gaik Oganian [2] and Jury Bublikov [1]

1 IDTI RAS, 127994 Moscow, Russia; yubu@rambler.ru
2 VTO Department, Moscow State Technological University STANKIN, 127994 Moscow, Russia; s.grigoriev@stankin.ru (S.G.); g.oganyan@stankin.ru (G.O.)
3 Mechanical and Mathematics Faculty, Belarusian State University, 220050 Minsk, Belarus; chigarevanatoli@yandex.ru
4 Materials Science and Metallurgy Shared Use Research and Development Center, National University of Science and Technology MISiS, 119049 Moscow, Russia; filippmilovich@mail.ru (F.M.); andreevn.misa@gmail.com (N.A.)
5 Department of Solid State Physics and Nanosystems, National Research Nuclear University MEPhI, 115409 Moscow, Russia; sitnikov_nikolay@mail.ru
* Correspondence: dr.a.veres@yandex.ru

Received: 28 August 2020; Accepted: 27 September 2020; Published: 28 September 2020

Abstract: The study proposes a model for the deposition of coatings, which takes into account the stochastic nature of the deposition process and is built considering the influence of the parallel convection and diffusion processes. The investigation has found that the dispersion of the motion direction of deposited particles in front of a substrate increases, which indicates a growth of the randomness in the trajectories of the particles being deposited. The obtained formulas indicate the fractal nature of the deposition process. During the formation of the multilayer coating structure, mismatched fractal structures of the layers overlap each other and thus the clustering effect is largely leveled out. The value of the nanolayer λ period has a significant influence on the fractal structure of the coating and the formation of feather-like hierarchical structures in it.

Keywords: hierarchical structure; multilayer PVD coating; stochastic process; convection and diffusion

1. Introduction

The process of the coating deposition is very complex and undoubtedly stochastic [1–7]. Accordingly, the modeling of the coating deposition process is also a complex and many-sided challenge, and in some cases, a probabilistic approach is required to meet this challenge [3–7]. Initially, one of the key tasks for the modeling of the processes under consideration was to predict a thickness of a coating, including a coating deposited on a complex-shape surface. To meet the above challenges, physical and geometric methods were applied, which made it possible to achieve the certain prediction accuracy. In particular, the studies proposed the coating thickness uniformity model [8,9], the cosine analytical and generic computational model [10], and the inverse power law relationship between thickness and source-to-substrate distance was considered [11]. Recent studies also applied true line-of-sight models, based on the inverse square law and also taking into account the effect of shadow masks [12]. In [13], the computational level set method was also applied for tracking topographic change of a surface.

The traditional approaches to the modeling of mass transfer processes, in particular, the coating deposition processes, have significant limitations when described in micro- and mesosystems [14,15]. In connection with the above, the processes are often modeled on micro- nano- and atomic levels, due to the integration of systems with a large number of particles, the motion of which is described, in particular, by equations of molecular dynamics of Newtonian mechanics. A significant problem of these methods is the existing limitation of the maximum number of particles in a system due to the limit of computing power.

Approaches of molecular dynamics are used to model the deposition process [16]. By simulating the interaction of individual atoms, this method makes it possible to predict the morphology, thermal and electronic properties, and also internal stresses of the coating being deposited. When the above method is applied, the forces of interatomic interaction are presented in the form of classical potential forces. The configuration sets obtained in the course of the calculations using the method of molecular dynamics are distributed in accordance with a certain statistical distribution function, for example, a one corresponding to microcanonical distribution. The exact information concerning the trajectories of system particles at large time intervals is not necessary to obtain results of a macroscopic (thermodynamic) nature [17]. The principals of molecular dynamics can be applied to simulate the coating deposition process using dynamic equations and pressure and temperature boundary conditions for substrate and coating surfaces, taking into account the results of the preliminary study of stresses in the coating.

In particular, to model the coating deposition process, the Particle-in-Cell Monte Carlo simulation method [3,4], with both modeling options of simpler 2D [5] and more complex 3D [6–8], requiring advanced computing power, is applied. During the comparison of 2D and 3D models, it was found that the main difference between two-dimensional and three-dimensional options of modeling was consideration of the effect of propagating plasma waves that had a significant influence on the properties of particle transport in discharges limited by a magnetic field. Two-dimensional modeling provides useful qualitative information about the general trends and mechanisms of magnetron discharges, while three-dimensional modeling makes it possible to quantitatively describe the electron transfer and current–voltage characteristics [18]. Based on the results of 2D modeling using the Monte Carlo algorithm, the formation of a coating with feather-like hierarchical structure was demonstrated in [19]. The results of the study focused on the development of a columnar coating structure have proved that the average distance between columns first increases sharply and then stabilizes due to cessation of the growth and regeneration of the structure. The Monte Carlo methods were applied for 2D modeling of the process of micropore formation in a coating, given the stochastic nature of the process [20]. Approaches of computational fluid dynamics were also applied, they give a much rougher description of the process, but at the same time contribute to a noticeable reduction in the amount of calculations, thus making it possible to introduce modeling in the 3D geometry with high-scale detailing [21,22]. A more detailed analysis of various methods for modeling the coating deposition process can be found in [23]. Thus, the most accurate and complete simulation of the deposition process should take into account the stochastic nature of this process and consider the molecular level, because a coating is formed precisely at this level. Several studies [24–26] consider a cluster, hierarchical structure of the coating, deposited through various methods, and the influence of the deposition process parameters on the formation of such structure. However, the influence of the parameters of the nanolayer coating structure on the formation of hierarchical structures in the coating requires an additional study.

2. Materials and Methods

The filtered cathodic vacuum arc deposition (FCVAD) [27–36] technology was applied to deposit the Ti–TiN–(Ti,Cr,Al)N coatings. A VIT-2 unit, (IDTI RAS—MSTU STANKIN, Moscow, Russia), equipped with two arc evaporators with a pulsed magnetic field and one arc evaporator with filtering of the vapor-ion flow, was used to deposit the coatings. In addition, the system was equipped with a source of pulsed bias voltage supply to a substrate, a dynamic gas mixing system for reaction gases, a system for automatic control of pressure in the chamber and a process temperature control system, and a system for stepless adjustment of planetary gear rotation [27–30,35–42]. The nanolayer structure of the coating provides high hardness at elevated temperatures and high oxidation resistance [27–31,37,38]. A three-layer structure of the coating was assumed [27–31].

The parameters for the process of coating deposition are presented in Table 1.

Table 1. Parameters of stages of the technological coating deposition process.

Process	p_N (Pa)	U (V)	I_{Ti} (A)	I_{Al} (A)	I_{Cr} (A)
Pumping and heating of vacuum chamber	0.06	+20	75	120	65
Heating and cleaning of products with gaseous plasma	2.0	100 DC/900 AC f = 10 kHz, 2:1	85	80	-
Deposition of coating	0.42	−800 DC	75	160	55
Cooling of products	0.06	-	-	-	-

Note: I_{Ti} = current of titanium cathode, I_{Cr} = current of Cr cathode, I_{Al} = current of aluminum cathode, p_N = gas pressure in chamber, U = voltage on substrate.

An analysis of the microstructure and nanostructure of the samples was studied using JEM 2100 (JEOL, Tokyo, Japan) high-resolution transmission electron microscope (TEM) with an accelerating voltage of 200 kV.

3. Results

3.1. Convection–Diffusion 2D Model for the Stochastic Coating Deposition Process

The creation of a wear resistant layer of the coating combines two parallel processes: convection and diffusion. Convection causes the rectilinear motion of particles from a cathode to the surface on which a coating is to be deposited, while diffusion gives rise to the consideration of random factors influencing particle trajectories, as a result of which the density of the substance being deposited becomes inhomogeneous, and the process represents the diffusion limited by the clustering or, alternatively, the clustering limited by the diffusion. A 2D model of the coating deposition process was considered, according to which the area where particles were deposited was plane (Figure 1). This assumption can be quite logically extended to a 3D model, because the flow of the particles moving through the fixed plane parallel to the 0XZ plane is quasi-isotropic. The typical linear size of the deposition area is about L. The particles are emitted from the line of $y = b$ and are deposited on the line of $y = a$.

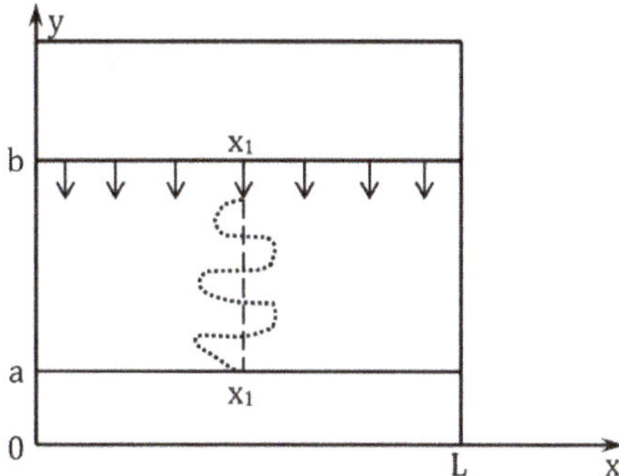

Figure 1. 2D model of the cross-section of the area perpendicular to the plane in which the particles emitted from the cathode in the direction of the substrate are being deposited.

In case if no diffusion occurs, a particle leaving a point with the coordinate (x_1, b) enters a point with the coordinate (x_1, a), while moving along a straight line connecting these points. In case of diffusion, a particle is moving from the point (x_1, b) to the point (x_1, a) along a random trajectory, and the probability of its getting into point (x_1, a) is not equal to 1 (Figure 2). Let it be assumed that at $t = 0$, the particle with the coordinate (x_1, b) begins to move along the straight line connecting it with the point with the coordinate (x_1, a). The one-dimensional process with $y = y(t)$ for constant $x = x_1$ is first considered.

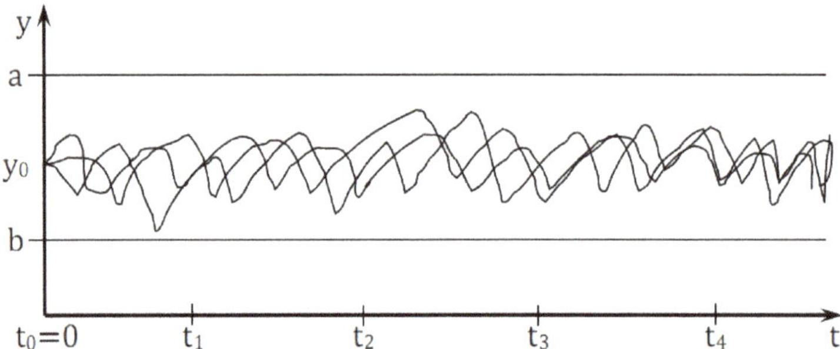

Figure 2. Diagram reflecting random particle trajectories (random walks) moving from the point (x_1, b) to the point (x_1, a).

Let at time $t = t_0$ the particle has coordinates of $y = y_0 (a < y_0 < b)$, $x = x_1$ $(0 < x_1 < L)$. The probability density *f(y)* and likelihood *P(y)* of this event are as follows:

$$f(y) = \delta(y - y_0), \quad \int_a^b f(y)dy = P(y = y_0) \tag{1}$$

When $t > t_0$, the coordinate $y = \xi(t)$ is a random time function, then its probability density can be written in the following form [43]:

$$f(y) = < \delta(y) > \tag{2}$$

where the brackets <> denote the averaging over realizations.

Let us assume that $\xi(t)$ satisfies a Langevin equation of the Brownian-motion type.

$$\begin{gathered} \tfrac{d\xi}{dt} = v(\xi, t) + f(\xi, t), \ \ \xi(0) = y_0 = \xi_0 \\ < f(\xi, t) > = 0, \ \ B(\xi, t, \xi', t') = < f(\xi, t) f(\xi', t') > = 2\delta(t - t') F(\xi, \xi', t) \end{gathered} \tag{3}$$

where $B(\xi, t, \xi', t')$ is a correlation function.

By (3), the differentiation of (2) with respect to t yields an equation for the probability density function of the particle's position:

$$\begin{gathered} \tfrac{\partial f_t}{\partial t} + \tfrac{\partial}{\partial y}\{[v(y, t) + A(y, t)] f_t(y) - \tfrac{\partial^2}{\partial y^2}[F(x, y, t) f_t(y)] = 0 \\ A(y, t) = \tfrac{\partial}{\partial y} F(y, y', t)\Big|_{y' = y} \end{gathered} \tag{4}$$

The Fokker-Planck-Kolmogorov differential Equation (4) should be solved with the initial condition (1). The physical meaning of the Equation (4) is in description of the stochastic processes for the motion of the coating's deposited particle in the gas phase from the cathode to the substrate. The coefficients in the equation at $f_t(y)$ are the particle transport and diffusion parameters, respectively. According to the Formulas (3) and (4), the equation terms with parameters A and F are specified by fluctuations of the field $f_t(y, t)$, affecting $v(\xi, t)$, which is a deterministic function describing the convective particle transport. If $f_t(y,t) = f(y)$, i.e., is constant in time, then A and F are also constant in time, which corresponds to the constant diffusion coefficient F and convection $A = 0$. Cases when $f(y,t)$ is non-stationary require a model specification.

The real process of the layer-by-layer formation of a coating is of stochastic nature due to the fact that the layer of particles which started their motion from a source plane does not remain planar, while the particles interact with each other and cannot simultaneously form microclusters during the deposition, with correlations formed between them. Such microconglomerates, being deposited on the substrate, will be considered as nuclei of clusters in the coating layer, which introduce randomness into the deposition process and make the coating structure heterogeneous, fractal. Depending on the mentioned random factors, the deposition time is a random variable for different particles.

Figure 3 exhibits an initial view of the surface of the Ti–TiN–(Ti,Cr,Al)N coating with a pronounced cluster structure.

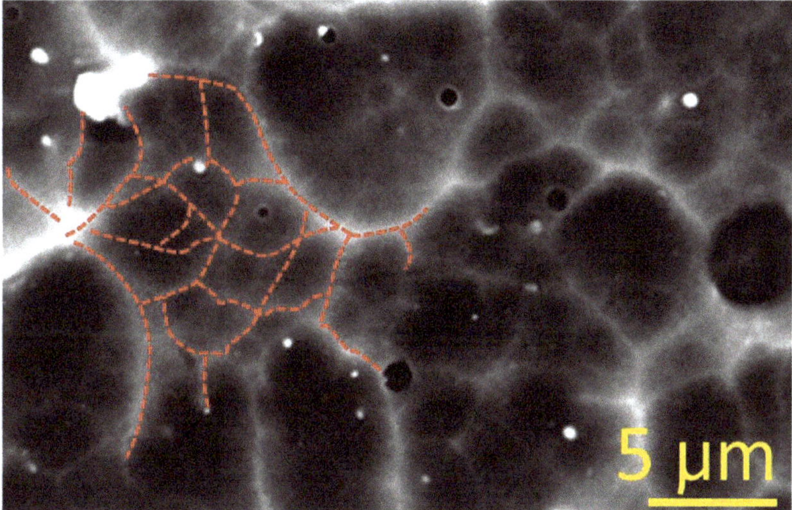

Figure 3. Cluster structure of the surface of the Ti–TiN–(Ti,Cr,Al)N coating after deposition.

The model of a random process reaching the boundaries of its variation area makes it possible to estimate the probability of this event and the average time required for that [43]. Let $P(y_0, t)$ denote the probability of the event that the random process will reach the boundary *a or b* with the initial condition of $P(y, o) = 0$ and the boundary conditions of $P(a,t) = P(b,t) = 1$, then $P(y_0, t)$ satisfies the equation [43]:

$$\begin{gathered} \frac{\partial P(y_0,t)}{\partial t} = K_1(y_0)\frac{\partial P}{\partial y_0} + \frac{1}{2}K_2(y_0)\frac{\partial^2 P}{\partial y_0^2} \\ K_1(y_0) = v(y_0) + \frac{AFF'}{4}\ K_2(y_0) = \frac{AF^2}{2},\ F' = \frac{dF}{dy_0} \end{gathered} \tag{5}$$

The drift coefficient K_1 and the diffusion coefficient K_2 are to be specified for every model. As is known [43], the Equation (5) can be solved only for some particular models. When the initial and boundary conditions are formulated, the probability to reach the boundary increases. Figure 4 exhibits in graphic form the probabilities P—to reach the boundary and $\overline{P}$—not to reach the boundary when the normalization condition is satisfied.

$$P + \overline{P} = 1 \tag{6}$$

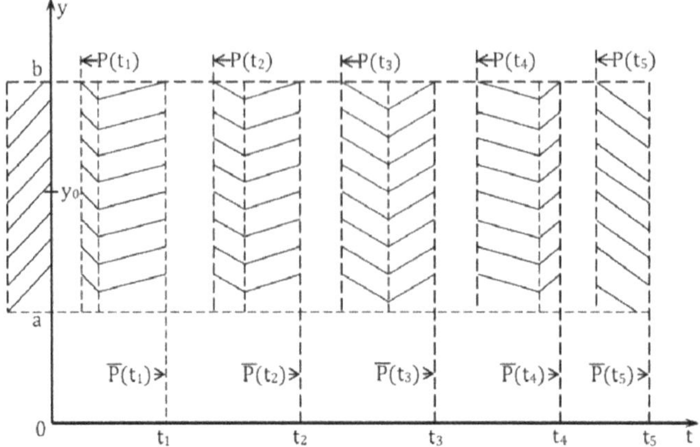

Figure 4. Graphic form depicting the probabilities P—to reach the boundary and $\overline{P}$—not to reach the boundary.

At Figure 4, $P + \overline{P} = 1$ is the normalization condition:

$$P_i = \frac{S_i}{S},\ \overline{P_i} = 1 - \frac{S_i}{S} = \frac{\overline{S_i}}{S},\ i = 1,\ 2,\ \ldots \tag{7}$$

S_i is the area of the rectangle $P(t_i)$,

$\overline{S_i}$ is the area of the rectangle $\overline{P}(t_i)$.

Thus, $P(t_i)$ describes a change in the rectangle, which area is numerically equal to the probability of reaching the boundary (of the substrate), while $\overline{P}(t_i)$ describes a change in the rectangle, which area is numerically equal to the probability of not reaching the boundary.

The time it takes for a particle to reach the substrate is a random variable over a set of realizations. Then to calculate the average time to reach $< t >$, a formula for mathematical expectation can be used as follows:

As far as $P(y_0, t) = 1 - \overline{P}(y_0, t)$, then

$$\begin{aligned} < t > = T = \int_0^\infty t\ \frac{\partial P(y_0,t)}{\partial t} dt = -\int_0^\infty t \frac{\partial \overline{P}}{\partial t} \partial t = \left\{ \frac{integration\ by\ parts}{t = K,\ \frac{\partial P}{\partial t}\ \partial t = \partial v} \right\} \\ = t\overline{P}|_0^\infty + \int_0^\infty \overline{P}(y_0, t) dt = \int_0^\infty \overline{P}(y_0, t) dt,\ as\ far\ as\ t\overline{P}(y_0, t)|_{t=\infty} \\ = 0,\ t\overline{P}(y_0, t)|_{t=0} = 0 \end{aligned} \tag{8}$$

Given (5), an equation for $T(y_0, t)$ can be derived from (7), as follows:

$$< t > = T = \int_0^\infty t \frac{\partial P(y_0, t)}{\partial t} dt = K_1 \int_0^\infty t \frac{\partial P}{\partial y_0} dt + K_2 \int_0^\infty t \frac{\partial^2 P}{\partial {y_0}^2} dt \tag{9}$$

Integration by parts:

$$\int_0^\infty t \frac{\partial P}{\partial t} dt = tP|_0^1 - \int_0^\infty P(y_0, t) \partial t = -1 \tag{10}$$

$$\int_0^\infty t \frac{\partial P}{\partial y_0} dt = \frac{d}{dy_0} \int_0^\infty tP dt = \frac{dT}{dy_0} \tag{11}$$

$$\int_0^\infty t\frac{\partial^2 P}{\partial {y_0}^2}dt = \frac{d^2}{dy^2}\int_0^\infty tPdt = \frac{d^2 T}{dy^2} \tag{12}$$

Through adding, we get:

$$K_2(y_0)\frac{d^2 T}{dy_0} + K_1(y_0)\frac{dT}{dy_0} + 1 = 0 \tag{13}$$

The boundary conditions for (13) are:

$$T = 0 \; at \; y_0 = a \; or \; y_0 = b \tag{14}$$

The form of solution for (13) depends on the form of the coefficients $K_1(y_0)$, $K_2(y_0)$, i.e., the form of the equation of motion. If the drift and diffusion coefficients are constant, then the source of particles coincides with the y-axis, i.e., $y_0 = 0$, and the real and imaginary substrates are located symmetrically: $a = -b$, then

$$T = \frac{b}{K_1} th\left(\frac{K_1}{K_2} b\right) \tag{15}$$

Function graph (15) is depicted in Figure 5.

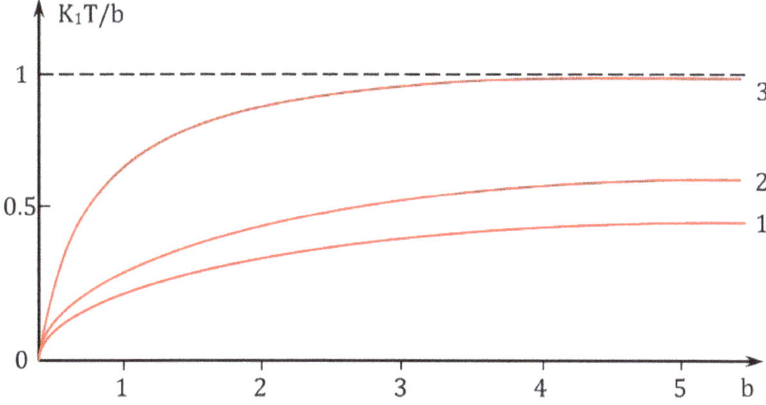

Figure 5. Dimensionless relationship for T and b at various values of K_1/K_2. 1: $K_1/K_2 = 10$, 2: $K_1/K_2 = 1$, 3: $K_1/K_2 = 0.1$

Figure 5 depicts the dimensionless relationship for T and b at various values of K_1/K_2.

b is the distance from the cathode to the substrate surface (in conventional units).

Figure 5 proves that the larger the drift coefficient K_1 is, the shorter time to deposition is, while the larger the diffusion coefficient K_2 is, the longer the time to deposition is. A model of coating synthesis during the layer-by-layer deposition of particles on a substrate was considered. Under the ideal conditions, a layer of particles emitted by a plane source is deposited on a substrate as a layer, and then the layer thickness $h(t)$ is a deterministic function. Under the real conditions, the coating thickness is a random function, as mentioned above, and it can be represented as follows:

$$H(t) = h(t) + v(t), \; < v > = 0, \; < v(t)v(t_1) > = \frac{N_0}{2}\delta(t - t_1) \tag{16}$$

where $v(t)$ is a random function of white-noise type.

After cooling down, a layer transforms, and the layer thickness at the time t is a certain random process, represented in the following form:

$$H_{\mathrm{KOH}}(t) = h_{\mathrm{KOH}}(t) + v_{\mathrm{KOH}}(t) \tag{17}$$

The coating synthesis process should end when $H_{\mathrm{KOH}}(t) = Q$ is reached. The probability that the $H_{\mathrm{KOH}}(t)$ function would reach the given thickness Q during the time t was considered. The quality of the coating is determined by the fact that its thickness in the final state should be a given constant value. The probability $P(y_0, t)$ of reaching the boundary Q is found by the formula:

$$P(y_0,t) = 1 - \overline{P}(y_0,t) = 1 - \int_{-\infty}^{Q} \overline{f}(y, y_0, t)dy \tag{18}$$

where $\overline{f}(y_0, y, t)$ satisfies the Equation [43]

$$\frac{\partial \overline{f}}{\partial t} = \alpha \frac{\partial}{\partial y}\left[y\overline{f}\right] + \frac{\alpha^2 N_0}{4}\frac{\partial^2 \overline{f}}{\partial y^2} \tag{19}$$

with boundary conditions:

$$f(y_0, Q, t) = f(y_0, a, t) = \overline{f}(y_0, b, t) = 0 \tag{20}$$

The initial condition is as follows:

$$\overline{f}(y_0, y, t_0) = \delta(y - y_0) \tag{21}$$

If $h(t) = Q$, i.e., a layer of the required thickness is deposited to the substrate, then the probability $P(y_0, t)$ has the following form:

$$\begin{gathered} P(y_0,t) = 2\left[1 - \Phi\left(\frac{Q - y_0 e^{-\alpha t}}{\sigma(t)}\right)\right] \\ \sigma^2(t) = \frac{N_0\alpha}{4}\left(1 - e^{-2\alpha t}\right) \end{gathered} \tag{22}$$

We considered Φ as the probability integral (standard Gaussian distribution function) [44].

A model of fluctuations in the direction of the deposited particle motion was considered. As a result of such fluctuation, a particle can be deposited not in the intended place, and that can lead to non-uniform density of the coating. In the $0xy$ plane, the position of a particle is characterized by the values of $x(t)$, $y(t)$. Instead of the parameter t, a parameter s is introduced, defined as follows:

$$s = \int V dt \tag{23}$$

where V is velocity of the particle motion.

At each point, particles deviate from the rectilinear trajectories of the motion along straight lines parallel to the y-axis (Figure 6).

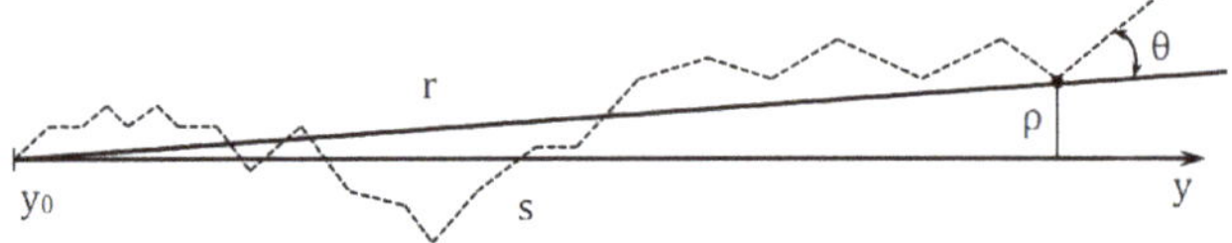

Figure 6. Deviation from the rectilinear trajectories of the motion along straight lines parallel to the y-axis.

The probability density $f(\theta|s,0)$ satisfies the following equation [45]:

$$\sin\theta\frac{\partial f}{\partial\theta} = \frac{B}{2}\frac{\partial}{\partial\theta}\left(\sin\theta\frac{\partial f}{\partial\theta}\right)$$
$$B = \lim_{\Delta t\to 0}\varepsilon^2/\Delta t,\ \varepsilon^2 = r^2 < (\Delta\varphi)^2 >,\ < \Delta r.r\Delta\varphi > \tag{24}$$
$$x = r\cos\varphi,\ y = r\sin\varphi$$

The value of B is related to the correlation function of the particle deposition rate

$$\psi_V(\rho) = < V(r)V(r') > - < V(\rho)^2 > \tag{25}$$

through the formula as follows:

$$B = -\frac{-2}{< V^2 > \int_0^\infty \frac{\partial}{\partial\rho}\left(\rho^2\frac{\partial\psi_v}{\partial\rho}\right)\frac{d\rho}{\rho}} \tag{26}$$

The solution to the Equation (26) is being sought with the initial condition of:

$$f(\theta|0,0) = \delta(\theta)/2\pi\sin\theta \tag{27}$$

and looks as follows:

$$f(\theta|s-s_1,0) = \frac{1}{4\pi}\sum_{n=0}^{\infty}(2n+1)P_n(\cos\theta)e\frac{-n(n+1)}{2}B(s-s_0) \tag{28}$$

where $P_n(\cos\theta)$ are Legendre polynomials [45].

At $s\to\infty$:

$$\lim_{s\to\infty} f(\theta|s-s_0,0) = \omega_1(\theta) = \frac{1}{4\pi} \tag{29}$$

It is convenient to describe the condition for a change in the particle velocity vector through $\cos\theta$, then $\cos\theta = P_1(\theta)$ and the average value of the direction cosine of the velocity vector is calculated by the following formula:

$$< \cos\theta >= \int_0^\pi \cos\theta V\sin\theta d\theta \tag{30}$$

The following differential equation $< \cos\theta >$ can be derived from (24):

$$\frac{d < \cos\theta >}{ds} = -B < \cos\theta > \tag{31}$$

The solution of Equation (30) with the boundary condition of $< \cos\theta >= 1$ at $s = 0$ is of the form:

$$< \cos\theta >= e^{-Bs} \tag{32}$$

At $s\to\infty$, $< \cos\theta >= 0$, and in accordance with (31), that means at a sufficiently large distance from a source to a substrate, all orientations of the particle motion are equally probable. For short distances:

$$< \cos\theta >= 1 - \frac{\theta^2}{2} = 1 - Bs \tag{33}$$

which follows that:

$$\theta^2 = 2Bs \tag{34}$$

The mean square of the particle deviation ρ from the initial (rectilinear) direction along the y-axis is calculated by the following formula [46]:

$$\langle\rho\rangle = \frac{4s}{3B} - \frac{2}{B^2}\left(1 - e^{-Bs}\right) + \frac{2}{2B^2}\left(1 - e^{-3Bs}\right) \tag{35}$$

It follows from (35) that when $Bs \ll 1$, then:

$$\sqrt{\langle\rho^2\rangle} = \sqrt{\frac{B}{6}}s^{3/2} \tag{36}$$

The mean square distance in a straight line to the particle being deposited is determined as follows:

$$\langle r^2\rangle = \frac{2s}{B} - \frac{2}{B^2}\left(1 - e^{-Bs}\right) \tag{37}$$

The Formula (37) at $Bs \ll 1$ is reduced to the formula for the average square of displacement of a heavy molecule in light gas, obtained by Smoluchowski [47–50]:

$$\langle r^2\rangle = s^2\left(1 - \frac{Bs}{3}\right) \tag{38}$$

The fractional degree in the Formula (35) indicates that the deposition of coating particles is described by the fractal dynamics.

An ideal model of coating deposition represents an emission of particles from a certain conventionally planar surface of a cathode with the same velocity (temperature) along a normal to a surface (plane). The particles move along the parallel trajectories and are being layer-by-layer deposited on the surface due to the complex planetary motion of the substrate surface (Figure 7a). The process is described by the convection equation.

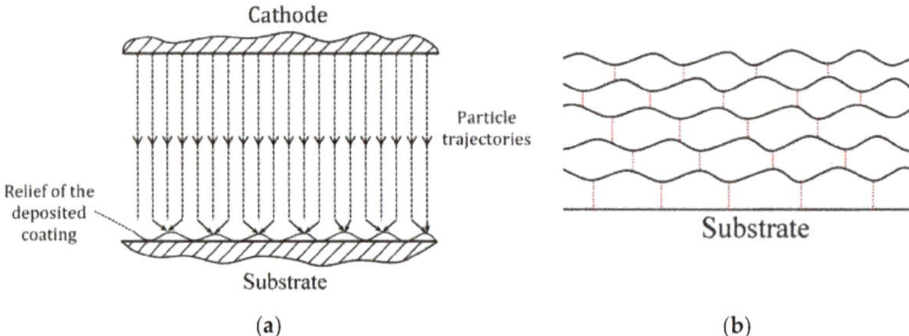

Figure 7. Diagram for (**a**) the formation of the relief of the coating being deposited; (**b**) the predicted cracking for the multilayer coating.

Under the real conditions, while emitted, particles demonstrate velocity fluctuations in value and direction, and during the particle motion, the process becomes even more irregular. The cloud has density fluctuations, and diffusion of particles occurs, as a result of which the density and hence the stiffness and strength of the layer being deposited fluctuate over the surface area. Furthermore, the very surface of the substrate is in the process of the complex planetary motion (rotation). As a result, at the initial stages of the coating destruction, damages and then cracks arise in the places of the deposited coating, in which the density is lower [28–30,49]. In terms of mathematics, this is described by a model of Brownian motion of a particle (Equation (3)), and the probability density satisfies the Equation (4).

The analysis of the resulting model indicates that the very coating deposition process implies the formation of a fractal structure, which further contributes to the formation of stress concentration areas, which, in turn, stimulate the process of cracking. At the initial stages of the coating destruction, damages and then cracks arise in the places of the deposited coating, in which the density is lower. During the formation of the multilayer coating structure, mismatched fractal structures of the layers overlap each other, and thus the clustering effect is largely leveled out. In turn, the above reduces the likelihood of cracking in the coating structure and makes it possible to predict an increase in the performance properties of the coating [27–31,51–53] (see Figure 7b).

3.2. Influence of the Nanolayer Period λ on the Cluster and Hierarchical Coating Structure

The study was focused on the influence of such an important coating parameter as the nanolayer period λ on a cluster and hierarchical structure using an example of the Ti–TiN–(Ti,Cr,Al)N coating with a nanostructured wear-resistant layer (Figure 8). For a coating with λ = 10 nm (Figure 8a), there is a formation of hierarchical feather-like structures, modeled and described, in particular, in [19,20,24–26] for various techniques of the coating deposition. While a coating is being deposited, there in an obvious increase in the cluster size. The studies conducted earlier [52,54,55] also detected an increase in grain size. With an increase in the nanolayer period up to λ = 38 nm (Figure 8b), the feather-like structure becomes less pronounced, and at λ = 160 nm (Figure 8c), no grain growth is detected with an increase in the coating thickness, while the deposition of a coating begins "from scratch" within each nanolayer period. No feather-like structure was also detected for a coating with an extremely long nanolayer period (λ = 1100 nm, Figure 8d). A column structure, without any noticeable increase in crystal size, is formed inside a nanolayer.

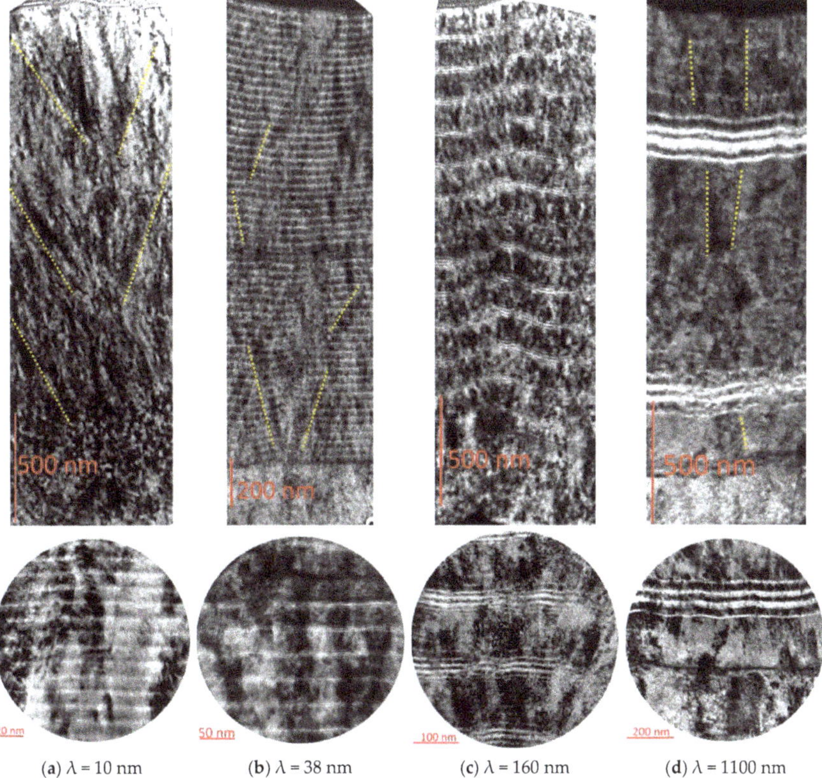

Figure 8. Formation of the cluster structure of the Ti–TiN–(Ti,Cr,Al)N coatings depending on the nanolayer period λ. (**a**) λ = 10 nm; (**b**) λ = 38 nm; (**c**) λ = 160 nm; (**d**) λ = 1100 nm.

4. Conclusions

- The obtained formulas prove that due to diffusion, particles can participate in collisions with other particles during the motion, and as a result, the velocity decreases and a particle enters a substrate at an angle, which may lead to a coating inhomogeneity over the area and the appearance of the surface microrelief, which also worsens the resistance and reliability of the coating. The Equation (16) is stochastic for various deposition trajectories.
- The Formula (32) indicates that if the distance from a cathode to a substrate is large enough, then there are a lot of particle in front of the substrate, and those particles move perpendicular to the initial direction of the motion (parallel to the substrate). The Formula (34) suggests that the dispersion of the motion direction in front of the substrate increases, which indicates an increase in randomness in the trajectories of the particles being deposited. The Formula (36) proves the fractal nature of the process of particle deposition on a substrate.
- During the formation of the multilayer coating structure, mismatched fractal structures of the layers overlap each other and thus the clustering effect is largely leveled out (averaged). In turn, the above reduces the likelihood of cracking in the coating structure and makes it possible to predict an increase in the performance properties of the products with such coatings.
- The nanolayer period λ has a significant influence on the fractal structure of the coating. When λ is low (less than 38 nm), with an increase in the coating structure, a pronounced feather-like structure is formed, and grains grow. With an increase in λ up to 160 nm and higher, no feather-like structure is formed, and no grain growth is detected during the coating deposition.

Author Contributions: Conceptualization, A.V. and S.G.; methodology, A.V. and A.C.; formal analysis, F.M. and G.O.; investigation, F.M., N.S., J.B., and N.A.; resources, S.G.; data curation, A.V., J.B., and A.C.; writing—original draft preparation, A.V. and A.C.; writing—review and editing, A.V.; project administration, A.V. and S.G.; funding acquisition, S.G. All authors have read and agreed to the published version of the manuscript.

Funding: This research was funded by Ministry of Science and Higher Education of the Russian Federation, Grant No. 0707-2020-0025.

Conflicts of Interest: The authors declare no conflict of interest.

References

1. Ghafouri-Azar, R.; Mostaghimi, J.; Chandra, S.; Charmchi, M. A stochastic model to simulate the formation of a thermal spray coating. *J. Therm. Spray Technol.* **2003**, *12*, 53–69. [CrossRef]
2. Grujicic, M.; Lai, S.G. Multi-length scale modeling of chemical vapor deposition of titanium nitride coatings. *J. Mater. Sci.* **2001**, *36*, 2937–2953. [CrossRef]
3. Birdsall, C.K. Particle-in-Cell charged-particle simulations, plus Monte Carlo collisions with neutral atoms, PIC-MCC. *IEEE Trans. Plasma Sci.* **1991**, *19*, 65–85. [CrossRef]
4. Kwok, D.T.K.; Cornet, C. Numerical simulation of metal plasma immersion ion implantation (MePIIID) on a sharp cone and a fine tip by a multiple-grid particle-in-cell (PIC) method. *IEEE Trans. Plasma Sci.* **2006**, *34*, 2434–2442. [CrossRef]
5. Bultinck, E.; Bogaerts, A. Particle-in-cell/monte carlo collisions treatment of an Ar/O_2 magnetron discharge used for the reactive sputter deposition of TiO_x films. *New J. Phys.* **2009**, *11*, 103010. [CrossRef]
6. Siemers, M.; Pflug, A.; Melzig, T.; Gehrke, K.; Weimar, A.; Szyszka, B. Model based investigation of Ar+ ion damage in DC magnetron sputtering. *Surf. Coat. Technol.* **2014**, *241*, 50–53. [CrossRef]
7. Cansizoglu, H.; Yurukcu, M.; Cansizoglu, M.F.; Karabacak, T. Investigation of physical vapor deposition techniques of conformal shell coating for core/shell structures by Monte Carlo simulations. *Thin Solid Film.* **2015**, *583*, 122–128. [CrossRef]
8. Fancey, K.S. A coating thickness uniformity model for physical vapour deposition systems: Overview. *Surf. Coat. Technol.* **1995**, *71*, 16–29. [CrossRef]
9. Fancey, K.S. A coating thickness uniformity model for physical vapour deposition systems: Further analysis and development. *Surf. Coat. Technol.* **1998**, *105*, 76–83. [CrossRef]

10. Fuke, I.; Prabhu, V.; Baek, S. Computational model for predicting coating thickness in electron beam physical vapor deposition. *J. Manuf. Process.* **2005**, *7*, 140–152. [CrossRef]
11. James, A.S.; Matthews, A. A simple model for the prediction of coating thickness uniformity from limited measured data. *Surf. Coat. Technol.* **1993**, *61*, 282–286. [CrossRef]
12. De Matos Loureiro da Silva Pereira, V.E.; Nicholls, J.R.; Newton, R. Modelling the EB-PVD thermal barrier coating process: Component clusters and shadow masks. *Surf. Coat. Technol.* **2017**, *311*, 307–313. [CrossRef]
13. Baek, S.; Prabhu, V. Technical paper: Simulation model for an EB-PVD coating structure using the level set method. *J. Manuf. Process.* **2009**, *11*, 1–7. [CrossRef]
14. Frenkel, D.; Smit, B. *Understanding Molecular Simulation*, 2nd ed.; Academic Press: Cambridge, MA, USA, 2002; p. 638.
15. Moarrefzadeh, A. Simulation and modeling of physical vapor deposition (PVD) process. *WSEAS Trans. Appl. Theor. Mech.* **2012**, *7*, 106–111.
16. Westkämper, E.; Klein, P.; Gottwald, B.; Sommadossi, S.; Baumann, P.; Gemmler, A. A contribution of Molecular Dynamics simulation to sophisticated engineering of coating processes applied to PVD DC sputter deposition. *Surf. Coat. Technol.* **2005**, *200*, 872–875. [CrossRef]
17. Aksenova, E.V.; Kshevetskiy, M.S. *Computational Methods for Studying Molecular Dynamics*; St. Petersburg State University Publishing House: Petersburg, Russia, 2009; p. 50.
18. Pflug, A.; Siemers, M.; Melzig, T.; Schäfer, L.; Bräuer, G. Simulation of linear magnetron discharges in 2D and 3D. *Surf. Coat. Technol.* **2014**, *260*, 411–416. [CrossRef]
19. Liu, M.-J.; Zhang, M.; Zhang, X.-F.; Li, G.-R.; Zhang, Q.; Li, C.-X.; Li, C.-J.; Yang, G.-J. Transport and deposition behaviors of vapor coating materials in plasma spray-physical vapor deposition. *Appl. Surf. Sci.* **2019**, *486*, 80–92. [CrossRef]
20. Yoshiya, M.; Wada, K.; Jang, B.K.; Matsubara, H. Computer simulation of nano-pore formation in EB-PVD thermal barrier coatings. *Surf. Coat. Technol.* **2004**, *187*, 399–407. [CrossRef]
21. Gidalevich, E.; Goldsmith, S.; Boxman, R.L. Modeling of nonstationary vacuum arc plasma jet interaction with a neutral background gas. *J. Appl. Phys.* **2001**, *90*, 4355–4360. [CrossRef]
22. Kubečka, M.; Obrusník, A.; Zikán, P.; Jílekjr, M.; Vencels, J.; Bonaventura, Z. Predictive simulation of antenna effect in PVD processes using fluid models. *Surf. Coat. Technol.* **2019**, *379*, 125045. [CrossRef]
23. Pinto, G.; Silva, F.; Porteiro, J.; Míguez, J.; Baptista, A. Numerical simulation applied to PVD reactors: An overview. *Coatings* **2018**, *8*, 410. [CrossRef]
24. Zhang, X.F.; Zhou, K.S.; Deng, C.M.; Liu, M.; Deng, Z.Q.; Deng, C.G.; Song, J.B. Gas-deposition mechanisms of 7YSZ coating based on plasma spray-physical vapor deposition. *J. Eur. Ceram. Soc.* **2016**, *36*, 697–703. [CrossRef]
25. Li, J.; Li, C.-X.; Chen, Q.-Y.; Gao, J.-T.; Wang, J.; Yang, G.-J.; Li, C.-J. Super-Hydrophobic surface prepared by lanthanide oxide ceramic deposition through PS-PVD process. *J. Therm. Spray Technol.* **2017**, *26*, 398–408. [CrossRef]
26. Hass, D.D.; Yang, Y.Y.; Wadley, H.N.G. Pore evolution during high pressure atomic vapor deposition. *J. Porous Mater.* **2010**, *17*, 27–38. [CrossRef]
27. Vereshchaka, A.A.; Vereshchaka, A.S.; Mgaloblishvili, O.; Morgan, M.N.; Batako, A.D. Nano-scale multilayered-composite coatings for the cutting tools. *Int. J. Adv. Manuf. Tech.* **2014**, *72*, 303–317. [CrossRef]
28. Vereschaka, A.; Tabakov, V.; Grigoriev, S.; Sitnikov, N.; Oganyan, G.; Andreev, N.; Milovich, F. Investigation of wear dynamics for cutting tools with multilayer composite nanostructured coatings in turning constructional steel. *Wear* **2019**, *420–421*, 17–37. [CrossRef]
29. Vereschaka, A.; Tabakov, V.; Grigoriev, S.; Sitnikov, N.; Milovich, F.; Andreev, N.; Bublikov, J. Investigation of wear mechanisms for the rake face of a cutting tool with a multilayer composite nanostructured Cr–CrN–(Ti,Cr,Al,Si)N coating in high-speed steel turning. *Wear* **2019**, *438–439*, 203069. [CrossRef]
30. Vereshchaka, A.S.; Vereshchaka, A.A.; Kirillov, A.K. Ecologically friendly dry machining by cutting tool from layered composition ceramic with nano-scale multilayered coatings. *Key Eng. Mater.* **2012**, *496*, 67–74. [CrossRef]
31. Vereschaka, A.; Tabakov, V.; Grigoriev, S.; Aksenenko, A.; Sitnikov, N.; Oganyan, G.; Seleznev, A.; Shevchenko, S. Effect of adhesion and the wear-resistant layer thickness ratio on mechanical and performance properties of ZrN–(Zr,Al,Si)N coatings. *Surf. Coat. Technol.* **2019**, *357*, 218–234. [CrossRef]

32. Volosova, M.A.; Grigor'ev, S.N.; Kuzin, V.V. Effect of titanium nitride coating on stress structural inhomogeneity in oxide-carbide ceramic. Part 4. Action of heat flow. *Refract. Ind. Ceram.* **2015**, *56*, 91–96. [CrossRef]
33. Metel, A.; Bolbukov, V.; Volosova, M.; Grigoriev, S.; Melnik, Y. Equipment for Deposition of thin metallic films bombarded by fast argon atoms. *Instrum. Exp. Tech.* **2014**, *57*, 345–351. [CrossRef]
34. Metel, A.S.; Grigoriev, S.N.; Melnik, Y.A.; Bolbukov, V.P. Characteristics of a fast neutral atom source with electrons injected into the source through its emissive grid from the vacuum chamber. *Instrum. Exp. Tech.* **2012**, *55*, 288–293. [CrossRef]
35. Metel, A.; Bolbukov, V.; Volosova, M.; Grigoriev, S.; Melnik, Y. Source of metal atoms and fast gas molecules for coating deposition on complex shaped dielectric products. *Surf. Coat. Technol.* **2013**, *225*, 34–39. [CrossRef]
36. Grigoriev, S.N.; Sobol, O.V.; Beresnev, V.M.; Serdyuk, I.V.; Pogrebnyak, A.D.; Kolesnikov, D.A.; Nemchenko, U.S. Tribological characteristics of (TiZrHfVNbTa) N coatings applied using the vacuum arc deposition method. *J. Frict. Wear* **2014**, *35*, 359–364. [CrossRef]
37. Vereschaka, A.A.; Grigoriev, S.N.; Vereschaka, A.S.; Popov, A.Y.; Batako, A.D. Nano-scale multilayered composite coatings for cutting tools operating under heavy cutting conditions. *Procedia CIRP* **2014**, *14*, 239–244. [CrossRef]
38. Vereschaka, A.A.; Volosova, M.A.; Grigoriev, S.N.; Vereschaka, A.S. Development of wear-resistant complex for high-speed steel tool when using process of combined cathodic vacuum arc deposition. *Procedia CIRP* **2013**, *9*, 8–12. [CrossRef]
39. Grigoriev, S.; Melnik, Y.; Metel, A. Broad fast neutral molecule beam sources for industrial-scale beam-assisted deposition. *Surf. Coat. Technol.* **2002**, *156*, 44–49. [CrossRef]
40. Grigoriev, S.N.; Melnik, Y.A.; Metel, A.S.; Panin, V.V. Broad beam source of fast atoms produced as a result of charge exchange collisions of ions accelerated between two plasmas. *Instrum. Exp. Tech.* **2009**, *52*, 602–608. [CrossRef]
41. Metel, A.S.; Grigoriev, S.N.; Melnik, Y.; Bolbukov, V.P. Broad beam sources of fast molecules with segmented cold cathodes and emissive grids. *Instrum. Exp. Tech.* **2012**, *55*, 122–130. [CrossRef]
42. Metel, A.S.; Grigoriev, S.N.; Melnik, Y.A.; Prudnikov, V.V. Glow discharge with electrostatic confinement of electrons in a chamber bombarded by fast electrons. *Plasma Phys. Rep.* **2011**, *37*, 628–637. [CrossRef]
43. Miklashevich, I.A.; Chigarev, A.V.; Korsunsky, A.M. Variational determination of the crack trajectory in inhomogeneous media. *Int. J. Fract.* **2001**, *111*, L29–L34. [CrossRef]
44. Brownlee, K.A. *Statistical Theory and Methodology in Science and Engineering*; Wiley: New York, NY, USA, 1965.
45. Miklashevich, I.A.; Chigarev, A.V. Equation of the crack front with allowance for the metric properties of the material. *Russ. Phys. J.* **2002**, *45*, 1159–1164. [CrossRef]
46. Lévy, P. *Processus Stochastiques et Movement Brownien*; Villars, G., Ed.; Academic Press: Cambridge, MA, USA, 1965.
47. Chigarev, A.; Polenov, V.; Shirvel, P. Simulation of tsunami effect by seismic wave propagation in hypoplastic medium at vicinity of free boundary. *Univ. J. Mech. Eng.* **2018**, *6*, 9–20. [CrossRef]
48. Polenov, V.S.; Chigarev, A.V. Propagation of waves in an inhomogeneous viscoelastic medium with initial stresses. *Pmm J. Appl. Math. Mech.* **1994**, *58*, 563–568. [CrossRef]
49. Chigarev, A.; Chigarev, J. Expansion of wave rays and fronts in media with inhomogeneous structure. *Univers. J. Mech. Eng.* **2018**, *6*, 76–95. [CrossRef]
50. Zhao, C.; Yang, C. Nonlinear Smoluchowski velocity for electroosmosis of power-law fluids over a surface with arbitrary zeta potentials. *Electrophoresis* **2010**, *31*, 973–979. [CrossRef]
51. Norris, J.R. Smoluchowski's coagulation equation: Uniqueness, nonuniqueness and a hydrodynamic limit for the stochastic coalescent. *Ann. Appl. Probab.* **1999**, *9*, 78–109. [CrossRef]
52. Vereschaka, A.; Tabakov, V.; Grigoriev, S.; Sitnikov, N.; Andreev, N.; Milovich, F. Investigation of wear and diffusion processes on rake faces of carbide inserts with Ti–TiN–(Ti,Al,Si)N composite nanostructured coating. *Wear* **2018**, *416–417*, 72–80. [CrossRef]
53. Vereschaka, A.; Tabakov, V.; Grigoriev, S.; Sitnikov, N.; Milovich, F.; Andreev, N.; Sotova, C.; Kutina, N. Investigation of the influence of the thickness of nanolayers in wear-resistant layers of Ti–TiN–(Ti,Cr,Al)N coating on destruction in the cutting and wear of carbide cutting tools. *Surf. Coat. Technol.* **2020**, *385*, 125402. [CrossRef]

54. Vereschaka, A.; Aksenenko, A.; Sitnikov, N.; Migranov, M.; Shevchenko, S.; Sotova, C.; Batako, A.; Andreev, N. Effect of adhesion and tribological properties of modified composite nano-structured multi-layer nitride coatings on WC-Co tools life. *Tribol. Int.* **2018**, *128*, 313–327. [CrossRef]
55. Vereschaka, A.; Grigoriev, S.; Sitnikov, N.; Milovich, F.; Aksenenko, A.; Andreev, N. Investigation of performance and cutting properties of carbide tool with nanostructured multilayer Zr–ZrN–($Zr_{0.5}$,$Cr_{0.3}$,$Al_{0.2}$)N coating. *Int. J. Adv. Manuf. Technol.* **2019**, *102*, 2953–2965. [CrossRef]

MDPI
St. Alban-Anlage 66
4052 Basel
Switzerland
Tel. +41 61 683 77 34
Fax +41 61 302 89 18
www.mdpi.com

Coatings Editorial Office
E-mail: coatings@mdpi.com
www.mdpi.com/journal/coatings

www.ingramcontent.com/pod-product-compliance
Lightning Source LLC
LaVergne TN
LVHW070114170726
843515LV00007B/1689

* 9 7 8 3 0 3 6 5 2 4 5 2 8 *